AF360844

Characterization of Welded Joints

Characterization of Welded Joints

Editor

Francisco J. G. Silva

MDPI • Basel • Beijing • Wuhan • Barcelona • Belgrade • Manchester • Tokyo • Cluj • Tianjin

Editor
Francisco J. G. Silva
ISEP—School of Engineering,
Polytechnic Institute of Porto
Portugal

Editorial Office
MDPI
St. Alban-Anlage 66
4052 Basel, Switzerland

This is a reprint of articles from the Special Issue published online in the open access journal *Metals* (ISSN 2075-4701) (available at: https://www.mdpi.com/journal/metals/special_issues/characterization_joints).

For citation purposes, cite each article independently as indicated on the article page online and as indicated below:

LastName, A.A.; LastName, B.B.; LastName, C.C. Article Title. *Journal Name* **Year**, *Article Number*, Page Range.

ISBN 978-3-03943-248-6 (Hbk)
ISBN 978-3-03943-249-3 (PDF)

Contents

About the Editor

Francisco J. G. Silva has a PhD in Mechanical Engineering from the Faculty of Engineering, University of Porto (Portugal), an MSc in Mechanical Engineering—Materials and Manufacturing Processes from the Faculty of Engineering, University of Porto, and a BSc in Mechanical Engineering—Industrial Management from the School of Engineering, Polytechnic of Porto (Portugal). He also has a postgrad qualification in Materials and Manufacturing Processes from the Faculty of Engineering, University of Porto. He is currently Head of the master's degree in Mechanical Engineering at School of Engineering, Polytechnic of Porto, a position that he has occupied since 2014. He was also Head of the bachelor's degree in Mechanical Engineering at Superior School of Industrial Studies and Management, Polytechnic of Porto, from 2003 to 2006. He has supervised several PhD students at the Faculty of Engineering, University of Porto, as well as more than 130 students at the School of Engineering, Polytechnic of Porto. Moreover, he has co-supervised more than 45 MSc students at the School of Engineering, Polytechnic of Porto and the Faculty of Engineering, University of Porto. He has more than 170 papers published in peer-reviewed journals with high impact factors, and has presented more than 125 papers at international conferences. He is a member of the Editorial Board of the scientific journal Coatings (MDPI) and Solids (MDPI). He is also Editor-in-Chief of the Journal of Coating Science and Technology (LifeScience Global) and Co-Editor-in-Chief of the Journal of Research Updates in Polymer Science (LifeScience Global). He has reviewed more than 400 papers for reputed journals with high impact factors, such as Journal of Cleaner Production, Composites Part A, Wear, Surface and Coatings Technology, Metals, Materials, Coatings, and Journal of Materials—Design and Applications, amongst many others. He is a member of the scientific committees of some international conferences, such as FAIM—Flexible Automation and Intelligent Manufacturing, among others. He has conducted several Special Issues in MDPI journals, such as Coatings and Metals. Moreover, he has conducted some national and international research projects involving other academic institutions and companies. He has a strong linkage to the industry, due to his past link to the industrial sector, having been the founder, owner and manager of a company in the electric sector for 18 years. Most recent papers: Ronny M. Gouveia, Francisco J. G. Silva, Eleonora Atzeni, Dušan Sormaz, Jorge Lino Alves, António Bastos Pereira (2020). Effect of scan strategies and use of support structures on surface quality and hardness of L-PBF AlSi10Mg parts, Materials, 13(10), E2248. doi: 10.3390/ma13102248.Vitor F. C. Sousa and Francisco J. G. Silva (2020), Recent Advances on Coated Milling Tool Technology - A Comprehensive Review, Coatings, 10, 235. doi:10.3390/coatings10030235.Vitor F. C. Sousa and Francisco J. G. Silva (2020), Recent Advances in Turning Processes Using Coated Tools-A Comprehensive Review, Metals, 10, 170. doi:10.3390/met10020170.Francisco J. G. Silva, A. P. Pinho, A. B. Pereira and O. C. Paiva (2020). Evaluation of Welded Joints in P91 Steel under Different Heat-Treatment Conditions, Metals, 10, 99; doi:10.3390/met10010099.R. Barros, Francisco J. G. Silva, Ronny M. Gouveia, Abdollah Saboori, Giulio Marchese, Sara Biamino, Alessandro Salmi and Eleonora Atzeni (2019), Stress Analysis Before and After Heat Treatment Laser Powder Bed Fusion of Inconel 718: Residual, Metals 9, 1290; doi:10.3390/met9121290. Most recent book: Francisco J. G. Silva, Ronny M. Gouveia (2020), Cleaner Production—Toward a better future, Springer Nature Switzerland, ISBN: ISBN: 978-3-030-23164-4.

Preface to "Characterization of Welded Joints"

The Characterization of Welded Joints book follows the intense research that continues to be carried out around a manufacturing process which is as important in the context of metalworking as it is in welding. Although welding brings together a set of processes that are quite old, significant developments have emerged in new processes and variants to existing processes, as a result of the continuous research that a wide range of researchers around the world continue to develop. In addition, the development of new materials, both in the field of metallic alloys and polymeric materials, induces the need to develop new investigations for the optimization of parameters and the weldability of these new materials, leading to numerous investigations in the area of the characterization of welded joint. This book brings together an interesting set of contributions that make it possible to understand, in a clear way, some phenomena directly related to certain welding processes, focusing on the processes and their parameters, their surroundings, such as heat treatments, and materials. About a third of the book is dedicated to developments around friction stir welding, probably one of the most investigated processes today, and still with a large potential to be investigated. Since welding is a process capable of generating numerous defects, some of which may constitute serious threats to the integrity of the welded assembly, the in-depth study of process/material binomials is extremely useful. Being a book created based on chapters developed by different groups of authors, the knowledge brought becomes even more effective, due to the fact that these contributions are brought by true specialists in each area, and the work has been reviewed by experts in the same area. Thus, the necessary conditions were created so that this book could bring high added value in terms of knowledge to the entire scientific community, with special attention to PhD students in this area and to professors who regularly teach these themes and seek a continuous update of their knowledge in this area. The authors hope that their contribution can be a real asset in increasing the knowledge of those who study and research in this field.

Francisco J. G. Silva
Editor

Article

Microstructure Evolution and Mechanical Properties of Titanium/Alumina Brazed Joints for Medical Implants

Hong Bian [1,2,3], Xiaoguo Song [1,2,3,*], Shengpeng Hu [3], Yuzhen Lei [3], Yide Jiao [3], Shutong Duan [3], Jicai Feng [1,2] and Weimin Long [4]

[1] School of Materials Science and Engineering, Harbin Institute of Technology, Harbin 150001, China; whenstreaming163@163.com (H.B.); feng_jicai@163.com (J.F.)
[2] State Key Laboratory of Advanced Welding and Joining, Harbin Institute of Technology, Harbin 150001, China
[3] Shandong Provincial Key Lab of Special Welding Technology, Harbin Institute of Technology at Weihai, Weihai 264209, China; husp@hitwh.edu.cn (S.H.); lei.yuzhen@163.com (Y.L.); harrisjiao@163.com (Y.J.); shutong0724@163.com (S.D.)
[4] State Key Laboratory of Advanced Brazing Filler Metals and Technology, Zhengzhou Research Institute of Mechanical Engineering, Zhengzhou 450001, China; brazelong@163.com
* Correspondence: xgsong@hitwh.edu.cn; Tel./Fax: +86-631-5678454

Received: 7 May 2019; Accepted: 31 May 2019; Published: 3 June 2019

Abstract: Medical titanium and alumina (Al_2O_3) bioceramic are widely utilized as biomaterials. A reliable brazed joint of titanium and alumina was successfully obtained using biocompatible Au foil for implantable devices in the present study. The interfacial microstructure and reaction products of titanium/Au/Al_2O_3 joints brazed under different conditions were investigated by scanning electron microscopy (SEM), energy dispersive spectroscopy (EDS), and X-ray diffraction (XRD). In this study, the typical interfacial microstructure of the titanium/Au/Al_2O_3 joint was titanium/Ti_3Au layer/TiAu layer/$TiAu_2$ layer/$TiAu_4$ layer/Au + granular $TiAu_4$ layer/TiO_x phase/Al_2O_3 ceramic. With increasing brazing temperature or holding time, the thicknesses of Ti_3Au + TiAu + $TiAu_2$ layers adjacent to the titanium substrate increased gradually. Shear tests indicated that the joint brazed at 1115 °C for 3 min exhibited the highest shear strength of 39.2 MPa. Typical fracture analysis displayed that the crack started at the Al_2O_3 ceramic and propagated along the interface of $TiAu_2$ and $TiAu_4$ reaction layers.

Keywords: brazing; titanium; alumina; interfacial microstructure; mechanical properties

---

1. Introduction

Titanium and its alloys have been intensively investigated and applied for biomedical applications because of their excellent biocompatibilities, mechanical properties, and corrosion resistances [1–6]. Applications have included dental implants, craniomaxillofacial implants, implants for artificial joint replacement and spinal components, internal fixation plates and screws, and housings for ventricular-assist devices and pacemaker cases [7–9]. Alumina, a ceramic with outstanding physical, chemical, and mechanical performances, has attracted great interest in industrial applications such as biomaterials, aerospace, nuclear power, automobiles, and electronics [10–13]. With excellent advantages in chemical stability, wear resistance, and biocompatibility, alumina has been a preferable orthopedic implant material used in dental and bone replacements as well as coatings for metallic materials [9,14–16]. Utilization of metal–ceramic composites for biomedical applications, including implantable pacemakers, retinal implants, and microstimulators, has dramatically increased in recent years [17]. To extend the practical utilization of metal–ceramic composite components, biocompatible metal–ceramic joints are desirable for implantable medical devices [17,18].

Nevertheless, ceramic–metal joints with sufficient mechanical integrity are difficult to achieve because of the large differences in physical and mechanical properties between ceramics and metals such as coefficient of thermal expansion (CTE), chemical composition, and modulus of elasticity (MOE) [10,19]. In the brazing of ceramics to metals, a main problem that needs to be solved is the poor wettability of liquid brazing alloy on ceramics. Active brazing is a promising approach that introduces active elements such as Ti, Zr, Ni, or V into brazing alloys [20–23], which significantly enhances wettability and spreading of liquid metal on ceramics by metallurgical bonding [24].

Ti is a typical active element that promotes wetting and adhesion. The interfacial reactions on the metal/alumina interface have been investigated using various Ti-containing metal alloys such as CuAg-Ti/alumina [21,25,26], AgCu-Ti/alumina [27,28], CuSn-Ti/alumina [29], NiPd-Ti/alumina [30], SnAgCu-Ti/alumina [31], and NiTiZr/alumina [24]. Voytovych et al. identified the existence of M_6X-type compounds and titanium oxides (TiO_x on a metal/Al_2O_3 ceramic interface) whose chemical compositions were believed to be greatly dependent on the activity of Ti [25]. Decrease in the activity of Ti in the system leads to the formation of titanium oxides with higher oxidation on the interface, resulting in a higher final contact angle. Similar conclusions were made by Kritsalis et al. after studying an NiPd-Ti/alumina system [30].

It has been widely reported that alumina could be brazed to different metals with Ti-containing filler alloys [10,21,32–37]. The types of titanium oxides that form in an Al_2O_3/Kovar joint using Ag-Pd/Ti filler are found to be affected by the thickness of the Ti layer, and the joint strengths are influenced by the thicknesses of the reaction layer and residual Ti layer [32]. Xin et al. investigated the reaction products on the Ti film/Al_2O_3 interface for an Al_2O_3/Kovar joint and suggested that a competitive reaction mechanism existed in the system. At temperatures lower than 1057 °C, Ti reacts with Al from the decomposition of Al_2O_3, resulting in the formation of Ti_3Al. As the reaction proceeds, TiO precipitates out from a Ti solid solution as the O concentration rises above the solubility in Ti and Ti_3Al. For test temperatures higher than 1057 °C, Ti directly reacts with O from Al_2O_3 to generate TiO and Ti_3Al [10]. Simultaneously, the type of Ti oxide depends on the activity of Ti in the reaction layer, which could be decreased by the interaction between Ti and Ni from Kovar substrate to form Ni_3Ti, resulting in a shift of reaction product from TiO to Ti_2O_3 or Ti_3O_5 [35]. This is also observed by other investigations [38–40].

The other challenge in brazing ceramics to metals is the thermal stress generated on the metal–ceramic interface resulting from CTE mismatch between them as the joint cools to room temperature. The addition of pure gold, which is biocompatible, is desired because it can release thermal stresses by plastic deformation.

In this study, reliable brazing of Al_2O_3 ceramic to medical titanium alloy was achieved using pure gold foil. Detailed investigations on the effects of brazing temperature and dwelling time on microstructure evolution and mechanical properties were conducted. Mechanical properties were analyzed from microhardness data for different phases as well as shear strength of the joints.

2. Experimental Materials and Methods

Commercial polycrystalline Al_2O_3 ceramic with a purity of 99.5% (Shanghai Unite Technology Co., Ltd., Shanghai, China) was processed into 5 mm × 5 mm × 5 mm cubes. The dimensions of pure α-titanium (Kunshan Bitaite Metal Products Co., Ltd., Kunshan, China) used herein were approximately 20 mm × 10 mm × 5 mm. Figure 1a shows the SEM image of Al_2O_3 in back-scattered electron (BSE) mode, and Figure 1b displays the metallograph of titanium, indicating the equiaxed structure of the α-Ti alloy. Au foil of purity 99.99% (KYKY Technology Co., Ltd., Beijing, China) with a thickness of 50 μm was used.

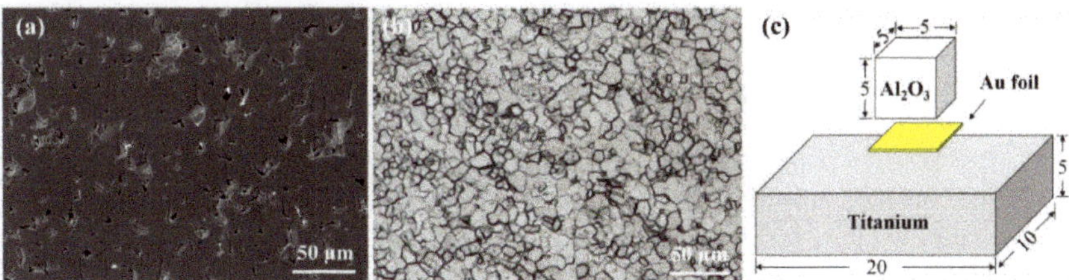

Figure 1. Microstructures of substrates and schematic diagram of brazing assembly. (**a**) BSE image of Al$_2$O$_3$ ceramic, (**b**) metallographic figure of α-Ti alloy, and (**c**) brazing assembly (mm).

To obtain titanium/Au/Al$_2$O$_3$ brazed joints, the joining surface of titanium was ground to a grit of 3000 with emery paper. The Al$_2$O$_3$ ceramic, Au foil, and α-titanium were all cleaned with acetone in an ultrasonic bath for 15 min, and then they were assembled as a sandwich structure, as described in Figure 1c. The atmosphere of the vacuum furnace was maintained at 1.3×10^{-3} Pa during the brazing process. The furnace was firstly heated to 1000 °C for 10 min at a rate of 20 °C/min then to the brazing temperatures at a rate of 10 °C/min. Afterwards, in order to investigate the impact of brazing temperature on the microstructures and mechanical properties of the brazed joints, the brazing specimens were held for 1 min at different brazing temperatures. Finally, the specimens were cooled down to 300 °C at a rate of 5 °C/min and then to room temperature spontaneously in the furnace. To investigate the effect of holding time on the microstructures and mechanical properties of the brazed joints, the brazing specimens were kept for different holding times at 1115 °C. About 10 specimens were prepared in the same condition for each parameter.

The cross-sections of titanium/Au/Al$_2$O$_3$ brazed joints were characterized using SEM (MERLIN Compact, ZEISS, Stuttgart, Germany) equipped with an energy-dispersive X-ray spectrometer (EDS, Octane Plus, EDAX, Mahwah, NJ, USA). Shear tests were conducted on at least six specimens at room temperature at a constant strain rate of 1 mm/min using a universal testing machine (Instron 5967, Instron, Boston, MA, USA). Average values and deviations of shear strengths were calculated from five specimens after removing outliers for each parameter. After the shear test, the fractures of titanium/Au/Al$_2$O$_3$ brazed joints were analyzed by three fractured specimens selected randomly using SEM and X-ray diffraction (XRD, JDX-3530M). To further evaluate mechanical properties of the reaction products in the joint, the hardness and elastic modulus across the joints were measured using a nanoindenter (G200, Agilent, Santa Clara, CA, USA).

3. Results and Discussion

3.1. Typical Interfacial Microstructure of the Titanium/Au/Al$_2$O$_3$ Joint

Vacuum brazing of titanium alloy and Al$_2$O$_3$ ceramic was achieved using Au filler foil. Figure 2 shows the typical interfacial microstructure and the main element distribution of the titanium/Au/Al$_2$O$_3$ joint brazed at 1115 °C for 1 min. As shown in Figure 2a, according to the different contrasts of each phase by SEM in the backscattered electron mode, the joint could be classified into four continuous reaction zones (zone I to IV), and zone V adjacent to the Al$_2$O$_3$ substrate consisted of a white phase dispersed with some light grey granular phases.

Figure 2a also shows variations of elemental concentration of Ti, Au, Al, and O along the white dashed line. The concentration profile of element Ti showed a stepwise decrease from titanium substrate to Al$_2$O$_3$ ceramic, with a noticeable enrichment on the metal/Al$_2$O$_3$ interface. Meanwhile, the main distribution of element Au in the seam exhibited a stepwise increase from titanium to Al$_2$O$_3$ ceramic, and it displayed a minute amount in titanium substrate. The elements of Al and O were mainly distributed in Al$_2$O$_3$ ceramic. These results were consistent with the corresponding elemental distribution in the typical joint displayed in Figure 2b–e.

The interdiffusion of Ti and Au indicated that the active element Ti diffused from the titanium substrate and spread in the brazing seam, and it eventually accumulated adjacent to the surface of Al_2O_3 ceramic. Meanwhile, Au melted and diffused into the titanium substrate during the brazing process.

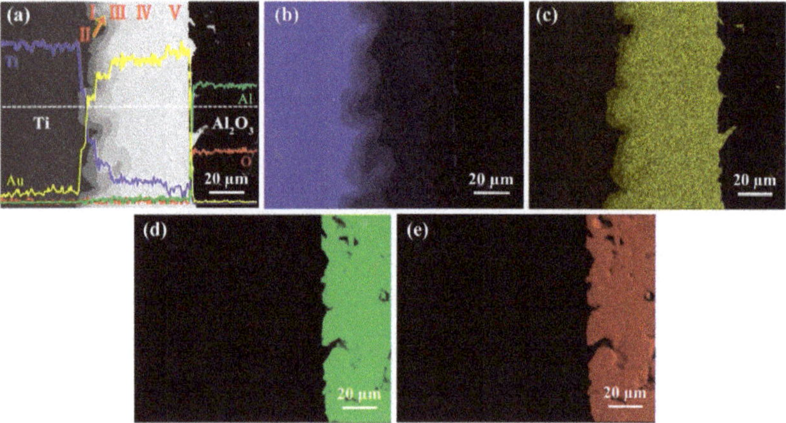

Figure 2. Microstructure and corresponding element distribution of the titanium/Au/Al_2O_3 joint brazed at 1115 °C for 1 min. (**a**) Microstructure and elemental distribution maps of (**b**) Ti, (**c**) Au, (**d**) Al, and (**e**) O.

In order to investigate microstructure characteristics of the titanium/Au/Al_2O_3 joint in detail, Figure 3 shows the microstructures of zones I–V under high magnification. EDS data showing elemental concentrations and possible phases at each spot are listed in Table 1. The characteristic areas of I to VI adjacent to titanium are shown in Figure 3a under high magnification. The characteristic areas of VI and V next to Al_2O_3 ceramic are displayed in Figure 3b under high magnification. According to the EDS results shown in Table 1 and the Ti–Au binary phase diagram illustrated in Figure 4 [41], the reaction layers that formed from the titanium substrate to Al_2O_3 substrate were a Ti_3Au phase (Spot A), a TiAu phase (Spot B), a $TiAu_2$ phase (Spot C), a $TiAu_4$ phase (Spot D), and an Au phase (Spot E) containing $TiAu_4$ particles, respectively.

From the above analysis, it was proposed that during brazing, active element Ti spread and accumulated on the metal/Al_2O_3 ceramic interface, which could be deduced from the reaction with Al_2O_3 and the formation of TiO_x [25,30,31,42,43]. However, TiO_x was hard to observe with SEM and EDS owing to its limited thickness. Apart from reacting with Al_2O_3, the dissolved Ti in the brazing seam also reacted with molten Au, forming Ti–Au intermetallic compounds (IMCs).

To sum up, the typical interfacial microstructure of the titanium/Au/Al_2O_3 joint brazed at 1115 °C for 1 min consisted of titanium/Ti_3Au layer/TiAu layer/$TiAu_2$ layer/$TiAu_4$ layer/Au + granular $TiAu_4$ layer/TiO_x phase/Al_2O_3 ceramic.

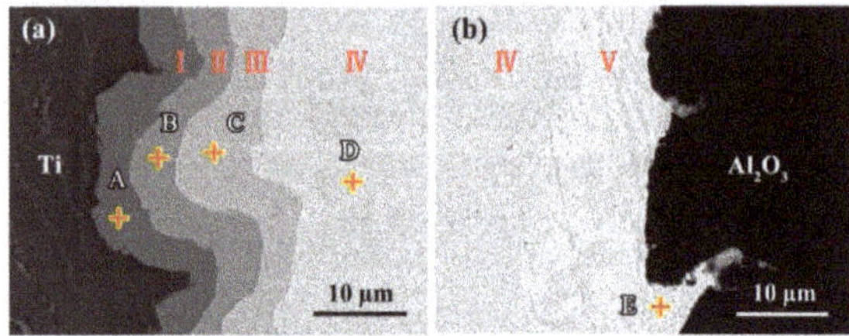

Figure 3. Microstructure of the titanium/Au/Al_2O_3 joint at a high magnification: (**a**) the titanium/brazing seam interface; (**b**) the brazing seam/Al_2O_3 interface.

Table 1. Energy dispersive spectroscopy (EDS) results of the spots marked in Figure 3 (at. %).

Spot	Ti	Au	Al	O	Possible Phase
A	72.64	23.63	0.02	3.71	Ti_3Au
B	48.44	45.89	0.06	5.61	TiAu
C	32.80	63.14	0.03	4.03	$TiAu_2$
D	18.44	73.80	0.04	7.72	$TiAu_4$
E	3.38	86.15	0.56	9.91	Au

In order to illuminate the formation mechanism of the typical interfacial microstructure and different Ti–Au IMCs in the titanium/Au/Al_2O_3 joint, the Ti–Au binary system was studied using the phase diagram shown in Figure 4 [41]. The complex interfacial microstructural morphology and arrangement of various intermetallic compounds (IMCs) generated during the brazing process were strongly dependent on the brazing temperature. The brazing process of titanium to Al_2O_3 ceramic can be deduced as follows. According to the Ti–Au binary phase diagram (Figure 4), it was observed that element Au melted to the liquid phase when the temperature exceeded 1064 °C. The active element Ti diffused from titanium substrate and dissolved into liquid Au gradually. As shown in Figure 4, marked by the red line at 1115 °C, with an increasing Ti concentration in the liquid phase, Ti began to react with molten Au to form $TiAu_4$ IMC by the peritectic reaction of Au(L) + Ti → $TiAu_4$. Thus, a continuous $TiAu_4$ layer formed in the brazing seam and inhibited the interdiffusion of Ti and Au. Because of the decreasing concentration gradient of Ti from titanium to the $TiAu_4$ layer, the Ti_3Au, TiAu, and $TiAu_2$ layers simultaneously formed between titanium and the $TiAu_4$ layer. During the cooling process, $TiAu_4$ particles and the Au phase directly precipitated in the remnant liquid phase by the eutectic reaction of L → Au + $TiAu_4$ adjacent to the ceramic substrate. Finally, the typical microstructure of titanium/Au/Al_2O_3 joint with five reaction zones was obtained in the brazing seam, as illustrated in Figure 3.

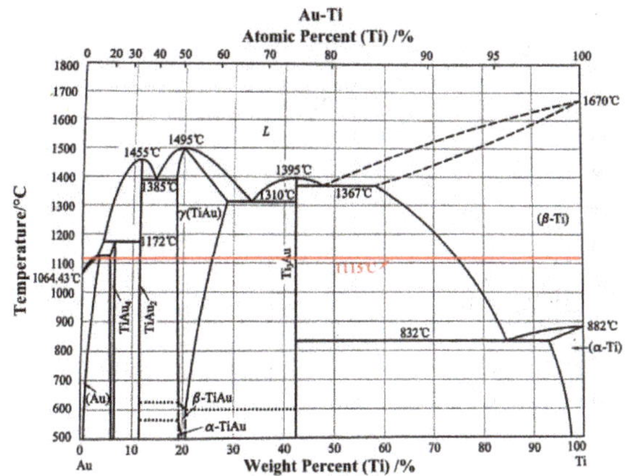

Figure 4. Ti–Au binary phase diagram.

3.2. Effects of Processing Parameters on the Microstructure of the Titanium/Au/Al_2O_3 Joint

Figure 5 displays the microstructure evolution of the joints brazed at different temperatures for 1 min. Brazing temperatures had a significant effect on the interfacial microstructure, and the thicknesses of reaction layers were measured and illustrated in Figure 5f. With increasing temperature, the thicknesses of Ti_3Au + TiAu + $TiAu_2$ layers (zone I–III) adjacent to the titanium substrate increased gradually (Figure 5a–e). The thickness of the $TiAu_2$ layer (zone III) increased first and then decreased, and the maximum thickness of 17.4 µm was obtained under 1115 °C. Meanwhile, as the brazing temperature increased, the thickness of the Au layer with granular $TiAu_4$ (zone V) next to Al_2O_3 ceramic notably decreased.

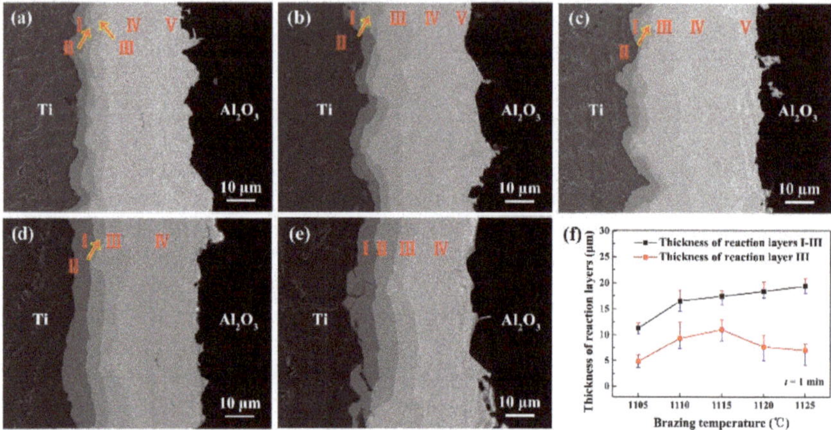

Figure 5. Microstructure of the titanium/Au/Al_2O_3 joint brazed at different brazing temperature for 1 min: (**a**) 1105 °C, (**b**) 1110 °C, (**c**) 1115 °C, (**d**) 1120 °C, (**e**) 1125 °C, and (**f**) the thicknesses of Ti-Au layers (zone I–III).

Figure 6 shows the microstructure evolution of the joints brazed at 1115 °C for different holding times. The microstructure of the joints changed significantly with the prolongation of holding time, and the thickness of the reaction layers were measured and illustrated in Figure 6d. As shown in Figure 6a–c, with the prolongation of holding time from 1 to 5 min, the thicknesses of Ti_3Au + TiAu + $TiAu_2$ layers (zone I–III) increased. The thickness of the $TiAu_2$ layer (zone III) increased first and then decreased, and the maximum thickness was obtained for a holding time of 3 min. Meanwhile, as the holding time increased, the thickness of the Au layer with granular $TiAu_4$ (zone V) next to Al_2O_3 ceramic did not change significantly.

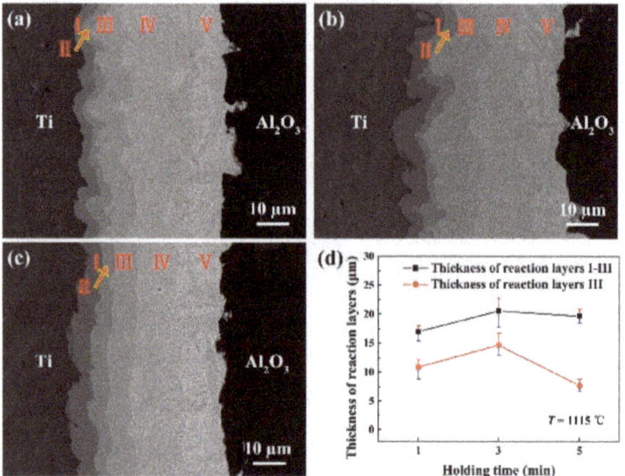

Figure 6. Microstructure of the titanium/Au/Al_2O_3 joint brazed at 1115 °C for different holding times: (**a**) 1 min, (**b**) 3 min, (**c**) 5 min, and (**d**) the thicknesses of Ti-Au layers (zone I–III).

Based on the above analyses, brazing temperature and holding time, which affected the dissolution of Ti from titanium substrate, had significant effects on the microstructure evolution of the joints.

A conceptual model was established and illustrated in Figure 7 to show the evolution of the microstructure. The reaction process could be classified into three stages. As shown in Figure 7a, during the brazing process when the temperature was above the melting point of Au, Au foil first converted into liquid. Then, Ti dissolved into molten Au under the driving force of the concentration gradient, and it reacted with Au to form the $TiAu_4$ layer between the Ti substrate and Au (Figure 7b). Finally, the Ti_3Au, TiAu, and $TiAu_2$ layers simultaneously formed between titanium and the $TiAu_4$ layer along the concentration gradient of Ti. During the cooling process, $TiAu_4$ particles and the Au phase directly precipitated because the residual element Ti was present in the remnant liquid phase adjacent to the ceramic substrate (Figure 7c). When brazing temperature or holding time increased, the mutual diffusion of Ti and Au became more sufficient. As a result, the thicknesses of Ti_3Au + TiAu + $TiAu_2$ layers increased gradually, especially the $TiAu_2$ layer. Meanwhile, the Au phase containing $TiAu_4$ particles reduced, as shown in Figure 7d. With the further increase of brazing temperature or holding time, the diffusion of Au was adequate, and the mount of Ti was sufficient. Ti_3Au and TiAu layers increased, resulting in the decreased thickness of the $TiAu_2$ layer. It was notable that the $TiAu_4$ layer almost occupied the brazing seam next to the ceramic (Figure 7e).

It has been widely reported that TiO_x could be generated on the interface of Ti containing metal and Al_2O_3 [25,30,32,35,38–40]. The limited thickness of TiO_x and many other compounds in metal–ceramic interfaces led to a decreased accuracy in the identification of titanium oxides [20,44]. As brazing temperature or holding time rose, more TiO_x phases formed adjacent to Al_2O_3 (Figure 7c–e).

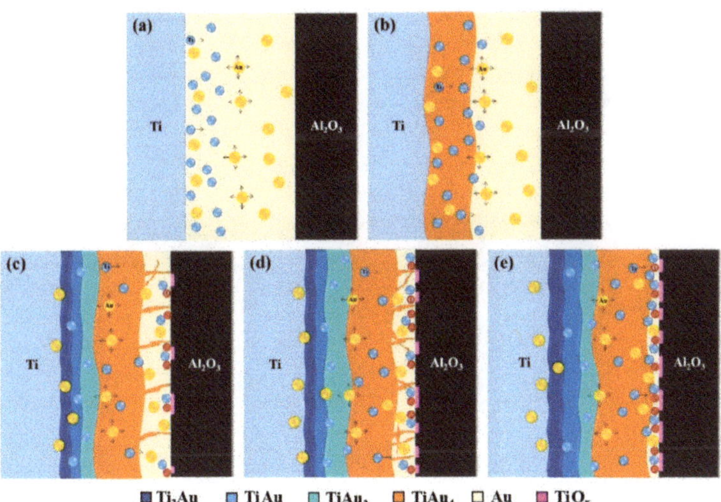

Figure 7. Schematic of the microstructure evolution for the titanium/Au/Al_2O_3 joint. (**a**) Mutual diffusion of Ti and Au; (**b**) formation of the $TiAu_4$ layer; (**c**) formation of reaction layers of the joint; and (**d**) and (**e**) growth of reaction layers with the increase of brazing temperature or holding time.

3.3. Mechanical Properties and Fracture Morphology of Titanium/Au/Al_2O_3 Joints

In order to evaluate the effect of brazing temperature and holding time on the mechanical properties of brazed joints, the shear strength of the joints was tested at room temperature, as shown in Figure 8. As shown in Figure 8a, when the joints were brazed at different temperatures varying from 1105 to 1125 °C for 1 min, the shear strength of the joints increased first and then decreased. The maximum average shear strength of 20.3 MPa was obtained when the joints were brazed at 1105 °C for 1 min.

As shown in Figure 8b, the shear strength of the joints firstly increased and then decreased when the joints were brazed at 1115 °C for different holding times prolonged from 1 to 5 min. The maximum value of shear strength reached 39.2 MPa when the holding time was 3 min, which was about twice that of the joints brazed at 1115 °C for 1 min.

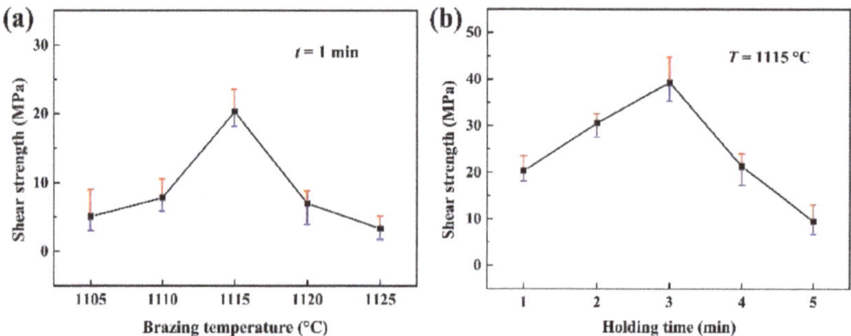

Figure 8. Effect of (**a**) brazing temperature and (**b**) holding time on shear strength of titanium/Au/Al$_2$O$_3$ joints.

Fracture analysis was conducted using an optical microscope, SEM, and XRD to investigate the fracture location and fracture path of the titanium/Au/Al$_2$O$_3$ joints brazed at 1115 °C for 1 min. As shown in Figure 9a,b, Al$_2$O$_3$ ceramic was observed on the fracture surface of the titanium side. Figure 9d shows the crack was initiated at the Al$_2$O$_3$ ceramic and propagated into the brazing seam via the Au/Al$_2$O$_3$ interface during the shear test. The magnified SEM image of Figure 9d is shown in Figure 9e. When the crack propagated into the brazing seam, the joints fractured along the interface of TiAu$_2$ and TiAu$_4$ reaction layers (the interface of zone III and zone IV). The joints brazed at 1115 °C for 1 min fractured in the brittle mode. To further investigate the fracture location, reaction phases on the fracture surface of the titanium side were identified using XRD analysis, as shown in Figure 9c. It was evident that the fracture surface of the titanium side consisted of Au, Al$_2$O$_3$, and TiAu$_2$, which corresponded to the fracture path analyses of Figure 9d,e.

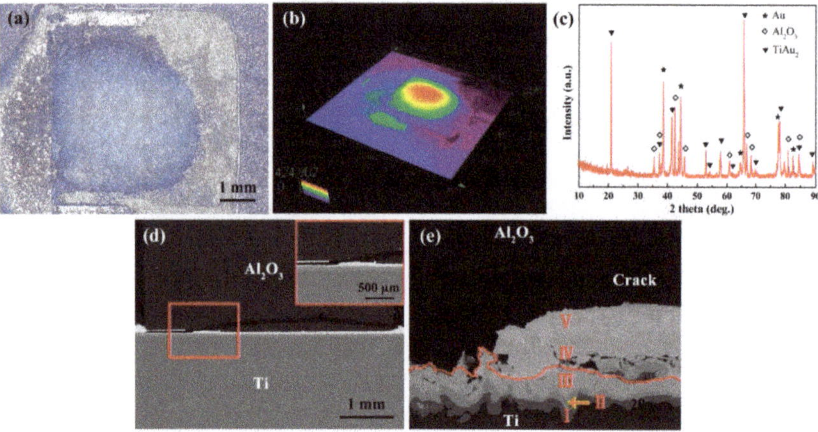

Figure 9. Fracture analysis of titanium/Au/Al$_2$O$_3$ joints brazed at 1115 °C for 1 min after the shear test. (**a**) Fracture surface of Ti alloy side, (**b**) 3D image of (**a**), (**c**) XRD pattern of (**a**), (**d**) fracture path of the Ti/Au/Al$_2$O$_3$ joint, and (**e**) high-magnification image of (**d**).

Figure 10 displays the fracture analyses of the joints brazed at different parameters. It was observed that two types of fracture patterns existed after the shear test. In the first fracture pattern, significantly flat fracture surfaces were clearly observed, and the joints fractured along the Au/Al$_2$O$_3$

interface during the shear tests when brazed at 1105 °C for 1 min and 1115 °C for 5 min (Figures 9c and 10a). XRD analyses of the fracture surface on the titanium side displayed detectable phases, including Au and Al_2O_3, which in turn supported the above analysis of the first fracture pattern. Meanwhile, as shown in Figure 10b, a second type of fracture pattern was observed when the joints were brazed at 1115 °C for 3 min, identical with that brazed at 1115 °C for 1 min. In the second type, the fracture started at the Al_2O_3 ceramic and propagated along the interface of $TiAu_2$ and $TiAu_4$ reaction layers, which was confirmed by the existence of Au, Al_2O_3, and $TiAu_2$ in the XRD result.

Variations in shear strength were significant count on the microstructure evolution of the joint. The increase of brazing temperature and holding time can promote the diffusion of active Ti from the titanium substrate and aggregation adjacent to Al_2O_3 ceramic. When the brazing temperature was lower (or the holding time was shorter), the diffusions of Ti and Au were limited, and the reaction layer of TiO_x was extremely thin as the weakest position of the bonding. Therefore, the shear strength of the joints was quite low, and the joint fractured along the Au/Al_2O_3 interface. With the increase of brazing temperature (or the prolongation of holding time), the TiO_x layer thickened, which could improve the metallurgical bonding between brazing alloy and ceramic. Therefore, the shear strength of the joints increased. Fractures occurred at the Au/Al_2O_3 interface and fragile Ti–Au reaction layers. When the brazing temperature further increased (or holding time was further prolongated), there was a drop in shear strength, which could be attributed to two factors: the over-thickened TiO_x layer and the higher stresses resulting from the increased temperature or changed microstructure of Ti–Au IMCs layers in the brazing seam. Based on the above analyses, it can be concluded that a suitable thickness of the TiO_x layer adjacent to ceramic had crucial influence on the shear strength of the joints.

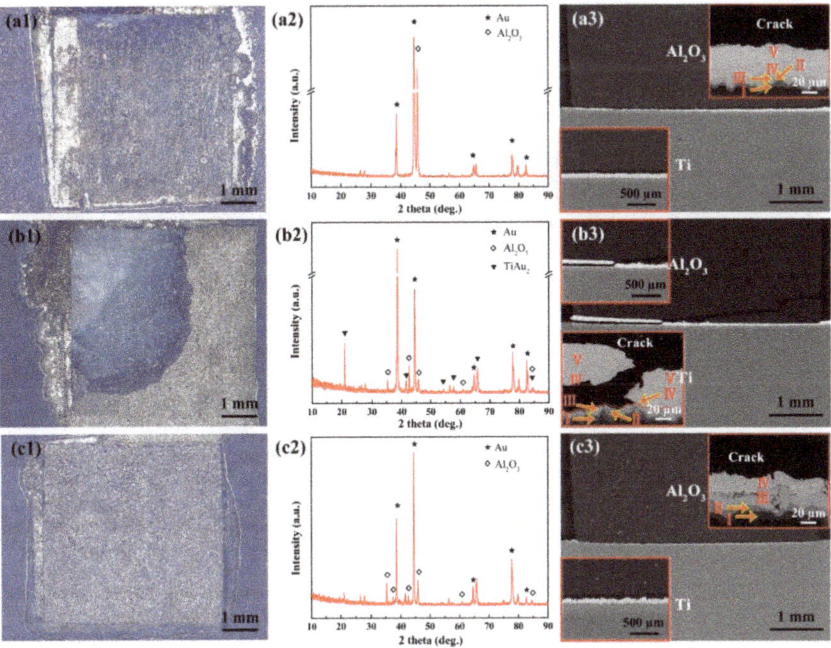

Figure 10. Fractographs and XRD patterns of titanium/Au/Al_2O_3 joints brazed at different parameters after the shear test. (**a1**–**a3**) 1105 °C for 1 min, (**b1**–**b3**) 1115 °C for 3 min, and (**c1**–**c3**) 1115 °C for 5 min.

Figure 11 shows nanoindentation test results, which displayed the variation in hardness and elastic modulus of reaction phases for the joint brazed at 1115 °C for 3 min. As shown in Figure 11a, the highest hardness (9.9 GPa) and elastic modulus (165.0 GPa) across the joint was found in the Ti_3Au

layer, while the Au phase showed the lowest hardness (2.7 GPa) and elastic modulus (115.0 GPa). In order to reveal elastic and plastic behaviors of reaction phases across the joint, typical loads versus depth curves are illustrate in Figure 11b. The deformation process of reaction phases could be divided into elastic-plastic loading and purely elastic unloading. It was apparent that the Au phase possessed the lowest elastic recovery of 14.1%, which recovered 69 nm of the total indentation depth (488 nm). These results showed that the deformation behavior of the Au phase was primarily plastic compared to other phases, which could be beneficial to release residual stress caused by CTE mismatch.

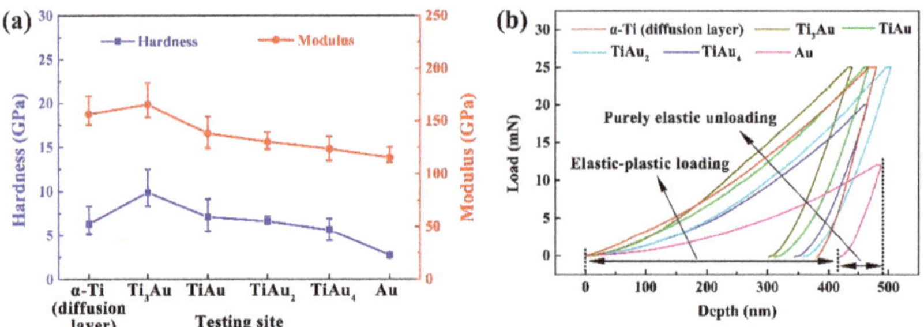

Figure 11. (**a**) Hardness and elastic modulus distribution across the joint interface; (**b**) typical load versus depth curves.

4. Conclusions

In this study, a reliable brazing joint of medical titanium and alumina bioceramic was successfully obtained using Au foil. The microstructure, reaction products, and shear strength of titanium/Au/Al_2O_3 joints were studied. The conclusions are summarized as follows:

(1)	The typical interfacial microstructure of the titanium/Au/Al_2O_3 joint was titanium/Ti_3Au layer/TiAu layer/$TiAu_2$ layer/$TiAu_4$ layer/Au + granular $TiAu_4$ layer/TiO_x phase/Al_2O_3 ceramic.

(2)	Brazing temperature displayed significant effects on the microstructure evolution and mechanical properties of brazed joints. With the increase of brazing temperature, the mutual diffusion of Ti and Au was enhanced, and the thickness of Ti_3Au + TiAu + $TiAu_2$ layers adjacent to the titanium substrate increased gradually. Meanwhile, the thickness of the Au layer with granular $TiAu_4$ next to Al_2O_3 ceramic notably decreased. The TiO_x phase, which promoted metallurgical bonding between the brazing alloy and Al_2O_3 ceramic, could increase as more Ti reacts with Al_2O_3. The shear strength of the joints increased first and then decreased. When the brazing temperature was 1115 °C, a maximum shear strength was obtained as a result of the TiO_x layer with a suitable thickness. Similar effects of holding time on microstructure evolution and mechanical properties were also observed, and the maximum shear strength was obtained for a holding time of 3 min.

(3)	Shear tests indicated that the joint brazed at 1115 °C for 3 min exhibited the highest shear strength of 39.2 MPa. Typical fracture analysis displayed that the crack started at the Al_2O_3 ceramic and propagated along the interface of $TiAu_2$ and $TiAu_4$ reaction layers.

Author Contributions: Conceptualization, X.S. and W.L.; Formal analysis, H.B.; Investigation, H.B. and S.D.; Project administration, J.F.; Resources, Y.L.; Supervision, X.S.; Validation, S.H.; Visualization, Y.J.; Writing—original draft and revising, H.B.; Writing—review and editing, all the authors; Writing—revising, S.H. and W.L.

Funding: This project is supported by the National Natural Science Foundation of China (Grant Nos. 51775138, U1537206 and U1737205), and the Key Research & Development program of Shandong Province (No. 2017GGX40103).

Conflicts of Interest: The authors declare no conflict of interest.

References

1. Geetha, M.; Singh, A.; Asokamani, R.; Gogia, A. Ti based biomaterials, the ultimate choice for orthopaedic implants–A review. *Prog. Mater. Sci.* **2009**, *54*, 397–425. [CrossRef]
2. Li, L.-H.; Kong, Y.-M.; Kim, H.-W.; Kim, Y.-W.; Kim, H.-E.; Heo, S.-J.; Koak, J.-Y. Improved biological performance of Ti implants due to surface modification by micro-arc oxidation. *Biomaterials* **2004**, *25*, 2867–2875. [CrossRef] [PubMed]
3. Symietz, C.; Lehmann, E.; Gildenhaar, R.; Krüger, J.; Berger, G. Femtosecond laser induced fixation of calcium alkali phosphate ceramics on titanium alloy bone implant material. *Acta Biomater.* **2010**, *6*, 3318–3324. [CrossRef] [PubMed]
4. Shukla, A.; Balasubramaniam, R.; Shukla, A. Effect of surface treatment on electrochemical behavior of CP Ti, Ti–6Al–4V and Ti–13Nb–13Zr alloys in simulated human body fluid. *Corros. Sci.* **2006**, *48*, 1696–1720. [CrossRef]
5. Ozdemir, Z.; Ozdemir, A.; Basim, G. Application of chemical mechanical polishing process on titanium based implants. *Mater. Sci. Eng. C* **2016**, *68*, 383–396. [CrossRef] [PubMed]
6. Revathi, A.; Borrás, A.D.; Muñoz, A.I.; Richard, C.; Manivasagam, G. Degradation mechanisms and future challenges of titanium and its alloys for dental implant applications in oral environment. *Mater. Sci. Eng. C* **2017**, *76*, 1354–1368. [CrossRef] [PubMed]
7. Chen, Q.; Thouas, G.A. Metallic implant biomaterials. *Adv. Mater. Sci. Eng. R. Rep.* **2015**, *87*, 1–57. [CrossRef]
8. Hussein, M.A.; Kumar, M.; Drew, R.; Al-Aqeeli, N. Electrochemical corrosion and in vitro bioactivity of nano-grained biomedical Ti-20Nb-13Zr alloy in a simulated body fluid. *Materials* **2018**, *11*, 26. [CrossRef]
9. Weng, Y.; Liu, H.; Ji, S.; Huang, Q.; Wu, H.; Li, Z.; Wu, Z.; Wang, H.; Tong, L.; Fu, R.K.; et al. A promising orthopedic implant material with enhanced osteogenic and antibacterial activity: Al_2O_3-coated aluminum alloy. *Appl. Surf. Sci.* **2018**, *457*, 1025–1034. [CrossRef]
10. Xin, C.; Liu, W.; Li, N.; Yan, J.; Shi, S.-Q. Metallization of Al_2O_3 ceramic by magnetron sputtering Ti/Mo bilayer thin films for robust brazing to Kovar alloy. *Ceram. Int.* **2016**, *42*, 9599–9604. [CrossRef]
11. Kar, A.; Mandal, S.; Venkateswarlu, K.; Ray, A.K. Characterization of interface of Al_2O_3–304 stainless steel braze joint. *Mater. Charact.* **2007**, *58*, 555–562. [CrossRef]
12. Li, J.; Pan, W.; Yuan, Z.; Chen, Y. Titanium metallization of alumina ceramics by molten salt reaction. *Appl. Surf. Sci.* **2008**, *254*, 4584–4590. [CrossRef]
13. Sakka, S.; Bouaziz, J.; Ben Ayed, F. Sintering and mechanical properties of the alumina–tricalcium phosphate–titania composites. *Mater. Sci. Eng. C* **2014**, *40*, 92–101. [CrossRef] [PubMed]
14. Smargiassi, A.; Bertacchini, J.; Checchi, M.; Cavani, F.; Ferretti, M.; Palumbo, C. Biocompatibility Analyses of Al_2O_3-Treated Titanium Plates Tested with Osteocyte and Fibroblast Cell Lines. *Biomedicines* **2017**, *5*, 32. [CrossRef] [PubMed]
15. Zhao, J.; Pan, N.; Huang, F.; Aldarouish, M.; Wen, Z.; Gao, R.; Zhang, Y.; Hu, H.-M.; Shen, Y.; Wang, L.-X. Vx3-Functionalized Alumina Nanoparticles Assisted Enrichment of Ubiquitinated Proteins from Cancer Cells for Enhanced Cancer Immunotherapy. *Bioconjug. Chem.* **2018**, *29*, 786–794. [CrossRef] [PubMed]
16. Chevalier, J.; Gremillard, L. Ceramics for medical applications: A picture for the next 20 years. *J. Eur. Ceram. Soc.* **2009**, *29*, 1245–1255. [CrossRef]
17. Jiang, G.; Mishler, D.; Davis, R.; Mobley, J.P.; Schulman, J.H. Zirconia to Ti-6Al-4V braze joint for implantable biomedical device. *J. Biomed. Mater. Part B Appl. Biomater.* **2005**, *72*, 316–321. [CrossRef]
18. Peytour, C.; Berthet, P.; Barbier, F.; Revcolevschi, A. Interface microstructure and mechanical behaviour of brazed TA6V/zirconia joints. *J. Mater. Sci. Lett.* **1990**, *9*, 1129–1131. [CrossRef]
19. Park, J.-W.; Mendez, P.F.; Eagar, T.W. Strain energy release in ceramic-to-metal joints by ductile metal interlayers. *Scr. Mater.* **2005**, *53*, 857–861. [CrossRef]
20. Bian, H.; Fu, W.; Lei, Y.; Song, X.; Liu, D.; Cao, J.; Feng, J. Wetting and low temperature bonding of zirconia metallized with Sn0.3Ag0.7Cu-Ti alloys. *Ceram. Int.* **2018**, *44*, 11456–11465.
21. Kozlova, O.; Braccini, M.; Voytovych, R.; Eustathopoulos, N.; Martinetti, P.; Devismes, M.-F. Brazing copper to alumina using reactive CuAgTi alloys. *Acta Mater.* **2010**, *58*, 1252–1260. [CrossRef]
22. Xiong, H.-P.; Chen, B.; Pan, Y.; Zhao, H.-S.; Mao, W.; Cheng, Y.-Y. A Cu-Pd-V System Filler Alloy for Silicon Nitride Ceramic Joining and the Interfacial Reactions. *J. Am. Ceram. Soc.* **2014**, *97*, 2447–2454. [CrossRef]

23. Wang, H.; Cao, J.; Feng, J. Brazing mechanism and infiltration strengthening of CC composites to TiAl alloys joint. *Scr. Mater.* **2010**, *63*, 859–862. [CrossRef]

24. Siegmund, P.; Guhl, C.; Schmidt, E.; Roßberg, A.; Rettenmayr, M. Reactive wetting of alumina by Ti-rich Ni–Ti–Zr alloys. *J. Mater. Sci.* **2016**, *51*, 3693–3700. [CrossRef]

25. Voytovych, R.; Robaut, F.; Eustathopoulos, N. The relation between wetting and interfacial chemistry in the CuAgTi/alumina system. *Acta Mater.* **2006**, *54*, 2205–2214. [CrossRef]

26. Kozlova, O.; Voytovych, R.; Eustathopoulos, N. Initial stages of wetting of alumina by reactive CuAgTi alloys. *Scr. Mater.* **2011**, *65*, 13–16. [CrossRef]

27. Ali, M.; Knowles, K.M.; Mallinson, P.M.; Fernie, J.A. Interfacial reactions between sapphire and Ag–Cu–Ti-based active braze alloys. *Acta Mater.* **2016**, *103*, 859–869. [CrossRef]

28. Liu, X.; Zhang, L.; Sun, Z.; Feng, J. Microstructure and mechanical properties of transparent alumina and TiAl alloy joints brazed using Ag-Cu-Ti filler metal. *Vacuum* **2018**, *151*, 80–89. [CrossRef]

29. Lin, C.-C.; Chen, R.-B.; Shiue, R.-K. A wettability study of Cu/Sn/Ti active braze alloys on alumina. *J. Mater. Sci.* **2001**, *36*, 2145–2150. [CrossRef]

30. Kritsalis, P.; Drevet, B.; Valignat, N.; Eustathopoulos, N. Wetting transitions in reactive metal/oxide systems. *Scr. Metall. Mater.* **1994**, *30*, 1127–1132. [CrossRef]

31. Kang, J.R.; Song, X.G.; Hu, S.P.; Liu, D.; Guo, W.J.; Cao, J.; Fu, W. Wetting and Brazing of Alumina by Sn0.3Ag0.7Cu-Ti Alloy. *Met. Mater. Trans. A* **2017**, *48*, 5870–5878. [CrossRef]

32. Zhu, W.; Chen, J.; Jiang, C.; Hao, C.; Zhang, J. Effects of Ti thickness on microstructure and mechanical properties of alumina–Kovar joints brazed with Ag–Pd/Ti filler. *Ceram. Int.* **2014**, *40*, 5699–5705. [CrossRef]

33. Fu, W.; Song, X.; Hu, S.; Chai, J.; Feng, J.; Wang, G. Brazing copper and alumina metallized with Ti-containing Sn0.3Ag0.7Cu metal powder. *Mater. Des.* **2015**, *87*, 579–585.

34. Cao, Y.; Yan, J.; Li, N.; Zheng, Y.; Xin, C. Effects of brazing temperature on microstructure and mechanical performance of Al$_2$O$_3$/AgCuTi/Fe–Ni–Co brazed joints. *J. Alloy. Compd.* **2015**, *650*, 30–36. [CrossRef]

35. Xin, C.; Yan, J.; Li, N.; Liu, W.; Du, J.; Cao, Y.; Shi, H. Microstructural evolution during the brazing of Al$_2$O$_3$ ceramic to kovar alloy by sputtering Ti/Mo films on the ceramic surface. *Ceram. Int.* **2016**, *42*, 12586–12593. [CrossRef]

36. Niu, G.; Wang, D.; Yang, Z.; Wang, Y. Microstructure and mechanical properties of Al$_2$O$_3$/TiAl joints brazed with B powders reinforced Ag-Cu-Ti based composite fillers. *Ceram. Int.* **2017**, *43*, 439–450. [CrossRef]

37. Valette, C.; Devismes, M.-F.; Voytovych, R.; Eustathopoulos, N. Interfacial reactions in alumina/CuAgTi braze/CuNi system. *Scr. Mater.* **2005**, *52*, 1–6. [CrossRef]

38. Selverian, J.H.; Ohuchi, F.S.; Bortz, M.; Notis, M.R. Interface reactions between titanium thin films and (1¯1 2) sapphire substrates. *J. Mater. Sci.* **1991**, *26*, 6300–6308. [CrossRef]

39. Meir, S.; Kalabukhov, S.; Frage, N.; Hayun, S. Mechanical properties of Al$_2$O$_3$\Ti composites fabricated by spark plasma sintering. *Ceram. Int.* **2015**, *41*, 4637–4643. [CrossRef]

40. Hayun, S.; Meir, S.; Kalabukhov, S.; Frage, N.; Zaretsky, E.; Subhash, G. Phase constitution and dynamic properties of spark plasma-sintered alumina-titanium composites. *J. Am. Ceram. Soc.* **2019**, *99*, 573–580. [CrossRef]

41. Murray, J.L. The Au-Ti (Gold-Titanium) system. *Bull. Alloy Phase Diagr.* **1983**, *4*, 278–283. [CrossRef]

42. Kar, A.; Mandal, S.; Ghosh, R.N.; Ghosh, T.K.; Ray, A.K. Role of Ti diffusion on the formation of phases in the Al$_2$O$_3$–Al$_2$O$_3$ brazed interface. *J. Mater. Sci.* **2007**, *42*, 5556–5561. [CrossRef]

43. Fu, W.; Hu, S.; Song, X.; Jin, C.; Li, J.; Zhao, Y.; Cao, J.; Wang, G. Effect of Ti content on the metallization layer and copper / alumina brazed joint. *Ceram. Int.* **2017**, *43*, 13206–13213. [CrossRef]

44. Lin, K.-L.; Singh, M.; Asthana, R. Characterization of yttria-stabilized-zirconia/stainless steel joint interfaces with gold-based interlayers for solid oxide fuel cell applications. *J. Eur. Ceram. Soc.* **2014**, *34*, 355–372. [CrossRef]

 *metals*

Article

Microstructure and Properties of Spot Welded Joints of Hot-Stamped Ultra-High Strength Steel Used for Automotive Body Structures

Zhixia Qiao [1,*], Huijun Li [2], Lianjin Li [3], Xiaoyu Ran [3] and Liwen Feng [3]

[1] Tianjin Key Laboratory of Refrigeration Technology, Tianjin University of Commerce, Tianjin 300134, China
[2] School of Mechanical, Materials, Mechatronics and Biomedical Engineering, University of Wollongong, Northfields Avenue, Wollongong, NSW 2500, Australia; huijun@uow.edu.au
[3] School of Mechanical Engineering, Tianjin University of Commerce, Tianjin 300134, China; lilianjin@163.com (L.L.); 13389905060@163.com (X.R.); fengliwen@163.com (L.F.)
* Correspondence: qzhxia@tjcu.edu.cn; Tel.: +86-1592-210-0905

Received: 1 February 2019; Accepted: 26 February 2019; Published: 2 March 2019

Abstract: Hot-stamped ultra-high strength steels have been widely used in automobile structural parts. Considering the high splash tendency in resistance spot welding due to their extremely high hardness, in this work, microstructural characteristics and mechanical performance of the resistance spot welded ultra-high strength steels are investigated. The results indicate that the interface between the nugget and heat-affected zone (HAZ) is the weakest zone where fractures initiate. In tensile shearing tests, a qualified spot welding joint failed with a button-shaped fracture. Welding defects would significantly decrease the load-carrying capacity and lead to interfacial fracture, except for a button-shaped fracture. In spot welding, it was found that a specific mid-frequency alternating current (AC) input mode, in which a 6 ms cooling cycle was inserted between every two neighboring current pulses, can avoid the splash in the spot welding of hot-stamped hardened steels.

Keywords: spot welding; hot-stamped hardened steel; microstructure; martensite; bainite

---

1. Introduction

Hot-stamped ultra-high strength steels have been widely used as automobile structural parts, due to the increasing requirements for safety and weight reduction of automobiles. The hot-stamped steel sheet is first heated to the austenite zone and then transferred quickly into deforming dies and hot pressed into structural components with required shapes, such as A-pillar, B-pillar, and bumpers, and finally is rapidly cooled to room temperature by flowing cooling water in dies. During the cooling process, essentially 100% martensite is produced, and the tensile strength can be up to 1500 MPa and the yield strength can reach 1100 MPa [1–3]. To avoid oxidation at high temperatures during hot stamping, the as-received steel sheets are usually pre-coated by Al-Si [4]. Resistance spot welding is one of the most important joining methods in automobile manufacture. However, the ultra-high strength steel exhibits a poor weld-ability during the spot welding process. Molten metal splash tends to occur in welding, which is related to the high hardness of the steel. On the one hand, the two overlapping steel sheets are hard to press together, and on the other, the contact resistance between the two ultra-high strength steel sheets is much higher than that of plain-carbon steels, which result in the faster melting of metal. Under the combined action of these two factors, splash is a common issue in the spot welding of ultra-high strength steel. Splash tends to cause many kinds of welding defects in the nugget and decreases the tensile capacity significantly. In order to optimize spot welding parameters of automobile steels, researchers have studied the influence of spot welding parameters, especially current, electrode force, and welding time, on the microstructure and mechanical properties of spot welding joints [5–7]. In recent years, alternating current (AC) spot welding machines with variable frequency have been developed and more widely used in resistance spot welding in the automobile industry [8–10]. In this case, except for the above mentioned three traditional parameters, some other parameters like frequency and pulse duration would influence the spot welding process, and thereby the mechanical properties of the joints. However, studies have seldom considered the effect of austenite pre-strain, which results from electrode force during welding on the phase transformation of the heating affected zone (HAZ) after spot welding. In this work, the damage mechanism of spot welding joints of hot-stamped strengthened steels was studied in detail, and then spot welding experiments were performed with different current input modes in order to optimize the parameters of resistance spot welding and decrease the splashing ratio in welding.

2. Experiment

Boron bearing steel with an Al-Si coating was used in the present study (GMW14400M-ST-S-HS1300T/950Y-MS). Rectangular specimens for spot welding tests with a dimension of $100 \times 25 \times 1.5$ mm^3 were cut from the hot-stamped hardened A-pillars or B-pillars. The chemical composition of the base steel was measured, and the results are shown in Table 1. The yield strength $R_{P0.2}$, tensile strength R_m, and elongation of the investigated hot-stamped hardened steel were determined as about 1100 MPa, 1500 MPa, and 5.4~7.7%, respectively. The microstructure of the hot-stamped hardened steel consists entirely of fine martensite, including martensite laths and plates, as shown in Figure 1. Microhardness of the hot-stamped hardened steel was determined as HV432.

Table 1. Chemical compositions of hot-stamped steel sheets (wt %).

C	Si	Mn	Cr	Ni	Mo	V	P	S	B	Fe
0.24	0.35	1.43	0.22	0.06	0.02	0.02	0.0317	0.0103	0.0025	base

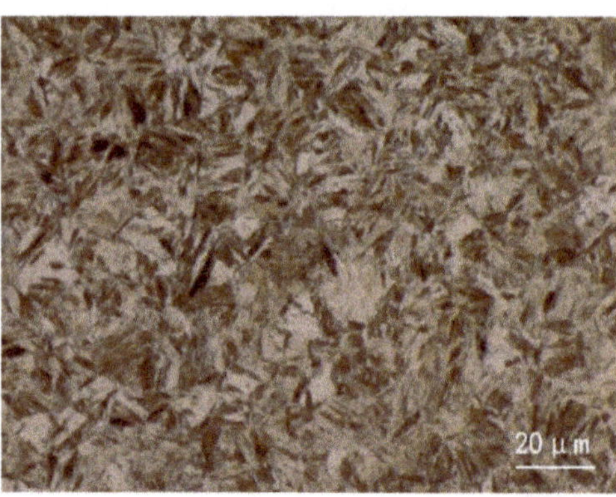

Figure 1. Optical microstructure of the hot-stamped hardened steel.

In order to clarify the phase transformation of the experimental steel after experiencing reheating in spot welding, its continuous cooling transformation (CCT) diagram was obtained by dilatometry measurements. The steels were heated to 1000 °C at a heating rate of 100 °C/s for 2 min, and then cooled to room temperature with different cooling rates.

In resistance spot welding experiments, DC (direct-current) and variable frequency AC (alternating-current) power controllers (Sunke Co. LTD, Tianjin, China) were used in this study. Three types of current modes were applied: (1) DC, with current size of 8~9 kA and welding time of 300~350 ms; (2) general mid-frequency AC, with frequency of 110 Hz, current size of 10~11 kA, and welding time of 20 cycles; (3) mid-frequency AC with cooling intervals (about 6 ms) between any two neighboring current pulses, with frequency of 110 Hz, current size of 11~12.7 kA, and welding time of 25~35 cycles. An electrode force of 550 kg was applied, and the diameter of the used electrode was 6 mm.

Specimens for microstructural examination were cut along a diameter of the spot welding nugget, as shown in Figure 2 by red dotted line of A-A. The specimen includes the whole cross section and it was embedded in epoxy resin with the cross section of the nugget at surface. One of the two cross sections was grounded and polished, and then etched with 3.5% nital reagent. Microstructural observations were conducted by optical microscopy (OM) (Leica DFC 450, Leica, Solms, Germany) and scanning electron microscopy (SEM) (Zeiss, Jena, Germany). Microhardness was determined to evaluate the property of different microstructural areas around the spot welding nuggets, by applying a load of 200 g. In order to examine the mechanical property of spot welding joints, tensile shear tests were performed using samples with dimensions as shown in Figure 2. Three samples were prepared for each welding condition and the average value of the three samples was used.

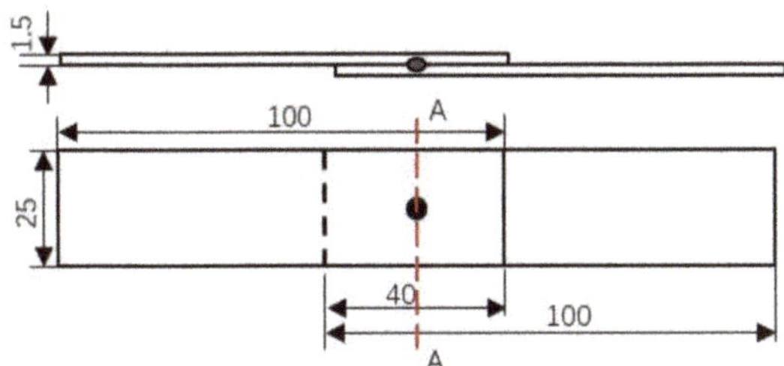

Figure 2. Specimen dimensions for tensile shear tests (unit: mm).

3. Results and Discussion

3.1. Continuous Cooling Phase Transformation

To elucidate the effect of the cooling rate on the phase transformation behavior of the experimental steel, the continuous cooling transformation diagram is established, as shown in Figure 3. Figure 4 shows the microstructures under different cooling rates. It can be seen that the critical cooling rate for martensite transformation was close to 17 °C/s, in which the microstructure consisted entirely of martensite, indicating the high hardenability of the steel. Under the slow cooling rate of 1.5 °C/s, as shown in Figure 4a, the microstructure consisted of polygonal ferrite and a small amount of pearlite, which was formed in the range 603~678 °C, as seen from Figure 3. With the cooling rate increasing to 3 °C/s, polygonal ferrite transformed to an acicular one, and a small amount of granular bainite could be found, as indicated by arrows in Figure 4b. The acicular ferrite nucleates at prior austenite grain boundaries and grows into the prior austenite grains. In the case of 3 °C/s, the phase transformation occurred in the temperature range 556~620 °C, as seen from Figure 3. As the cooling rate further increased to 8 °C/s, granular bainite tended to predominate in the produced microstructure, as shown in Figure 4c. As the cooling rate went higher than 17 °C/s, the product was completely martensite, as shown in Figure 4c. The martensite start and finish temperatures (M_s and M_f) were respectively determined as 438 and 339 °C.

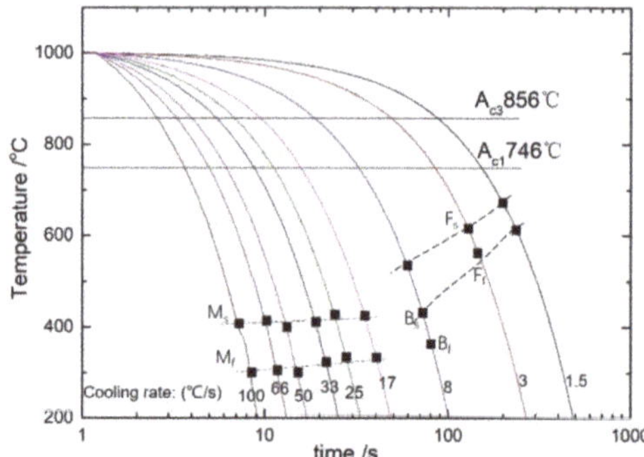

Figure 3. Continuous cooling transformation diagram of the experimental hot stamping steel.

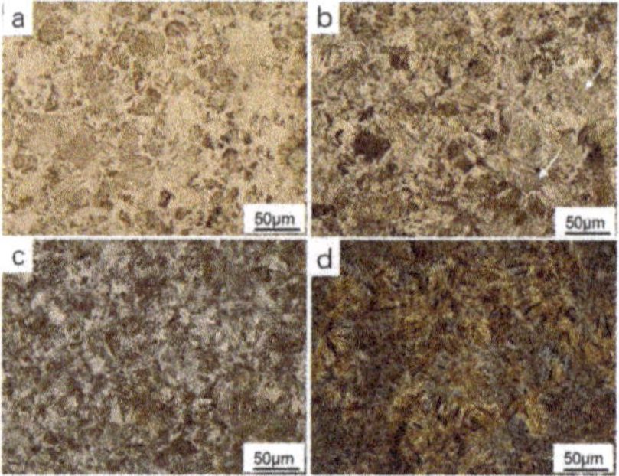

Figure 4. Optical microstructures under typical cooling rates: (**a**) 1.5 °C/s, (**b**) 3 °C/s, (**c**) 8 °C/s, (**d**) 17 °C/s.

3.2. Microstructures of a Spot Welded Joint

Figure 5 shows the low-magnification microstructure of a whole spot welding joint, which was obtained under mid-frequency AC with cooling intervals of 6 ms between any two neighboring current pulses (named AC-6), with a peak current of 12 kA, frequency of 110 Hz, and welding time of 35 cycles. In Figure 5, a whole nugget with a diameter of about 6 mm can be clearly seen, as shown by the yellow double-headed arrow. It is known that $4\sqrt{t}$ (*t* denotes sheet thickness) is the minimum requirement for the nugget diameter. Thickness of the sheet samples used in this work was 1.5 mm, and thus the $4\sqrt{t}$ should be 4.9 mm. The actually obtained diameter was much larger than the standard requirement for a spot welding nugget.

Figure 5. Low-magnification optical microstructure of a whole spot welding joint obtained at AC-6, with a peak current of 12 kA, frequency of 110 Hz, and welding time of 35 cycles.

The spot welded joint includes three regions: nugget, heat-affected zone (HAZ), and base metal. There were two micro areas, just like two white arc lines, on the left and right sides of the nugget in each steel sheet, just outside the upper and under indentations, marked by "e" in Figure 5. These areas exhibited as white due to the specific microstructure of acicular ferrite. Figure 6 presents the microstructures of different regions in Figure 5. The nugget was formed by solidification of melted base metal after spot welding, and it consisted of a columnar grain perpendicular to the surface of steel sheets. Due to the strong cooling effect resulting from the flowing water inside Cu electrodes, the nugget after spot welding was cooled at a rate much higher than the critical quenching rate, and its microstructure was completely composed of martensite, as shown in Figure 6a. Although the martensitic transformation in the nugget occurred within the large columnar austenite grains, martensite laths were relatively fine and dense. The widths of the columnar austenite grains were generally small and less than 50 µm, and most martensite laths form by crossing the width of the columnar grains, therefore, the lengths of the martensite laths were limited within the widths of the columnar austenite grains and were exhibited as fine and dense.

At the interface between the nugget and HAZ, i.e., the fusion zone, extremely coarse prior austenite grains can be distinguished, as shown in the lower right part of Figure 6b. The large-sized austenite grains were related to the higher austenitizing temperature, which was close to the melting point of the steel. This region would be the weakest part of the whole spot welding joint upon loading.

Just outside the fusion zone, the microstructure consisted of relatively fine martensite, as shown on the upper left part of Figure 6b. With the distance away from the nugget center increasing, the maximum heating temperature in welding was decreased, and the size of martensite laths was decreased, as shown in Figure 6c,d. From the fusion zone to the white arc line (marked by "e" in Figure 5), the microstructure was just composed of martensite.

Figure 6e and 6e' show the OM and SEM images of the microstructure in the white arc line in Figure 5. The microstructure in this area consisted of fine acicular ferrite. The width of the white arc line region was about 100 µm and it was a part of the HAZ of a spot welding joint. This region is suggested to be completely austenitized during the welding process, but the cooling rate was too slow to obtain the martensite, since it was relatively far away from the electrodes and cannot be effectively cooled by the flowing water inside the electrodes. It should be noted that the acicular ferrite in Figure 6e is much finer than that in Figure 4b. It was reported that the pre-strain imposed on austenite could affect the phase transformation behavior in subsequent cooling process [11]. Pre-strain can produce a considerable amount of crystal defects (dislocations) within austenite grains. These defects can serve as nuclei for ferrite, therefore ferrite can not only nucleate on grain boundaries but also within the austenite grains. Abundant nucleation sites within austenite grains result in the formation of fine acicular ferrite. When the pre-strain is absent, ferrite would nucleate only on grain boundaries and grow into the austenite grain, resulting in the formation of coarse acicular ferrite, as shown in Figure 4b. From Figure 5, the white arc lines were located just outside the electrode indentation, where the pre-strain accumulated significantly in prior austenite grains. Coupled with the lowest cooling rate, fine acicular ferrite was produced in this region. This area also exhibited the lowest hardness in the whole spot welding joint.

Regions outside the white arc line areas experienced a faster cooling rate due to better heat conduction into surrounding material, and the microstructures were composed of granular bainite, as shown in Figure 6f,g. In addition, with the distance away from the nugget increasing, the density of particles in granular bainite increased and the polygonal bainite matrix evolved to a lath-shaped one. Figure 6h shows the microstructure far away from the nugget, where the peak heating temperature was lower than Ac1 and the martensite in base metal was tempered to some extent.

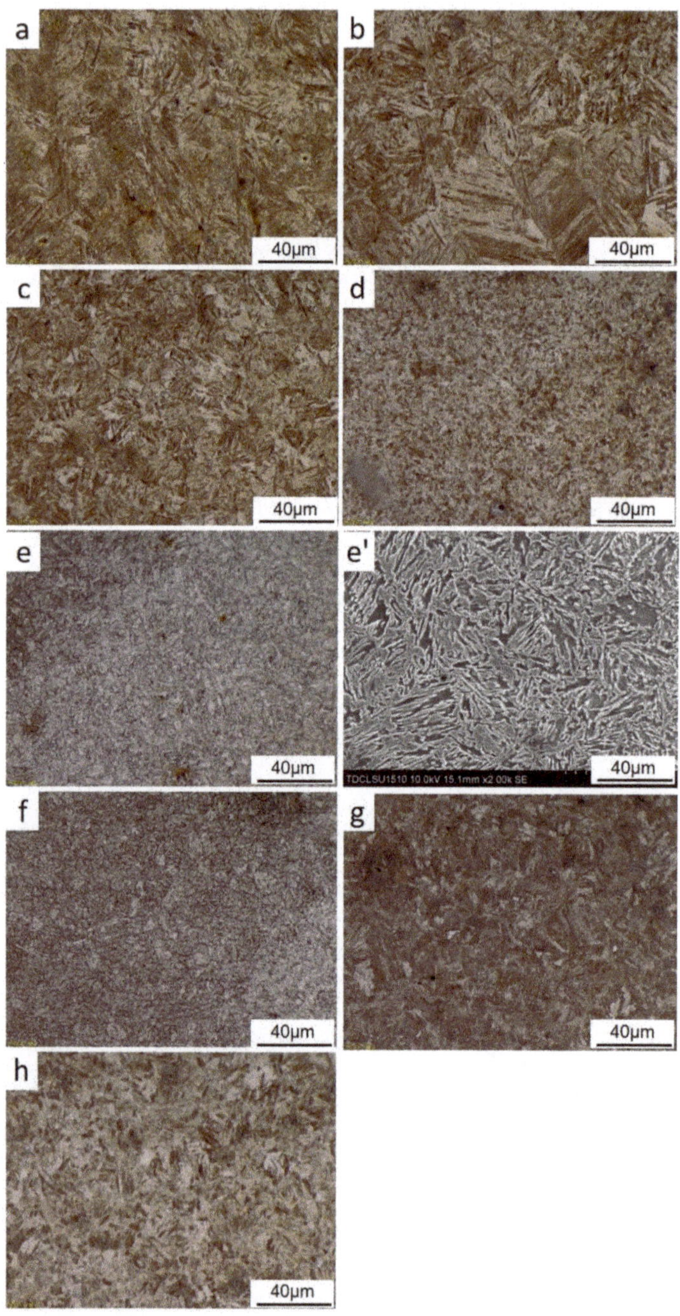

Figure 6. Microstructures of different micro regions in a spot welding joint, corresponding to (**a**)~(**e**) marked in Figure 5: (**a**) nugget, composed of fine martensite within columnar grains; (**b**) fusion zone, mix of coarse martensite and fine martensite; (**c**) martensite; (**d**) minute martensite; (**e**) acicular ferrite under an optical microscope; (**e′**) acicular ferrite under SEM; (**f**) granular bainite with polygonal matrix; (**g**) granular bainite with a lath-shaped matrix; (**h**) tempered martensite.

3.3. Micro-Hardness Distribution Across a Spot Welded Joint

Figure 7 shows the distribution of hardness along the white dashed line in Figure 5. The microhardness of the nugget is somewhat higher than that of the base metal, by about 30–50 HV. This resulted from the extra-high cooling effect of the electrodes on the phase transformation. The region just outside the nugget, which consisted of martensite, had a hardness slightly lower than that of the nugget. The lowest hardness throughout the welding joint in Figure 7 (about 310 HV) corresponds to the white arc line regions in Figure 5, which consist of acicular ferrite. Outward from the white arc line regions, the cooling rate increased rapidly and granular bainite was obtained, therefore, microhardness rose sharply. Microhardness tended to stabilize at about 450 HV as soon as the base metal appeared.

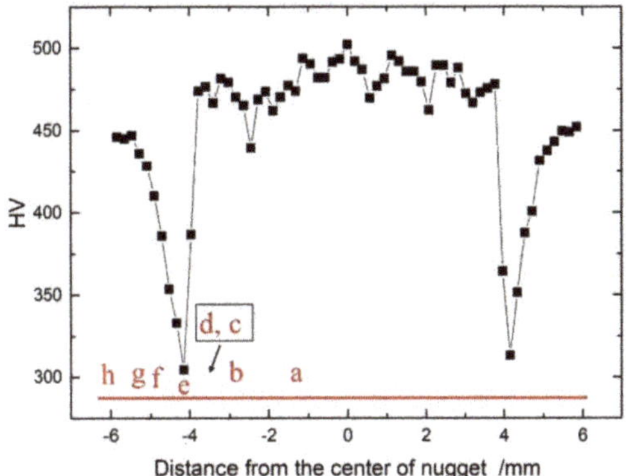

Figure 7. Microhardness measurement results on the typical spot welded joint in Figure 5; (**a**)–(**h**) correspond to the micro areas in Figure 5.

3.4. Tensile Shear Fracture

Spot welding joints exhibit two types of fracture modes: button-shaped fractures and interfacial fractures. A spot welded joint is generally considered to be qualified if a button-shaped fracture is produced when the applied load exceeds its capacity in tensile shearing tests. That is, the whole joint is integrally pulled away from one of the two overlapped sheets, leaving a hole in this sheet and a button on the other one. Figure 8 is an overall view of a failed specimen as a button fracture, obtained at AC-6, with a peak current of 12.7 kA, frequency of 110 Hz, and welding time of 35 cycles. In tensile shear tests, in order to discover the origin of cracks, a high-speed camera was used to take continuous photographs of the weld surface during loading. It was learned from the obtained continuous pictures that for any one of the two overlapped sheets, stresses concentrate significantly on the chuck-nearing side of the joint at an early loading stage, as shown by the two rectangles in Figure 8a. For the upper sheet in Figure 8a, stress concentrates at the right side of the joint and for the lower sheets, stress concentrates at the left side. When stress exceeds the strength of the joint, cracks originate at these stress concentration sites and subsequently fracture occurs. Once cracks appear on either side, the stress concentration on the other side would be relieved greatly, and afterwards the cracks further develop until fracture takes place at that side. From the low-magnification optical micrograph shown in Figure 8b, it is known that fracture originated on the boundary of the nugget, i.e., the fusion zone, which consisted of coarse martensite, as shown in Figure 6b. This indicates that the fusion zone is the weakest region throughout the weld joint, although the white arc line region is the softest region, consisting of acicular ferrite. Figure 8c shows a magnified SEM image of the fracture edge, clearly showing that the original fracture occurred on the boundaries of coarse grains. Later, tearing occurred along half of the nugget edge on the fractured sheet and the nugget was left on the other sheet, leaving a button. Under large compressive stress, the opposite side of the fracture originating point in the joint was deformed significantly, and the formed fiber-like structure is marked by white arrows in Figure 8d.

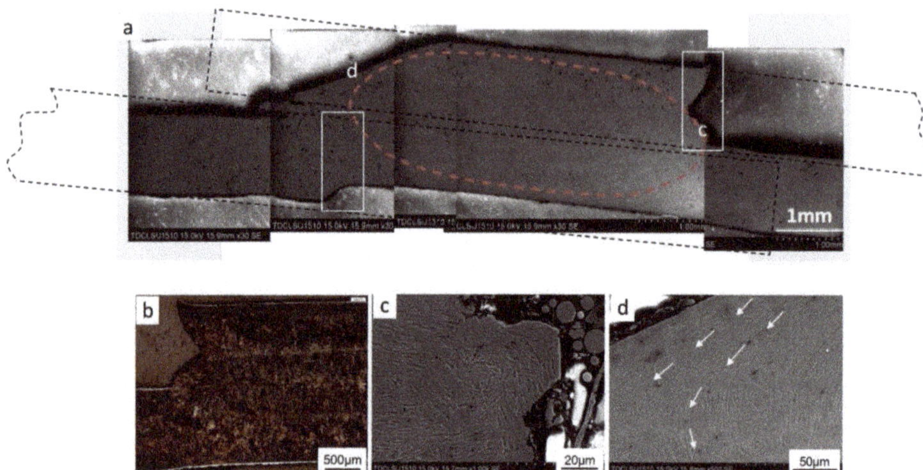

Figure 8. Button-shape fracture images, obtained at AC-6, with a peak current of 12.7 kA, frequency of 110 Hz, and welding time of 35 cycles: (**a**) overall view of a fractured joint, (**b**) optical image showing nugget, (**c**) SEM image of fracture edge at "c" point, (**d**) SEM image of fracture edge at "d" point.

A spot welding joint with the interfacial fracture mode is generally considered unqualified. In this case, the nugget was divided into two parts along the bonding surface of two overlapped sheets. The flat fracture surface is left on each sheet. In this case, the spot welding joint fails mainly under the action of shearing force. Figure 9 shows the typical morphology of an interfacial fracture surface in a failed specimen in tensile shear tests. It can be noted that many small cracks exist in the nugget, as shown in Figure 9a. Such small cracks tend to form in the spot welding process when splash occurred, in which shrinkage stresses were more likely to be produced. High-magnification observation on regions around the cracks shows no dimple produced after fracture, as marked by white circles in Figure 9b, indicating that brittle fracture occurred around cracks. Yet regions far away from the cracks exhibit plastic fracture characteristics, with a lot of dimples left. Some areas exhibit typical shearing fracture characteristics, with small dimples stretched along a uniform direction, as shown in Figure 9c,d, indicating that shearing stress was responsible for the fracture. In addition, inclusions and shrinkage defects in the nugget were also factors leading to interfacial fracture, because their appearance would decrease the effective loading area of the joint.

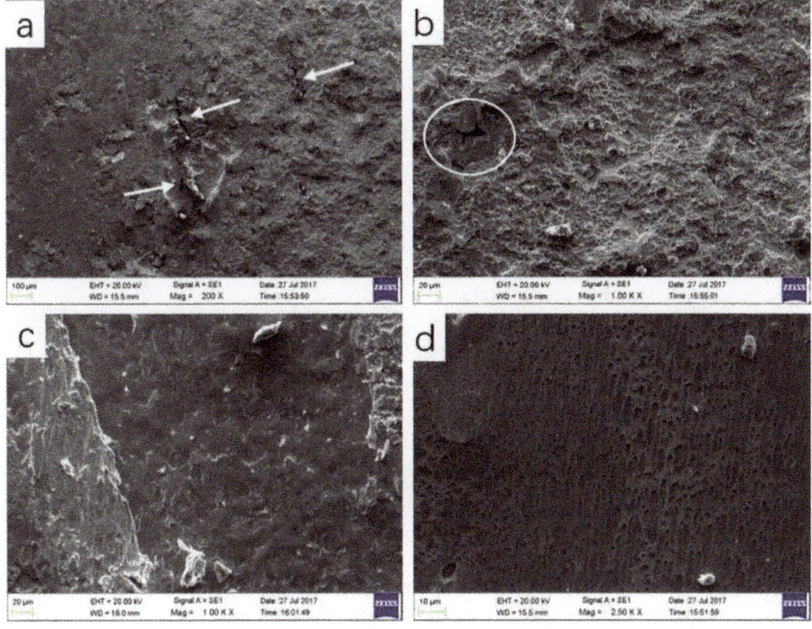

Figure 9. SEM images of an interfacial fractured specimen: (**a**) showing small cracks at low magnification, (**b**) showing brittle fracture around cracks and plastic fracture far away from cracks, (**c**) shearing surface, (**d**) stretched dimples on shearing surface.

3.5. Parameter Optimization for Spot Welding of Hot-Stamped Hardened Steels

In case of spot welding of hot-stamped hardened steel, splash is an important problem demanding a solution. This is due to its ultra-high hardness, which requires a relatively large electrode force to keep the two overlapped sheets in close contact. Therefore, liquid metal tends more to be squeezed out from the molten nugget and thus splash occurs. In this case, a larger plastic metal ring around the molten nugget is beneficial for avoiding splash in spot welding. In order to optimize the technology of spot welding of hot-stamped hardened steel, three current modes were used in this investigation—direct current (DC), mid-frequency alternative current (AC), mid-frequency AC with a 6 ms cooling interval every other cycle (named AC-6)—as shown in Figure 10.

For the AC-6, the time interval was inserted between two neighboring cycles aiming to slow down the expanding rate of the molten metal nugget, and at the same time give more time for the plastic ring areas to expand outward. This is beneficial for the molten metal to be included in the plastic ring and prevent splash happening. Large enough plastic ring areas are important for avoiding splash in spot welding, which can ensure the molten metal is enclosed through close contact of the two overlapped sheets at the plastic areas under great pressure from electrode forces. The length of the time interval should be long enough to ensure the plastic ring expands faster than the molten metal nugget. If the length of the interval is too long, the nugget size would be too small and more welding time would be needed for a joint. After experimenting repeatedly, 6 ms was determined for the experimental ultra-high strength steel of 1.5 mm thickness.

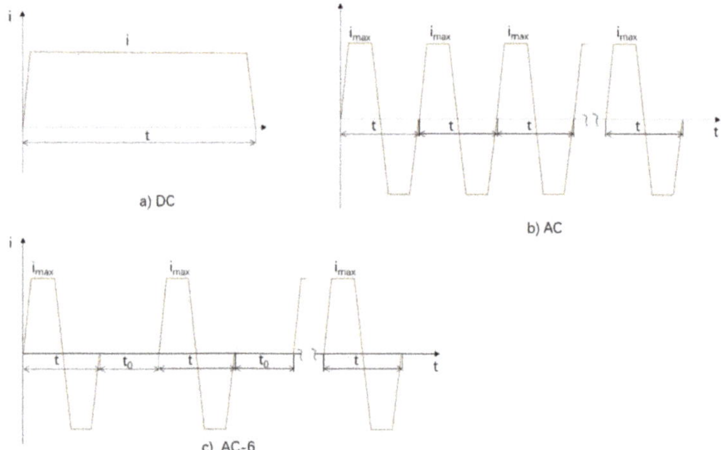

Figure 10. Current modes applied in the present investigation: (**a**) DC: direct current, (**b**) AC: mid-frequency alternating current, (**c**) AC-6: mid-frequency alternating current with a 6 ms cooling interval every other cycle.

Under the two spot welding conditions of AC (AC and AC-6), it was found that the frequency of 110 Hz was the most suitable parameter for the present steel through a series of orthogonal tests by changing current value and frequency. Therefore, frequency of 110 Hz was used here. Under each welding condition, the most suitable current was determined and used, as shown in Table 2. Spot welding test results are summarized in Table 2, showing the diameter of the nugget, splash ratio, and maximum bearable tensile load. It was AC-6 that can solve perfectly the splash problem in spot welding, and only 5% of weld joints were splashed.

Table 2. Spot welding test results using three different current modes.

Current Mode	i_{max}/kA	Virtual Current/kA	Weld Time/ms	Nugget Diameter/mm	Non-Splash Ratio	Max-Tensile Load/kN
AC-6	11	6.8	377	6.2	95%	22
AC	11	9.49	182	5.5	10%	19
DC	8.5	8.5	300	5.7	10%	13.5

It is known that a plastic metal ring around the nugget is important for avoiding splash during spot welding, which can encircle the molten nugget and prevent it from splashing out. Therefore, keeping a large enough plastic ring during welding is important for avoiding splash. At the end of heating, the liquid molten nugget has the largest volume, and at the same time, the surrounding metal is in austenite condition and exhibits good ductility. Under the large electrode force, the austenite zone is extruded significantly, leading to the two overlapped sheets compressed against each other closely. Therefore, the HAZ with a large width is more favorable in order to avoid splash in spot welding. The distance from edge of the nugget to the nearest white arc line was measured for the three welding conditions using different current modes, as shown in Figure 11. The results, as shown in Figure 11d, indicate that the distance from edge of the nugget to the nearest white arc line is the largest under the AC-6 condition, which proves that the austenite zone around the nugget plays an important role in avoiding splash.

From Table 2, it is noted that the welded joints obtained under the AC-6 condition exhibit the highest mechanical property (maximum tensile load). In tensile shearing tests, the welded joints under AC-6 always fractured with a button-shaped mode, yet interfacial fractures may occur in samples obtained under AC and DC conditions. Microstructural examination of the welded joints obtained under three different current modes was carried out. Defects like shrinkage holes, micro cracks, and Al-Si inclusions were found in DC and AC specimens, as shown in Figure 12a–d. These defects probably result from the splash during the welding process. Splash leads to the reduction of liquid metal in the nugget, resulting in the solidification shrinkage holes and cracks. In addition, some molten Al-Si coating was pushed back into the liquid nugget when splash occurred and led to Al-Si inclusions in the nugget, as shown in Figure 12a,c, which significantly decreases the load-carrying capacity of the spot welding joint. Microstructural examinations on the spot welding joint of the AC-6 specimen showed that the microstructure in the nugget was in good condition, and there were no obvious weld defects.

It can be concluded from the above results that applying AC with 6 ms intervals between two neighboring cycles is a good method to address the splash problem in spot welding of hot-stamped hardened steels, and thus increase evidently the tensile shearing property.

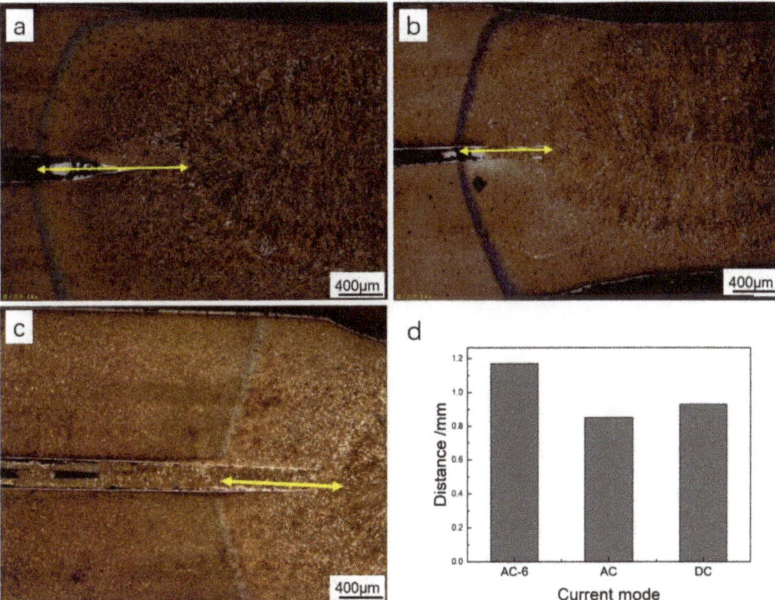

Figure 11. Comparison of plastic zone size under three conditions with different current modes: (**a**) AC-6, (**b**) AC, (**c**) DC, (**d**) results summarization.

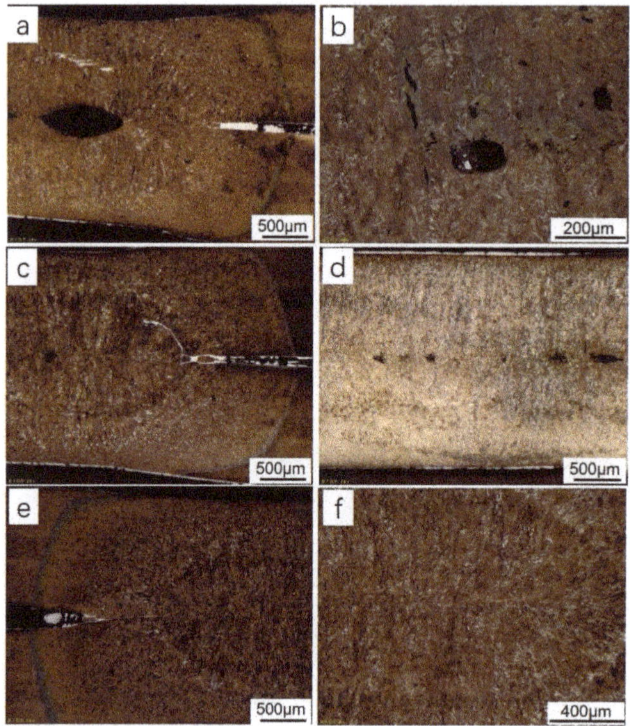

Figure 12. Various defects produced in spot welded specimens applying AC and DC, showing shrinkage holes, Al-Si inclusions, and cracks: (**a**) and (**b**) AC condition, (**c**) and (**d**) DC condition, (**e**) and (**f**) AC-6 condition.

4. Conclusions

The microstructure of spot welded joints of hot-stamped hardened steel was studied by establishing a CCT diagram and microstructural examination. Nugget microstructure completely consisted of martensite within columnar crystals and HAZ was composed of coarse martensite, fine martensite, acicular ferrite, granular bainite, and tempered martensite, in the order from nugget edge to outside. The nugget exhibited the highest microhardness and the acicular ferrite zone was the softest area.

The boundary between the nugget and the coarse martensite zone was the weakest area and fracture was prone to occur in the boundary when the tensile shearing load exceeded its limit. Therefore, for a qualified spot welding joint, a button-shape fracture was generally obtained. If a joint has a lot of welding defects, i.e., shrinkage holes, cracks, and Al-Si inclusions, interfacial fracture tends to occur, which is characterized by a shearing fracture.

Splash is liable to produce various defects in a spot welded nugget, resulting in interfacial fracture of the joint. The splash problem in the spot welding of hot-stamped hardened steel was solved perfectly by using a specific mid-frequency AC current input mode, in which a 6 ms cooling cycle was inserted between every two neighboring current pulses. Under traditional mid-frequency AC and DC heat input modes, splash easily occurs in spot welded joints, resulting in defects in the nugget, and significantly decreasing the load-carrying capacity of welded joints.

Author Contributions: Writing—original draft preparation, Z.Q. and H.L.; investigation, L.L.; formal analysis, X.R. and L.F.

Funding: This research was funded by the Key Technologies R & D Program of Tianjin (Grant No. 18YFZCGX00050), the National Natural Science Foundation of China (Grant No. 51204121), and the Natural Science Foundation of Tianjin City (Grant No. 16JCTPJC48900).

Conflicts of Interest: The authors declare no conflict of interest.

References

1. Merklein, M.; Wieland, M.; Lechner, M.; Bruschi, S.; Ghiotti, A. Hot stamping of boron steel sheets with tailored properties: A review. *J. Mater. Process. Technol.* **2016**, *228*, 11–24. [CrossRef]
2. Karbasian, H.; Tekkaya, A.E. A review on hot stamping. *J. Mater. Process. Technol.* **2010**, *210*, 2103–2118. [CrossRef]
3. Yi, H.L.; Ghosh, S.; Bhadeshia, H.K.D.H. Dual-phase hot-press forming alloy. *Mater. Sci. Eng. A* **2010**, *527*, 4870–4874. [CrossRef]
4. Ighodaro, O.L.; Biro, E.; Zhou, Y.N. Comparative effects of Al-Si and galvannealed coatings on the properties of resistance spot welded hot stamping steel joints. *J. Mater. Process. Technol.* **2016**, *236*, 64–72. [CrossRef]
5. Jong, Y.-S.; Lee, Y.-K.; Kim, D.-C.; Kang, M.-J.; Hwang, I.-S.; Lee, W.-B. Microstructural Evolution and Mechanical Properties of Resistance Spot Welded Ultra High Strength Steel Containing Boron. *Mater. Trans.* **2011**, *52*, 1330–1333. [CrossRef]
6. Choi, H.-S.; Park, G.-H.; Lim, W.-S.; Kim, B.-M. Evaluation of weldability for resistance spot welded single-lap joint between GA780DP and hot-stamped 22MnB5 steel sheets. *J. Mech. Sci. Technol.* **2011**, *25*, 1543. [CrossRef]
7. Lu, Y.; Peer, A.; Abke, T.; Kimchi, M.; Zhang, W. Subcritical heat affected zone softening in hot-stamped boron steel during resistance spot welding. *Mater. Des.* **2018**, *155*, 170–184. [CrossRef]
8. Ding, L.; Min, H. Simulink and design of mid-frequency inverter resistance spot welding control system. *Electr. Weld. Mach.* **2015**, *45*, 26–31.
9. Brezovnik, R.; Cernelic, J.; Petrun, M.; Dolinar, D.; Ritonja, J. Impact of the switching frequency on the welding current of a spot-welding system. *IEEE Trans. Ind. Electron.* **2017**, *64*, 9291–9301. [CrossRef]
10. Son, J.-I.; Im, Y.-D. Intelligent Controller Implementation for Decreasing Splash in Inverter Spot Welding. *IEICE* **2009**, *92*, 1708–1712. [CrossRef]
11. Yogo, Y.; Kurato, N.; Iwata, N. Investigation of Hardness Change for Spot Welded Tailored Blank in Hot Stamping Using CCT and Deformation-CCT Diagrams. *Metall. Mater. Trans. A* **2018**, *49*, 2293–2301. [CrossRef]

Article

Evaluation of Welded Joints in P91 Steel under Different Heat-Treatment Conditions

Francisco José Gomes Silva [1,2,*], **António Pedro Pinho** [1], **António Bastos Pereira** [2] and **Olga Coutinho Paiva** [1]

[1]   ISEP—School of Engineering, Polytechnic of Porto, 4200-072 Porto, Portugal; avp@isep.ipp.pt (A.P.P.); omp@isep.ipp.pt (O.C.P.)

[2]   TEMA—Centre for Mechanical Technology and Automation, Department of Mechanical Engineering, University of Aveiro, Campus de Santiago, 3810-193 Aveiro, Portugal; abastos@ua.pt

*   Correspondence: fgs@isep.ipp.pt; Tel.: +351-228340500

Received: 18 December 2019; Accepted: 7 January 2020; Published: 8 January 2020

Abstract: P91 steel has been of interest to many researchers over the past two decades. This interest is because this steel has very interesting characteristics for application in power plants, where it is common to have pipes that need to support steam at temperatures between 570 and 600 °C, and at pressures in the range of 170 to 230 bar. These working conditions are quite severe for most common steels, requiring increased high-temperature mechanical strength as well as high creep resistance. The manufacture of these pipes normally includes welding operations, which must preserve the main characteristics of this type of steel. This justifies the concern of the researchers to ensure the best welding conditions so that the preservation of the properties of these steels becomes possible. The present work intends to depict the best results obtained varying the heat-treatment conditions applied to weldments made on heat-resistant steel P91. This steel usually takes the designation SA 213 T91 (seamless tube) or SA 335 P91 (seamless pipe), according to ASME II, as well as the designation X10CrMOVNb9-1 according to EN 10216-2. The purpose of this study is to compare the behavior of pipe welding under different post-welding heat-treatment (PWHT) conditions. One of them is performed with thermal cycles (preheating, post-heating, and the post-weld heat treatment) in agreement with most construction codes and standard rules. The second one is performed without any thermal cycle before and after welding. Both welds were made by the same process, TIG (Tungsten Inert Gas, or GTAW—Gas Tungsten Arc Welding) in the horizontal position (2G according to ASME IX) and the same welding parameters. In order to evaluate the results obtained in the welds, microstructure analyses, hardness measurements, bending tests, and tensile tests at room and high temperature (600 °C) have been performed. Other tests were also carried out according to the quality procedures, such as visual, penetrant dye, and X-ray tests. Regarding the different strategies used in the heat treatments, the best results have been obtained using a strategy similar to the one currently in use and recommended by construction codes and steel manufacturers but excluding the phases' transformation time, and it was possible to observe that the tensile strength is impaired by about 2% to 9% at room and elevated temperatures, respectively; the elongation is reduced by 39% at room temperature but keeps a good performance at elevated temperature; the hardness profile is very similar at both temperatures; the microstructure presented is compatible with the requirements; and no cracking trend has been reported. Thus, a new strategy for the welding heat treatment of grade 91 steels was drawn, saving energy and processing time.

Keywords: P91 steel; heat-resistant steels; welding; heat treatment; PWHT; welds characterization; microstructures; tensile strength; hardness; heat-treatment processing time; sustainability

---

1. Introduction

Grade 91 steel was developed based on steels that emerged in the 1960s with typically 12% Cr content. The development of grade 91 was mainly due to creep problems, as 12% Cr steels traditionally failed when exposed to prolonged creep conditions such as those in power plants where these steels were used, exposing them to high pressures and temperatures of around 565 °C. Thus, steels with increased creep resistance have emerged. Grade 91 was originally developed by Oak Ridge National Laboratory in Tennessee, USA, typically consisting of 9% Cr and 1% Mo, which were initially called P9 steel presented as its main focus use in power plants [1]. Subsequently, this steel was studied and its composition was evolved through the addition of other elements, such as vanadium and Nb, and with controlled N content, thus giving rise to 91 steel grade. This new grade of steel substituted the P22 steel grade and can assume various designations, unfolding under designations such as SA 213 P91 (seamless tube) and SA 335 P1 (seamless pipe) according to ASTM, or X10CrMoVNb9-1 according to BS EN 10216-2 [2]. The P91 steel grade also responded to the demand for the increased efficiency of power plants, which now need to operate at higher temperatures in order to release lower amounts of CO_2 for the same volume of energy generated. Indeed, the latest composition of the P91 steel grade allows continuous working at temperatures in the range of 600 °C without being affected by creep phenomena, even under elevated stress conditions. Since welding of the various components that form part of the power plants is required, P91 steel has been the subject of numerous studies, most of which are briefly described below. The excellent properties of P91 steel also come from the careful distribution of fine Nb and V carbonitrides, which have a microstructure that can be changed during the welding process [3].

Several researchers have devoted their attention to the characterization of P91 steel welds made using different processes. In a recent study, Vidyarthy and Dwivedi [4] compared the use of TIG and A-TIG (Activated Flux TIG) processes (GTAW) on P91 steel welds, investigating the influence of CeO_2 and MoO_3-based activating flux on some factors that strongly condition welding, such as such as heat input, weld bead geometry, and angular distortion during single-pass execution. The use of A-TIG aims to overcome the productivity limitations imposed by the conventional TIG process, which is essentially geared to small thicknesses. The thin activated flow layer used in the A-TIG process substantially improves the process productivity [5]. Welding beads performed by both processes were also investigated, analyzing the resulting microstructure, mechanical strength, microhardness, and impact strength (Charpy). It was observed that for the same set of parameters used in both welding processes, the A-TIG process promoted an increase in the heat input transmitted to the joint due to the action of the activating fluxes, which resulted in an increase of 200% in the joint penetration when CeO_2 flux is used, and 300% when MoO_3 flux is used, compared to the conventional TIG process. In the analysis of welded joints, other benefits were also observed, such as less angular distortion, which dropped from 1.96° in conventional TIG welding to 0.78° using CeO_2 flux or 0.12° when using MoO_3 flux. The microstructure was predominantly dominated by martensite in the welding zone, and coarse precipitates such as $M_{23}C_6$ carbides were also observed in the primary austenite grain boundaries. The ultimate tensile strength (UTS) in welded samples increased by 2% related to the parent metal, and the microhardness increased as well. Conversely, impact resistance decreased in the A-TIG process compared to the conventional TIG process. Dhandha and Bandheka [5] also studied the A-TIG process applied to P91 steel using as fluxes Fe_2O_3, ZnO, MnO_2, and CrO_3. These authors also confirmed that penetration is always improved at least 100% using these kind of activated fluxes in that process, and a decrease in bead width was observed as well, as usually required. The best results were achieved using ZnO activated flux. The surface appearance/morphology of the weldments was considered as very good.

Marzocca et al. [3] studied P91 steel welds performed by means of the flux cored arc welding (FCAW) process, using two different rutilic filler metals (E91T1 and E91T1-G) and 80% Ar/20% CO_2 shielding gas. The main focus of that work was to study the resulting microstructure of the welded zone, using five welding passes to fill up the chamfer previously prepared and a heat input energy of 1.5 kJ/mm. $M_{23}C_6$ carbides were found in all zones, i.e., the parent metal (PM), fused zone (FZ), and heat-affected zone (HAZ). VN (Vanadium-based) precipitates were also observed in all zones but with a decreased size and greater dispersion. However, NbCN was only found in the PM and HAZ. In addition, trying to overcome the lack of productivity characteristic of the conventional TIG process, Krishnan et al. [6] used the pulsed gas arc welding (GMAW-P) process using a cored wire filler material to weld 12 mm thick P91 steel sheets in a single pass. The authors reported that the best welding results were obtained using 75° bevel aperture and 1.38 kJ/mm heat input, which corresponds to a welding speed of 320 mm/min and a current intensity of 270 A. Very interesting properties of mechanical strength (UTS = 812–849 MPa) and the impact strength of the weld bead (104–127 J) have also been reported. The deposition rate with flux cored wire was increased by about 42% compared to the use of a common solid wire, considering the same set of parameters and welding conditions. Other authors using the same process reported a significant decrease in the defects generated during the welding process, namely spatter, welding porosity, and lack of fusion decreasing, as well the weld width [7,8]. Other processes such as shielded metal arc welding (SMAW), submerged arc welding (SAW), and flux cored arc welding (FCAW) have also been tested to maximize the welding efficiency of P91 steels, but non-metallic inclusions have been observed in the weld beads, loss of toughness, and excessively high oxygen content in the welds, considering the studies conducted during the 2000–2009 decade [9,10]. However, a further study developed by Arivazhagan and Kamaraj [11] in 2015 is in line with other works published more recently, which reported a very low amount of fine microinclusions less than 2 µm in size, allowing toughness values around 47 J. However, this value can be improved by 15%–25% using 100% Argon instead 80% Ar/20% CO_2. An increase in the post-welding heat-treatment (PWHT) duration at 760 °C from 2 to 5 h has been reported as the main factor behind the 30% to 50% toughness improvement in welded joints.

Shanmugarajan et al. [12] used an autogenous laser beam in P91 steel welding but reported the presence of the delta ferrite phase in the weld beads, which was attributed to the amount of heat given to the joint. Kundu et al. [13] used an electron beam to weld P91 steel, but too high residual stresses were observed in welding of thicker thicknesses. In order to increase the efficiency in the GTAW process, Pai et al. [14] reached an increased filler metal deposition rate using it in preheated conditions, but in contrast, they observed a significant decrease in the joint toughness, which was attributed to the excessive heat passed to the joint in the process.

One of the main problems that has been worrying researchers is the possible drop in toughness and creep behavior due to the welding process. This has motivated several studies. El-Dosoky et al. [15] investigated the behavior of welded joints subjected to prolonged exposure at 600 °C under 120 MPa and 70 MPa loads, verifying that the creep resistance is conditioned by the fine grain of the heat-affected zone (FGHAZ). It has also been reported that welded samples have a higher creep rate than the parent metal mainly in the tertiary region, and that creep begins earlier in welded samples relative to the parent metal for the same load applied at 600 °C. Hyde et al. [16] studied the crack growth in welded P91 steel samples by creep crack growth tests at 650 °C, using compact strain (CT) test specimens, comparing the results obtained experimentally with simulations performed by the finite element method (FEM). The authors also confirmed a good correlation between the creep crack growth rates in the P91 parent metal and the cross-weld specimens for a given contour-integral (C*) [17], in which the crack growth rate was 10 times higher in the cross-welded CT specimens than those of the parent metal. Using the same type of specimens, Kumar et al. [18] reported similar conclusions, adding the idea that the HAZ enables the faster generation and growth of cracks to the detriment of the parent material and melted zone, creating the best conditions for the deviation of crack paths from those zones to the HAZ. In addition, using CT specimens of P91 steel welded and non-welded, Venugopol et al. [19] reached similar conclusions, using the fracture mechanics parameter C*. The conclusion drawn by these researchers, which was later confirmed by other works already cited here, emphasized that the creep crack growth rate is higher at the HAZ, which is especially true when lower C* values are reported. A similar study was recently conducted by Baral et al. [20], using non-welded P91 steel samples and welded samples of the same material, in a temperature range between 600 and 650 °C, and loads between 50 and 180 MPa. The observations made by these authors allowed ratifying the previous opinion of other researchers, in which the influence of the $Cr_{23}C_6$ coarse grains in the HAZ intercritical zone is clear, significantly conditioning the creep resistance, and that the minimum creep rate clearly follows Norton's power law. Trying to minimize the HAZ width, Divya et al. [21] carried out a comparative study between the laser welding (LW) of P91 steel welding, which leaves a HAZ of about 1 mm, with shielded metal arc welding (SMAW), which leaves a HAZ of about of 2.5 mm. The study was conducted to study the failure under creep conditions in the HAZ, which is commonly known as Type IV cracking. It was found that the microstructural damage induced by the LW is lower than in the case of the SMAW process. Although the width of the HAZ was effectively smaller in the case of LW, all specimens failed in the intercritical region of the HAZ. This shows that there is no benefit brought by LW in this case, because no significant improvements are brought to the Type IV cracking resistance. Wang et al. [22] performed PWHT at different temperatures on P91 welds, noting that there is a transition from Type IV cracking to Type I cracking in the HAZ intercritical region when moving from 600 to 840 °C treatment temperatures. In fact, the authors emphasize the idea that it is impossible to eliminate the HAZ intercritical region by PWHT if the temperature at which this treatment is performed does not exceed the Ac_1 critical temperature of the parent metal. The intercritical structure formed on the basis of transformed austenite grains and untransformed ferrite grains in the fused zone may be the source of creep failure, i.e., Type IV cracking. These conclusions emphasize the importance of the temperature at which the PWHT must be conducted. Thus, PWHT conditions deserve special attention.

Sharma et al. [23] recently studied the effect of PWHT on welded P91 steel, reporting that the best conditions encountered to maximize the mechanical strength of welded pipe would be a 2 h treatment at 760 °C. In this study, the TIG process with heated wire was used, which allowed a smaller heat passage to the welded joint, minimizing the HAZ. In 2014, Venkata et al. [24] reported that the maximum temperature at which a PWHT should be performed is 770 °C and should always be lower than the austenite starting temperature (Ac_1). In a study conducted in 2012 by Paddea et al. [25], the highest residual stresses (600 MPa) have been reported to be located near the outer boundary of the HAZ and toward the weld root in both PWHT and as-welded samples. As a result of these residual stresses, premature Type IV creep failures were observed in these 9–12% Cr (P92) steel welds. However, after PWHT, the residual stresses dropped to values around 50 MPa in the vicinity of the HAZ. Regardless of PWHT, the region where the highest level of residual stresses was measured has always been HAZ's intercritical region, which is the most vulnerable to Type IV cracking phenomena. Pandey et al. [26] performed creep tests at a temperature of 620 °C and loaded in the range of 150 to 200 MPa in multi-pass welded samples in P91 steel both in an as-welded condition and subjected to a set of heat treatments after welding. This set of treatments consisted of keeping the samples at 760 °C for 2 h with subsequent air cooling, followed by a re-austenitization treatment at 1040 °C for 1 h, and then further tempering at 760 °C again for 2 h, with a new air cooling process. It was reported by the authors that this treatment substantially increased the creep life of the samples, especially for a 150 MPa load. Laves phases have also been reported, as well as the change in the preferred failure location which, when in the as-welded condition, was caused by Type IV cracking in the HAZ intercritical region, but when subjected to the latter treatment, the breaking zone happened in the base material because the treatment conveniently unifies the structure along the samples. The same results are also reported by some of the same authors in [27]. Similar experiments carried out by a similar team of authors but using multi-pass shielded arc welded metal (SMAW) in P91 steel butt joints on 18 mm thick plate samples, which showed that the hydrogen taken to the joint in the welding process (6.21 mL/100 g) gave rise to hydrogen embrittlement, which is a situation that cannot be overcome by the sequence of treatments performed (identical to that described above). However, without such a large content of hydrogen present, the treatment produces a clear uniformity of the microstructure in the welded samples [28].

Several studies focusing on welding P91 steel with other materials, i.e., dissimilar joint, including P92 steel [29], PM2000 steel [30], AISI 304 stainless steel [31], and IASI 316L stainless steel [32], have also been carried out. The results are promising and some of the problems reported above when using P91/P91 joints seem to tend to be softened, depending on the welding process and conditions used.

Studies in recent years reveal how important it is to know how to properly weld and treat P91 steel welds. Thus, this work aims to deepen the previous studies by performing different PWHT cycles to welded P91 steel samples, analyzing the resulting microstructure, mechanical strength, and hardness.

2. Materials and Methods

2.1. Materials

In order to carry out the experimental work, a pipe with Ø48.3 mm and a thickness of 5 mm in SA 355 P91 steel provided by Vallourec Brazil (Belo Horizonte, Brazil) was selected. Its chemical composition and main mechanical properties provided by the supplier can be seen in Tables 1 and 2, respectively. The length of each sample was 150 mm, and the preparation (chamfer) was made in just one tip of the pipe. The preparation details can be seen in Table 3. Twenty-five samples have been produced under each set of conditions, in order to provide at least five samples for each kind of destructive test. The results of the tests carried out in this work present the average value followed by the standard deviation.

Table 1. Chemical composition of the parent metal SA 355 P91.

C	Si	Mn	P	S	Al	Cr	Ni	Mo	V	Cu	W	Nb	N
0.100	0.290	0.500	0.017	0.001	0.009	8.600	0.170	0.980	0.230	0.120	0.030	0.080	0.052

Table 2. Main mechanical properties of the parent metal SA 355 P91 (Data from Vallourec company).

$Rp_{0.2}$ (20 °C)	Rm (20 °C)	Elong. (20 °C)	$Rp_{0.2}$ (600 °C)	Rm (600 °C)	Hardness
521 MPa	693 MPa	25.2%	320 MPa	365 MPa	221 HV_{30}

Table 3. Chemical composition of the filler metal ER90S-B9.

C	Si	Mn	P	S	Cr	Ni	Mo	V	Cu
0.095	0.235	0.545	0.007	0.003	8.980	0.545	0.910	0.210	0.125

As filler metal, wire with Ø2 mm with the reference ER90S-B9 was used, which was provided by Electro Portugal, Lda. (Porto, Portugal); its chemical composition can be observed in Table 3. The mechanical properties of this filler metal follow the EN10204 standard. This filler metal was selected because it is recommended by the ASME manufacturing code for P91 steel.

2.2. Methods

Since the main objective of this work is to analyze the influence of different heat treatment cycles on P91 steel welded joints, the welding conditions were kept constant for producing the samples. Initially, the pipes were cut in a circular sawing machine Kaltenbach KKS 450 S, and the chamfer was performed in a conventional lathe Knuth V-Turn Pro (KNUTH Werkzeugmaschinen GmbH, Wasbek, Germany). After that, the samples were properly cleaned, avoiding the presence of chips. The welding process used in this work was GTAW in the position of 2G using a Kemppi (Kemppi, Lahti, Finland) Master TIG MLS 4000 multi-process power source welding machine. To perform the welding, the samples were properly aligned, and thermocouples were attached to the surface close to the weld joint. Following Kobelko's recommendation for this type of material (between 0.8 and 1.6 kJ/mm), a thermal delivery of 1.33 kJ/mm was selected, which conditioned the welding parameters. Good engineering practice recommends that a current rating of about 45 A per each millimeter diameter of the filler material should be used. Thus, a current intensity of 95 A was selected, as the diameter of the wire used as the filler metal was ±2 mm. The main parameters used in the welding process are shown in Table 4.

Table 4. Main parameters used in the welding process.

Parameter	Value	Parameter	Value
Process	GTAW	Current/Polarity	DC (Direct polarity)
Shielding gas	100% Ar$^+$	Current intensity	95 A
Gas flow	14 L/min	Voltage	14 V
Welding position	2G	Traveling speed	42 mm/min
Chamfer type	V	Thermal input	1.33 kJ/mm
Chamfer angle	60°	Filler metal diameter	2 mm
Root opening	3 mm	Tungsten electrode	Ø2.4 mm (EWCE-2)
Root face	1 mm	Number of passes	2

The voltage was adjusted automatically, following the characteristic curve of the welding machine. After the welder obtained the current intensity, the voltage, and the range of thermal delivery, the traveling speed was adjusted, taking into account the welding position and the thickness of the pipe to be welded. For calculating the traveling speed, the following expression for the thermal delivery was used:

$$Q = k \times \left(\frac{U \times I}{v}\right) \times 10^{-3} \tag{1}$$

where Q is the thermal delivery (kJ/mm), k is the thermal efficiency factor, U is the arc voltage (V), I is the current intensity (A), and v is the traveling speed (mm/s). This is valid for the first pass, because it is well known that for the second pass, a higher level of current can be used (+10%), as well as a lower traveling speed (−20%), in order to properly fulfill the chamfer. These new parameters for the second pass will result in higher thermal delivery, although it is controlled in order to prevent the appearance of welding defects.

For preheating, post-heating, and PWHT processes, a Weldotherm (Weldotherm, Essen, Germany) VAS 82-12 Digit 1000 Heat Treating Unit provided with 12-channel capacity was used, with six channels that were capable of using electrical resistances up to 135 A and another six channels that were capable of using electrical resistances up to 90 A.

Different thermal cycles have been applied to the welded samples in order to study the mechanical behavior of the joints. In Figure 1, it is possible to observe the different cycles selected, as well as the code used in this work, making the correlation with the subsequent analyses easier. Table 5 presents the main levels of temperature used in each set of heat treatments. The main argumentation behind the selection of the different thermal cycles imposed to the samples used in this study can be found below:

- P_APP_T00: these samples were used in the basic condition, i.e., without any kind of pre- or post-treatment. Although it is well known that this condition does not present the best conditions to be used in practice, it can be useful to know what happens when the best practices are not applied and compare the results with the other sets of heat treatment applied to the other sets of samples;

- P_APP_T01: the procedure used in this set of samples followed the rules recommended by the construction codes, standards, and some important manufacturers. This procedure uses all the advantageous settings and is considered as ideal for this kind of base material. The complete procedure includes preheating, welding, post-heating, transformation time, and finally, PWHT. Its main disadvantages are the long processing time as well as the energy consumed; furthermore, it is not environmentally friendly or productive;

- P_APP_T02: this batch of samples has been produced to verify the influence of hydrogen (if inserted into the weld), and how it interferes with bonding, along with residual stresses in the weld and corrosion phenomena. In order to simulate this situation, preheating was used, but without post-heating treatment. At the end of the welding process, the samples have been protected with a ceramic fiber blanket, avoiding a fast cooling process until it reached room temperature. No phases' transformation time was used. The PWHT was applied just six months later;

- P_APP_T03: the purpose of the selection of these conditions is the same as that for P_APP_T02 but, in this case, a post-welding treatment was performed in order to eliminate any possible retained hydrogen. It is intended to compare this situation with the previous one, analyzing the effect of the post-welding treatment;

- P_APP-T08: the conditions used in this set of parameters are very similar with those of the P-APP_T01 set but, in this case, the time of transformation was changed. After cooling to room temperature, the samples have been subjected immediately to PWHT. The main purpose of this procedure is to save energy and time in the process.

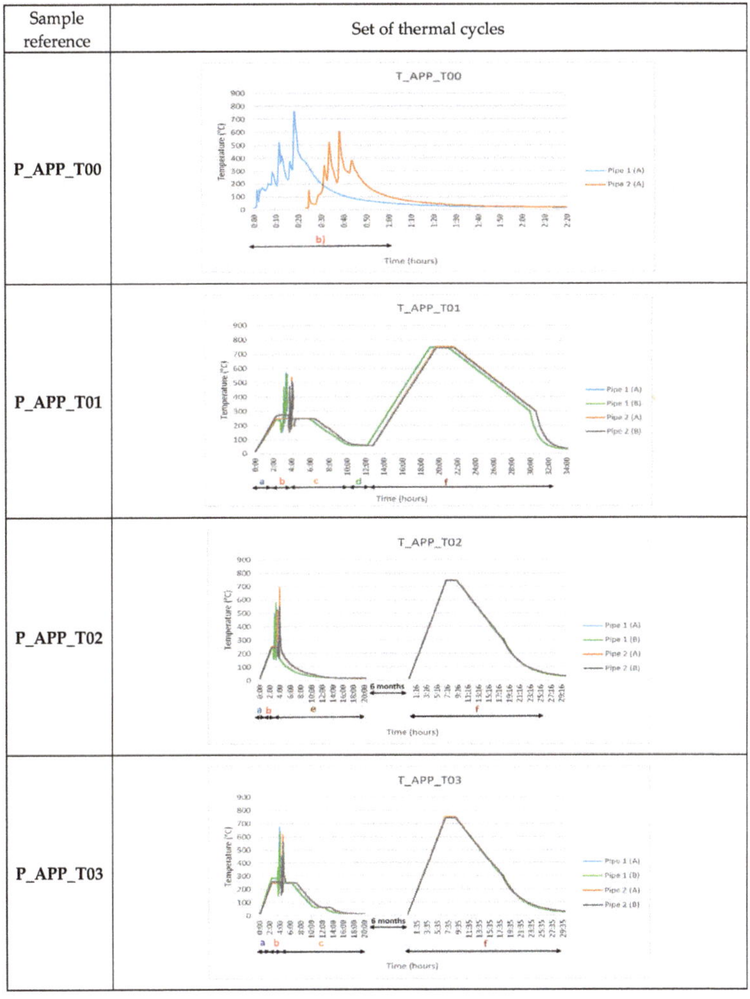

Figure 1. *Cont.*

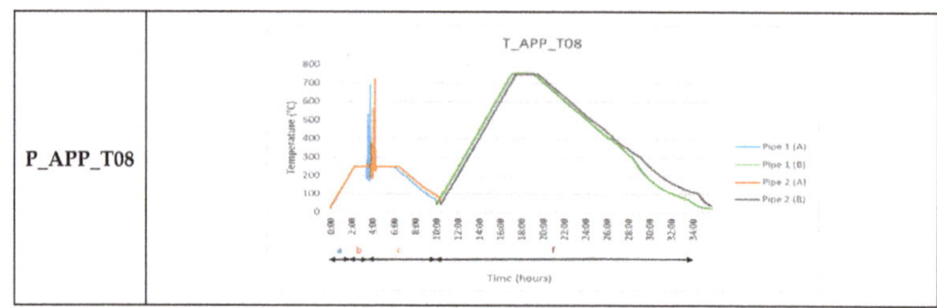

Figure 1. Different sets of thermal cycles used for each set of samples utilized in this work. Legend: (**a**) Pre-heating; (**b**) Welding; (**c**) Post-heating; (**d**) Transformation time; (**e**) Controlled cooling; (**f**) Post-welding heat-treatment (PWHT).

Table 5. Different thermal cycles applied to the P91 samples.

| Sample Code | Pre-Heating and Global PWHT | | | | | | Localized PWHT | | | | |
	Heating Rate (°C/h)	Temperature (°C)	Time After Welding (min)	Cooling Rate (°C/h)	Temperature (°C)	Time (min)	Heating Rate (°C/h)	Temperature (°C)	Time After Welding (min)	Cooling Rate (°C/h)	Temperature (°C)
P_APP_T00	As-received						As-welded				
P_APP_T01	100	250	120	50	60	120	100	750	120	50	300
P_APP_T02	100	250	Protected until reach room temperature				100	750	120	50	300
P_APP_T03	100	250	120	50	60	120	100	750	120	50	300
P_APP_T08	100	250	120	50	20	–	100	750	120	50	300

The purpose of this test is to verify to what extent hydrogen, if inserted into the weld, interferes with the joint together with the internal stress and corrosion in the weld. It is well known that the lack of tempering of this material after welding for extended periods can be detrimental. This is because, due to their late performance, the welds have a microstructure that is susceptible to internal stresses and often external stresses which, together with a certain level of hydrogen, can result in cold cracking, which is also known as hydrogen cracking. Apart from the presence of these agents, it is possible that welds may also be subject to corrosion, which may lead to stress cracking in addition to cold cracking. This was the main reason to include a waiting period of 6 months after the welding and before PWHT.

Figure 2 shows the setup used to perform the preheating process (a), as well as the welding process (b).

(**a**)　　　　　　　　　　　(**b**)

Figure 2. (**a**) Pipes preheating setup and (**b**) Pipe welding.

The characterization of the joints has been carried out following different techniques with a view to get important information about the properties achieved by the samples under each set of thermal cycles used. Thus, the characterization techniques used can be seen below, as well as the main reasons behind their selection.

- Liquid penetrant testing: used after the chamfer preparation;
- Visual testing: immediately before the welding process, checking the morphology of the welds and possible external defects;
- X-ray analysis: looking for internal defects;
- Hardness surface testing: checking for possible deviations regarding the expected values;
- Bending tests: trying to explore the appearance of eventual defects when loaded by bending;
- Tensile testing at 20 °C: checking the mechanical strength of the joint at room temperature;
- Tensile testing at 600 °C: checking the mechanical strength of the joint at high temperature;
- Hardness profile: analyzing the hardness along the cross-section of the welds;
- Chemical composition: trying to characterize the weld's composition;
- Electronic and optical microscopy: observing the resulting microstructure;
- Microhardness: trying to identify soft spots, where Type IV cracking usually appears.

After welding, the samples were immediately subjected to a first check in order to verify if the samples met the acceptance criteria defined in EN ISO 5817–Level C [33]. Afterwards, an additional visual inspection following ISO-17637:2003 standard [34] was performed, getting the corresponding approval. After that, samples were checked using liquid penetrant following the ISO 3452-1:2013 [35], looking for defects that had reached the surface. Since all the samples did not present surface signs of defects, they obtained the corresponding approval. Trying to detect internal defects, the samples also were subjected to X-ray inspection following the ISO 20769-2:2018 standard [36], using an ICM machine with 300 kVA, obtaining the corresponding approval as well. Indeed, some samples presented an excess of penetration, which is a defect that is acceptable, but there were no other defects such as root lack of penetration, porosities, or inclusions. The criteria used to approve the welds also followed ASME Code Sections I, V, VIII Div. 1 and 2, ASME B31.1 [37]. After these previous non-destructive tests, the samples were cut and machined in a vertical milling machine Baileigh VM-1054-3. A summary of these previous tests can be seen in Figure 3. After machining, it was necessary to smooth the weld until it is leveled with the base material, so that the entire test area was of the same thickness and section in order to remove the notch factor resulting from the border of the bead with the base material. This operation was performed for the tensile and bend samples.

In order to perform the other tests, different equipment was used, which is described as follows. Regarding the hardness assessment, three different tests were performed: hardness tests with portable equipment to carry out a first evaluation of the welded joints, a microhardness cross-section evaluation in order to assess the hardness reached in each zone of the joint, and finally another microhardness test allowing the detection of soft spots. Thus, the surface preparation was carried out using sandpaper provided with different grain sizes (80, 220, 310, 500, and 1000 mesh) until the Rz roughness parameter was lower than 5 μm. A Krautkramer–MIC 10 equipment provided with a MIC 205-A probe was used because it is portable and easy to handle, providing results accurate enough regarding the level required at this stage, which was just a general previous evaluation. The load used to perform these tests was 49 N (5 kgf) with a dwell time of 30 s, and a diamond Vickers indenter with 136° was used to perform the indentations. In order to obtain more accurate results, further microhardness tests were carried out using Shimadzu HSV-20 equipment, following the NP EN 1043-1 standard, using as the load 98.1 N (10 kgf) and a Vickers diamond indenter. However, to do that, it was necessary to improve the surface roughness through a new polishing process with diamond slurry of 1 μm over 10 min in order to decrease the Rz roughness to values under 2 μm. Two rows of indentations were produced as depicted in Figure 4. Finally, the opposite surface of the same samples was prepared following the same procedure as well, in order to identify soft spots close to the base material. Indeed, Newell [38] argues that these soft points are responsible for a decrease in creep resistance by about 20% compared to an unaffected base material and are formed mainly in the fine grain region of the HAZ. Using the same microhardness equipment and indenter, these tests have been carried out using just 1 kgf (HV_1) and a dwell time of 30 s, still following the ISO 6507:2018 standard [39]. Three rows of indentations were produced, one close to the weld root, another in the middle of the cross-section, and the last one close to the surface of the base material; all of the indentations were distanced 0.5 mm each other, which is the minimum distance recommended by the standard, avoiding the influence on the results among indentations. This value was selected based on a good compromise between the lowest value recommended for this effect and the small grain size usually observed close to the HAZ. In this case, just one sample was analyzed regarding each set of welding and heat-treatment conditions.

(a)

(b)

(c)

(d)

Figure 3. Different non-destructive test stages regarding the samples approval: (a) Liquid penetrant before welding; (b) Dimensional control; (c) Liquid penetrant after welding and cutting; (d) X-ray analysis.

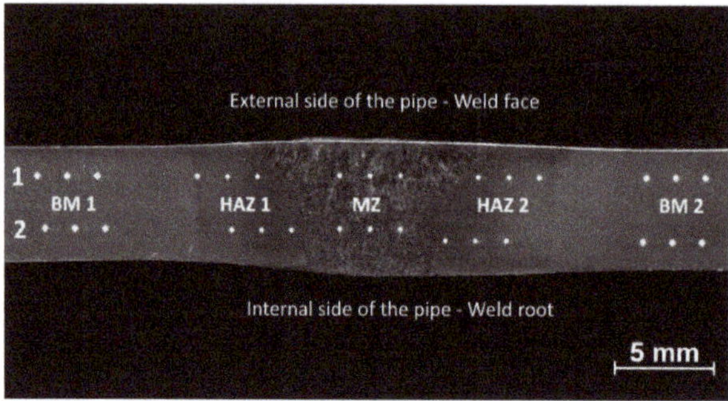

Figure 4. Example on how the microhardness indentations were produced across the cross-section.

In order to analyze the chemical composition of the weld, an optical emission spectrometer Spectro, model Spectrolab M8 was used. This test was performed trying to understand the composition of the weld and corresponding dilution. The values achieved are based on six analyses in different samples, and the results presented are average values with standard deviations lower than 5%.

The micrographic analysis aims to analyze the microstructure present in the samples, thus allowing verification of the grain size and its distribution. In order to perform these analyses, a Carl Zeiss (Carl Zeiss, Oberkochen, Germany) optical microscope model Axioskop 2 Mat was utilized, using different lens and magnifications. This analysis was also an important tool to verify the influence of thermal cycles on the steel microstructure, both in melted zone (MZ) and the HAZ. This optical microscopy technique is ideal for microstructure analysis, grain size analysis, particle analysis, and the identification of voids or cracks. For this purpose, the samples were prepared by sanding and polishing with 80 mesh sandpapers, followed by 220, 310, 500, and finally 1000 mesh. Then, the samples were polished sequentially with 3 µm and 1 µm diamond slurry, after which the samples were etched using a reagent called Vilella, which consists of hydrochloric acid, 5 mL; picric acid, 2 g; and ethyl or methyl alcohol, 100 mL. The etching time was 20 s. The test was performed according to BS EN 1321:1997—Destructive test on welds in metallic materials—Macroscopic and microscopic examination of welds [40].

Tensile tests allow a vast characterization of the materials, both in terms of mechanical strength and ductility. Moreover, this type of test can be performed at different temperatures, simulating hard work conditions that can be applied to the materials and joints in service. In this case, tensile tests were performed at room temperature, 20 °C, and at high temperature, 600 °C, from which yield strength, rupture and elongation data were collected. In order to increase the confidence in the values collected, three tests were carried out for each condition considered in this work. In order to perform the tensile tests, as Instron universal testing machine model 4208 was used, which was provided with a load cell of 300 kN. Tensile test specimens were taken from the weld cross-section and prepared according to ASME IX: 2015–QW 150 [41]. Regarding the tensile tests carried out at elevated temperature (600 °C), they were performed in a similar universal testing machine (Instron 8562) provided with a heated camera where the samples are kept at constant temperature during the tensile tests. The load cell used is a 100 kN cell. The samples were prepared following the ISO 6892-2:2018 standard [42].

Bending tests are usual in welded samples because they allow the assessment of the samples' behavior under very demanding work circumstances. In this work, bending tests were performed using a hydraulic CIATA press, model P-115/HP provided with a maximum compression capacity of 147 kN (15 tons). Samples were tested according to ASME IX–QW163 [43], with a bending angle of 180°, a bearing distance of 36 mm, and a spindle diameter of 24 mm. Samples were used in which the compressive effort was performed on the welding face and others were performed on the welding root to cover all situations. After the tests, the samples were analyzed using the penetrant liquid technique, which was previously described in this work.

3. Results and Discussion

3.1. Non-Destructive Test Results

Some of the initially performed tests were merely qualitative, thus providing information about whether the samples were ready to pursue the analysis or not. These tests were already described before in this work.

3.2. Hardness and Microhardness Analyses

The hardness tests performed allowed assessing the hardness profile across the joint, from the base material 1 (BM1) to the base material 2 (BM2), passing by the thermal-affected zones 1 and 2 (HAZ1 and HAZ2) and the melted zone (MZ). The microhardness values can be observed in Table 6, and the profiles corresponding to the different samples can be seen in Figure 5, resulting from the average values of five samples measured in the same places (15 indentations per sample), taking into account the central area of each sample. The calculated standard deviation can be observed in Table 6 as well. The letter B included in the label of samples P_APP_T02 and P_APP_T03 in Figure 5 represents the values obtained before the PWHT was carried out six months after the welding and corresponding heat treatments performed in the first stage of the samples' preparation. As expected, the hardness values before PWHT are higher in the HAZ and MZ, because the martensite is not tempered yet, being extremely hard in that state. After PWHT, the values achieved are within the expected range. Since the sample P_APP_T00 was not subjected to heat treatments, its hardness is extremely high in the MZ, as well as in the HAZ. Thus, sample P_APP_T00 does not fulfill the properties required by the usual applications of this kind of material, and the procedure used in this sample can be considered inadequate. It is worth noticing that the remaining samples show hardness values within the usual range for this material; thus, at this stage, excluding the P_APP_T00 sample, the other samples can be considered for the next tests. Performing this test only lets us validate in a first moment, in the shop floor, the ability of the work pieces to advance to the next stages, because the accurateness of the hardness measuring equipment is ±40 HV. Thus, in order to evaluate the surface hardness of the samples with adequate accuracy, microhardness tests were performed using the microhardness measuring equipment previously described. The microhardness profile can be seen in Figure 6, showing clearly that sample P-APP_T00 presents a different hardness behavior resulting from the lack of heat treatments. In this case, the microhardness measurements were performed only after the PWHT treatment; thus, there are no curves corresponding to this situation, as shown in Figure 5. The values reported in the curves can be considered common for this kind of material and joint. The increased hardness reported in the MZ and HAZ is typical (250–280 HV_{10}) as well in these conditions, and the values registered are within the range of acceptable values (200–275 HV_{10}, but they can reach 300 HV_{10} without special concerns). Excluding the sample P_APP_T00, which presents a singular behavior due to the absence of heat treatments before the PWHT applied to this sample, the samples were mainly constituted of non-tempered martensite, which was hard and brittle. The pattern followed by the remaining samples is very similar among them, with microhardness values of around 230 HV_{10} in the base metal and slighting increasing across the HAZ, reaching steady values between 250 and 280 HV_{10} in the MZ. Obviously, sample P_APP_T00 is not acceptable, reaching microhardness values out of the range of acceptable values. This corroborates the unconditional need for applying heat treatments to this base material when welding is included in the manufacturing processes. According to ISO 18265 (Metallic materials—Conversion of hardness values), it is possible to estimate the ultimate strength through hardness values [34]. This is because both hardness and ultimate strength are indicators of the mechanical resistance of a metal to plastic deformation, and they are approximately proportional. However, this proportionality is not valid for all metals. For this case, it may be pointed out that the breaking stress in MPa is 3.13 times higher than the hardness in HV.

Table 6. Surface microhardness of the different samples (HV_{10}). HAZ: heat-affected zone, MZ: melted zone.

Samples Ref.	BM1	HAZ1	MZ	HAZ2	BM2
			HV_{10}		
P_APP_T00	207 ± 5	271 ± 12	356 ± 16	292 ± 8	211 ± 12
P_APP_T01	199 ± 17	198 ± 20	245 ± 12	208 ± 16	196 ± 14
P_APP_T02	212 ± 11	268 ± 14	327 ± 9	278 ± 13	214 ± 9
P_APP_T03	204 ± 8	287 ± 12	310 ± 11	288 ± 8	206 ± 10
P_APP_T08	195 ± 6	195 ± 7	212 ± 9	191 ± 5	193 ± 3

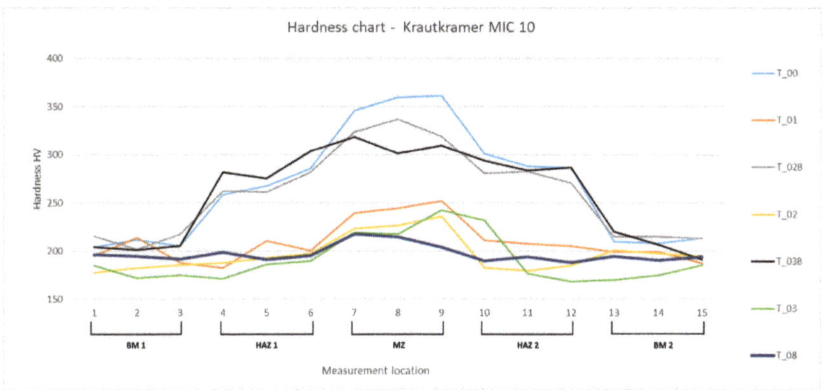

Figure 5. Hardness profile of the different samples across the welded and neighboring areas. The label B in some graphs represents the values obtained before the PWHT was carried out 6 months after the welding.

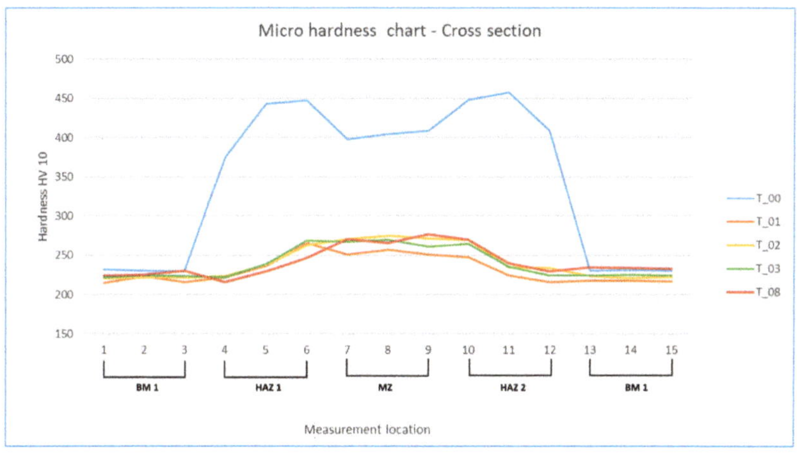

Figure 6. Microhardness profile of the different samples across the welded and neighboring areas.

The microhardness tests were also performed having as the main goal identifying soft spots in the cross-section, i.e., very small areas where the hardness is lower than in the base metal or in the fine grain region of the HAZ, which is also known as the intercritical region. The summary of the results by the crossed region can be seen in Table 7. The intercritical region is located in the connection zone of the HAZ, which is very close to the base material. This is where the so-called Type IV crack usually appears, which often occurs both at the manufacturing stage and during the life of the component.

Table 7. Cross-section microhardness of the different samples (HV_1).

Samples Ref.	BM (Left)	HAZ (Left)	MZ	HAZ (Right)	BM (Right)
			HV_{10}		
P_APP_T00	231 ± 2	422 ± 45	404 ± 12	439 ± 23	231 ± 4
P_APP_T01	219 ± 6	242 ± 19	253 ± 6	230 ± 15	218 ± 2
P_APP_T02	222 ± 2	242 ± 19	272 ± 6	247 ± 17	223 ± 2
P_APP_T03	224 ± 2	243 ± 21	266 ± 5	242 ± 18	225 ± 2
P_APP_T08	227 ± 3	231 ± 13	271 ± 6	247 ± 18	234 ± 2

Regarding the soft spot analysis, no special attention was paid to the P_APP_T00 samples, due to its non-heat-treated condition. Taking into attention the P_APP_T01 samples, lower microhardness values were identified in the range of 7.5 to 10 mm from the center line of the weld. Indeed, this is the most critical area in the weld cross-section because this is a HAZ that is close to the base material, which is an area that is well known as the region where the grain size is finer and the hardness can drop about 20 HV of the substrate's usual level. Regarding the samples P_APP_T01, some microhardness measurements provided values under 200 HV (187 HV, while the base material presents 220–230 HV) in the intercritical area, indicating that there are soft spots that can induce Type IV cracks. The samples P_APP_T02 and P_APP_T03 did not present values under 200 HV. Indeed, the lowest value obtained in both samples was 204 HV, showing that these samples do not present soft spots and are not prone to develop Type IV cracks in the intercritical area. The sample P_APP_T08 presented just one (one in 14) value below 200 HV (170 HV), which was located into the HAZ and close to the surface. Since all the other results were close to 220 HV in this sample, the result was not valued also because in the root and middle indentations rows, the same effect was not felt. Thus, it was considered that sample P_APP_T08 presents good conditions to avoid the development of Type IV cracks, presenting less concerns than sample P_APP_T01.

3.3. Welds Chemical Analysis

In order to check the final composition of the weld, optical spectroscopy was used to obtain the weld composition. However, as the chemical composition of the base material and filler metal are very similar, it would be expected that the weld follows a chemical composition very similar to the base material and filler metal. The results can be seen in Table 8, where the compositions related to Pipe 1, Pipe 2, and the filler metal have been taken from the information provided by the suppliers and the composition related to the weld face and root face were taken from analyses performed by optical spectroscopy. As can be observed, the chemical composition of the weld, both in the face and in the root, are very similar to the compositions provided by the suppliers to the base materials and filler metal. Thus, the dilution in the welding process will not significantly affect the weld composition.

Table 8. Chemical analysis of the materials used in the welding and the weld composition (wt %).

Material	Reference	C	Si	Mn	P	S	Al	Cr	Ni	Mo	V
Pipe 1	SA 213 T91	0.091	0.35	0.52	0.016	0.0011	0.006	8.71	0.28	0.95	0.21
Pipe 2	SA 213 T91	0.102	0.29	0.50	0.017	0.0012	0.009	8.60	0.17	0.98	0.23
Filler metal	ER 90S B9	0.092	0.22	0.54	0.008	0.0020	-	8.90	0.46	0.90	0.22
Weld face	-	0.095	0.29	0.55	0.007	0.0006	0.005	8.88	0.48	0.89	0.22
Weld root	-	0.091	0.30	0.53	0.010	0.0008	0.005	8.90	0.42	0.89	0.22

3.4. Microstructure Analysis

The microstructure analysis gives relevant information about the behavior of the different zones of the welds' cross-section. Thus, each sample was carefully analyzed, allowing a deep understanding of the phenomena produced by the different thermal cycles applied to the samples. A vast number of micrographs were collected, sorted, and analyzed, and the selection of the most representative micrographs regarding each sample and area of analysis are presented in Figure 7.

All the micrographs present microstructures consisting of tempered martensite, which gives the joint very good properties, excluding the P_APP_T00 sample. This presents a microstructure with martensite, but it is not tempered, because the PWHT was not performed. Samples P_APP_T01, P_APP_T02, and P_APP_T03 present very similar microstructures with identical grain sizes. Precipitates are found along the grain boundaries. The typical grain size in the HAZ is about 20–30 μm, and in the MZ it appears larger, with grains of 60–70 μm. In sample P_APP_T01, the grains appear more defined, and the martensite areas appear tempered as well. The sample P_APP_T08 presents a regular HAZ; however, the MZ presents an atypical structure relative to the other samples used in this work. This reveals that the creep behavior can limit the application of this kind of joint in classic applications of this kind of material, but this expectation needs to be confirmed.

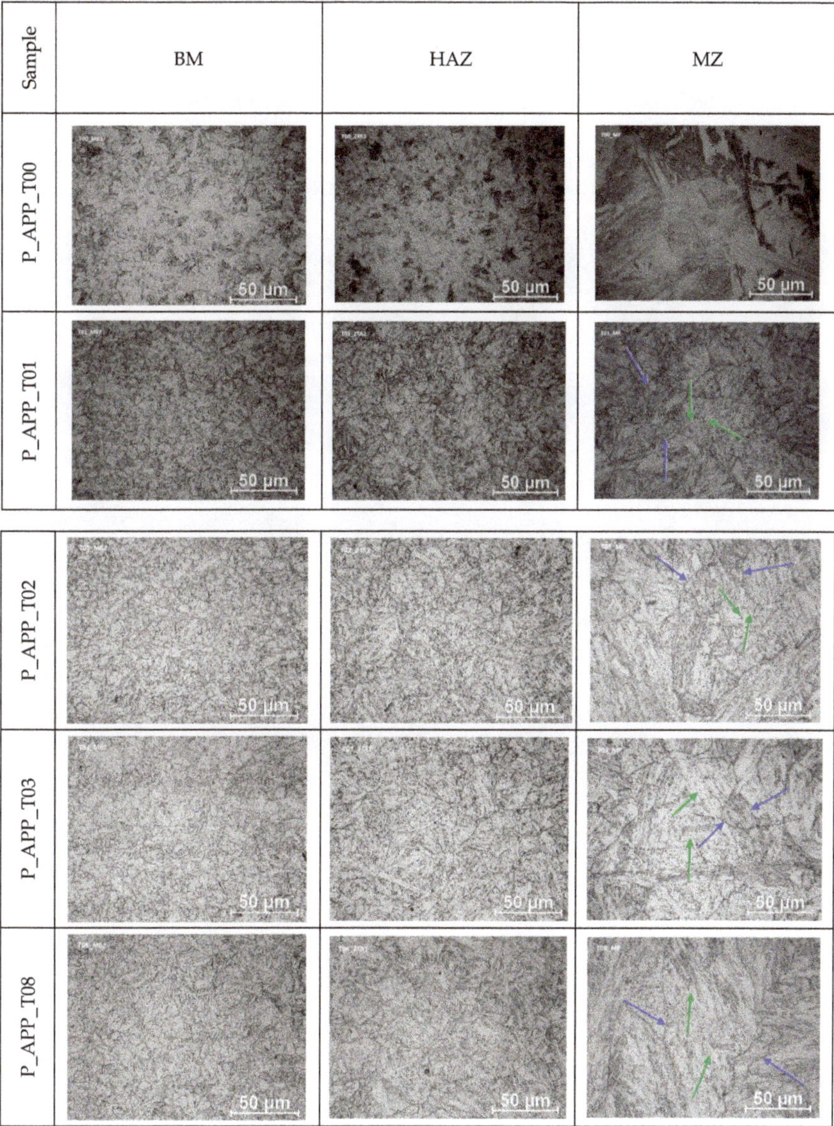

Figure 7. Micrographs of the different cross-section areas for samples subjected to different thermal cycles. Blue arrows highlight grain boundaries and green arrows highlight precipitates.

3.5. Tensile Tests at Room and High Temperature

The ASME IX: 2013–QW 150 [41] code refers to 585 MPa as the minimum yield strength to be achieved in welds of this kind of material; this is the value that is also required as a minimum for the base material. Regarding yield strength and elongation, the code does not specify any minimum value. ASME code B31.3—Process Piping [37] refers to minimum yield stress 413 MPa and ultimate tensile strength 586 MPa. Analyzing the rupture results, it was found that it always occurred in the base material at a considerable distance from the welding area, as can be seen in Figure 8. Moreover, all ruptures presented a ductile look, even when the elongation was relatively low. The values obtained in the tensile tests can be observed in Table 9, namely the yield strength, ultimate tensile strength, and elongation at room and high temperature. The results can be observed in Figures 9 and 10. The highest values are presented by sample P_APP_T00, because they were not subjected to heat treatments and PWHT. Thus, the yield strength and ultimate strength are the highest values, but the elongation presents the lowest values, although these were not significantly different from samples P_APP_T02, P_APP_T03, and P_APP_T08. Sample P_APP_T03 presents the lowest ultimate strength value at room temperature, and P_APP_T02 presents the lowest ultimate strength value at high temperature.

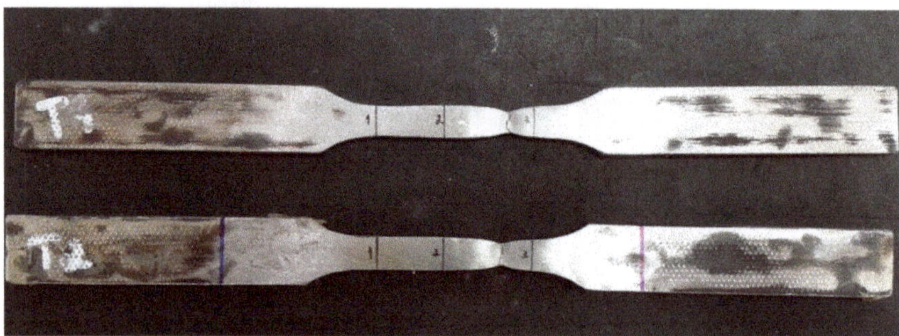

Figure 8. Aspect of the samples used in tensile tests after rupture. Line 2 indicates the welding zone.

Table 9. Yield strength, ultimate tensile strength, and elongation at room and high temperature.

Samples	Room Temperature (20 °C)			High Temperature (600 °C)		
	$Rp_{0.2}$	Rm	Elong.	$Rp_{0.2}$	Rm	Elong.
	MPa	MPa	%	MPa	MPa	%
P_APP_T00	559 ± 16	696 ± 15	14.5 ± 0.4	350 ± 12	405 ± 9	14.4 ± 0.3
P_APP_T01	526 ± 7	696 ± 3	24.8 ± 0.9	331 ± 6	377 ± 8	16.3 ± 0.8
P_APP_T02	511 ± 2	666 ± 1	16.5 ± 1.2	235 ± 3	239 ± 3	15.8 ± 0.9
P_APP_T03	509 ± 2	548 ± 1	17.6 ± 0.3	287 ± 2	345 ± 3	18.3 ± 0.4
P_APP_T08	515 ± 5	682 ± 8	15.1 ± 0.9	301 ± 5	362 ± 6	16.7 ± 0.8

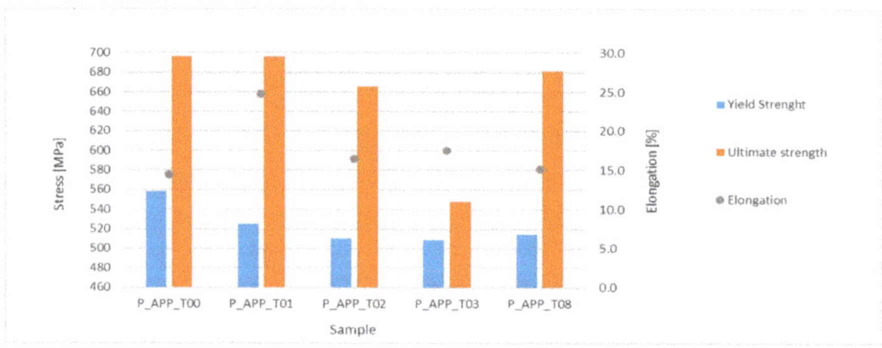

Figure 9. Results of the tensile tests performed at room temperature.

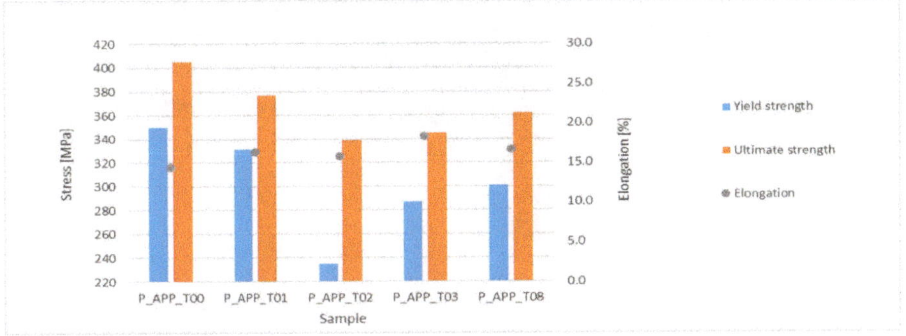

Figure 10. Results of the tensile tests performed at elevated temperature (600 °C).

Regarding the results obtained at elevated temperature, ASME B31.3—Process Piping code refers to the minimum yield strength of 71 MPa at a temperature of 600 °C [37]. In some codes, this value may be higher, requiring a minimum of 215 MPa. In any case, the results presented are above these minimum values for both codes cited. As in the tests carried out at room temperature, the samples also presented a ductile rupture clearly out of the MZ or HAZ.

Regarding Table 9, and excluding the sample P_APP_T00 which presents high risks of cracking in service due to its structure in the HAZ, and considering the sample P_APP_T01 as the model usually recommended by the main steel manufacturers and construction codes, the sample that presents the closest yield and ultimate strength both at room and elevated temperatures is sample P_APP_T08. The main drawback presented by this sample is the low elongation which, being higher than that in the samples P_APP_T02 and P_APP_T03, is relatively lower than sample P_APP_T01. Indeed, P_APP_T01 is the sample that presents a better relation between mechanical strength and ductility, mainly at room temperature. However, at elevated temperature, the elongation decreases substantially and reaches similar values to the other samples. Sample P_APP_T03 is the one that presents the best elongation at elevated temperature, but its mechanical resistance decreases more than that of the P_APP_T01 and P_APP_T08 samples. At this stage, samples P_APP_T02 and P_APP_T03 can be discarded, because they present a considerable drop in the mechanical resistance relative to the P_APP_T01 sample.

3.6. Bending Tests

After the bending tests have been completed, the samples were initially observed with the naked eye and subsequently subjected to penetrant liquid tests (Figure 11) to identify the possible existence of microcracks. In fact, no cracks or microcracks were found in any of the samples in either MZ, HAZ, or BM. Penetrant testing is not usual in industrial terms to verify this kind of joint, as cracks usually develop quite rapidly and are perfectly noticable to the naked eye. However, the penetrant liquid technique was used to confirm the absence of any kind of cracking in this work. All specimens described a perfect curvature, except for the sample P_APP_T00. These samples showed an atypical curvature, where it can be seen that the MZ offered such high bending resistance that the curvature took place in the BM. This effect is clearly depicted in Figure 12. Anyway, this sample does not present any crack or discontinuity. Given that the P_APP_T00 sample presented high mechanical strength and low elongation, and that its microstructure consisted essentially of non-tempered martensite, it would be expected that this samples would exhibit cracks when subjected to more intense deformation stresses. However, this was not the case. Thus, as a summary of these tests, it can be stated that all the samples are approved, but the P_APP_T00 sample shows greater resistance to deformation in the weld bead zone, transfering the deformation to an adjacent zone (HAZ and BM).

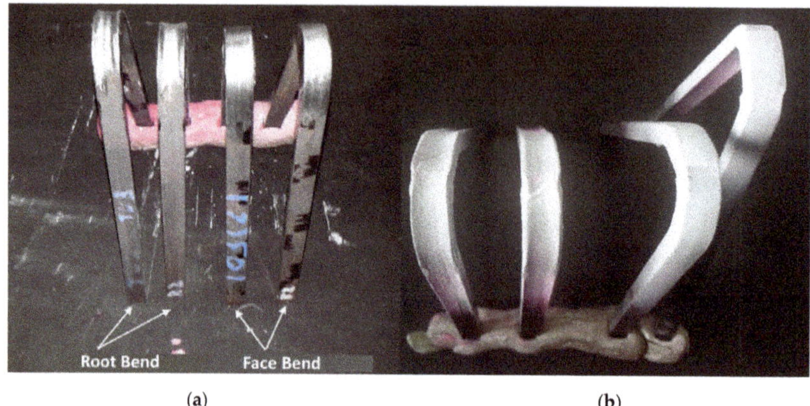

(a)   (b)

Figure 11. Samples subjected to bending tests: (**a**) samples bent in face and root surfaces, and (**b**) liquid penetrant technique applied to bent samples.

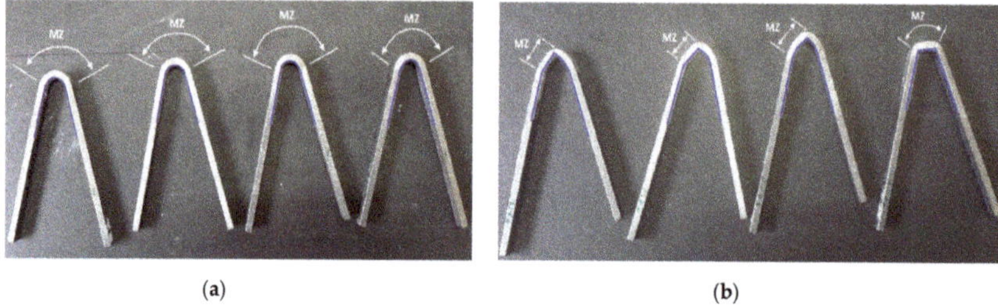

(a)   (b)

Figure 12. Bending tests corresponding to samples (**a**) P_APP_T08 and (**b**) P_APP_T00.

3.7. Discussion about the Results and Corresponding Thermal Cycles

After collecting all the results and analyzing them individually, it is now necessary to gather this information and relate it to the thermal cycles initially performed.

It remains clear that the welding process generates martensite in the weld and HAZ, increasing the mechanical resistance, but limiting the ductility, as can be seen in the tensile tests and optical microscopic analysis carried out on sample P_APP_T00. This is valid for both room temperature and elevated temperature, which creates limitations regarding the use of this kind of joint in their traditional applications, heat exchangers, due to creep phenomenon. The microhardness measured clearly confirms these results, showing an increase in hardness in the MZ and HAZ of these samples. However, regarding the bending tests, no cracks have been reported, but the high hardness presented in the weld moves the bending center from the weld to the base material. Thus, it is clear that martensite needs to be tempered after welding, and some heat treatments can help to decrease the stress accumulated in and close to the weld, improving the overall mechanical behavior of the joints.

Sample P_APP_T01 was prepared following the codes' recommendations, including preheating, post-heating, and PWHT. The procedure is time consuming and costly, but the results are excellent. The mechanical resistance remains within the range of values taken as acceptable for this kind of joint, and the ductility is not severely compromised, remaining as well in an acceptable value, although it is more affected at elevated temperatures than at room temperature. The microstructure shows the presence of tempered martensite, and the hardness values are lower than in the case of the P_APP_T00 samples. As expected, no cracks have been developed during or after the bending tests. Thus, the main disadvantage of this procedure is the time spent to perform it and the energy consumed, making this process less friendly to the environment.

Sample P_APP_T02 was prepared saving the post-heating treatment and keeping the pre-heating and PWHT. The phases' transformation time included in the P_APP_T01 sample was suppressed in this and the following thermal cycles. This thermal cycle gave interesting mechanical resistance values at room temperature, but it did harm the ductility a little. However, analyzing the mechanical resistance at elevated temperature, a severe decrease in the yield strength and ultimate strength can be observed, presenting as well the lowest elongation value. Thus, suppressing the post-heat-treatment, the properties at elevated temperature are severely affected. The grain seems to not be significantly affected by the thermal cycles in the HAZ, and regarding the MZ, it seems to be less affected than the other samples. Therefore, the absence of post-heat-treatment affects the properties at elevated temperature, which is drastic for this kind of material regarding the typical applications referred above. It is worth noting that no cold cracking effect was felt in the samples due to the waiting time imposed between the welding process and PWHT.

The P_APP_T03 sample followed the same principles of the P_APP_T02 sample, but in this case, a post-heat-treatment was included after the welding process and before the six-month waiting time. These samples presented the lowest ultimate strength values at room temperature, although the values relative to tensile at elevated temperature were improved relative to the similar P_APP_T02 sample. This is the main concern of these samples. It is also worth noting that these samples present a very good ductility behavior at room temperature and the best ductility at elevated temperature. At this point, it needs to be referred that the only difference between the procedure applied to these samples and the procedure used for sample P_APP_T01 is the inclusion of the six-month waiting time after the post-heat-treatment and before the PWHT. However, the inclusion of this waiting time was detrimental to the mechanical properties at room temperature, although other properties were improved, such as the ductility.

Regarding the P_APP_T08 sample, the thermal cycles used are very similar to the thermal cycles applied to P_APP_T01, having as the only difference suppression of the phases' transformation time. This suppression is important in terms of productivity and energy saving. The results obtained based on these samples allow concluding that there is a good balance between the mechanical resistance obtained and the ductility presented by these samples; both for room temperature and elevated temperature, the grain is a little more acicular than in the other samples, which is not translated in terms of hardness, which presents values extremely close to the average results obtained for samples P_APP_T01, P_APP_T02, and P_APP_T03. Moreover, regarding the bending tests, no cracking or its initiation has been detected. Thus, although these samples show a slight decrease in mechanical resistance and ductility, this is a good alternative to the thermal cycle P_APP_T01.

In summary, Table 10 gives a brief overview of the thermal cycles applied to each set of samples and the corresponding qualitative results, following the criteria pointed out at the bottom of Table 11. It is worth noting that samples P_APP_T08 and P_APP_T01 present slight differences in the procedure, as well as in the final results, not compromising the behavior of welds produced in this material in its main applications.

Regarding the main parameters provided by tensile tests, and in order to make a decision regarding the selection of the best strategy based on quantitative data, Table 11 was elaborated classifying from 1 to 4 (1 = worst value and 4 = best value) the different quantitative data obtained from that test. As can be seen in Table 11, after the recommended procedure corresponding to the P_APP_T01 sample, the best alternative is the P_APP_T08 strategy, with a sum of 6.1 points. This confirms the analysis previously performed.

Table 10. Summary of the thermal cycles applied and corresponding results by test type.

Samples	Thermal Cycles				Performed Tests					
	Preheating	Post-Heating	Time Transf.	PWHT	Hardness	Macro	Micro	Tensile RT	Tensile HT	Bend
T00	No	No	No	No	Not Acceptable	Acceptable	Not Acceptable	Good	Good	Good
T01	Yes	Yes	Yes	Yes	Very Good	Very Good	Very Good	Very Good	Good	Very Good
T02	Yes	No [1]	No	Yes [2]	Good	Very Good	Good	Acceptable	Not Acceptable	Very Good
T03	Yes	Yes	No	Yes [2]	Good	Very Good	Good	Not Acceptable	Good	Very Good
T08	Yes	Yes	No	Yes	Good	Acceptable	Acceptable	Good	Good	Very Good

(1) Protected cooling until reaching room temperature; (2) PWHT performed 6 months after welding; Rating Scale: Not Acceptable–Acceptable–Good–Very Good.

Table 11. Summary of the thermal cycles applied and corresponding results by test type.

Samples	Room Temperature (20 °C)							High Temperature (600 °C)							Pts Total (Room + High)
	Rp0.2 (MPa)		Rm (MPa)		A (%)		Pts Total	Rp0.2 (MPa)		Rm (MPa)		A (%)		Pts Total	
	Value	Pts	Value	Pts	Value	Pts		Value	Pts	Value	Pts	Value	Pts		
P_APP_T01	526	4	696	4	24.8	4	4.00	331	3	377	4	16.3	3	3.60	7.60
P_APP_T02	511	2	666	2	16.5	2	2.00	235	1	339	1	15.8	2	1.15	3.15
P_APP_T03	509	1	548	1	17.6	3	1.30	287	2	345	2	14.6	1	1.85	3.15
P_APP_T08	515	3	682	3	15.1	1	2.70	301	4	362	3	16.7	4	3.40	6.10

4. Conclusions

Grade 91 steel is usually applied in situations where the temperature and pressure are very demanding. The weldability of these steels presents well-known challenges, which are usually overcome using complex thermal cycles before and after the welding process. This work intended to test other thermal cycles with a view to shorten the treatment time, making the process more productive and more environmentally friendly. Thus, five different thermal cycles were drawn and tested: (a) welding without any pretreatment, post-treatment, or PWHT; (b) using the thermal cycle recommended by the construction codes and steel manufacturers; (c) using a pretreatment and the PWHT, using as well a six-month waiting period between welding and PWHT without phases' transformation time, in order to study the possible development of cold cracking; (d) using the previous strategy but including the post-welding treatment and; finally, (e) using the same strategy as that in (b) but excluding the phases' transformation time.

The hardness and microstructure assessed in P_APP_T00 indicate that this strategy cannot be used, because if the mechanical strength is even higher than that in the P_APP_T01 case, the risks of further cracking are higher, making this option not safe. The non-tempered martensite in the MZ and HAZ induces a clear hardness increase, which can be capable of inducing further cracks in service. Thus, this option was discarded due to these facts.

The last strategy (P_APP_T08) for thermal cycles around the welding process produced closer results in terms of mechanical properties at both room and elevated temperatures, comparing with the reference: P_APP_T01. A slight reduction in yield strength (2% at room temperature and 3.9% at elevated temperature) and ultimate strength is reported (2% at room temperature and 9% at elevated temperature), as well as a reduction of about 39% in the elongation at room temperature. However, at elevated temperature, the elongation is even a little bit higher than in P_APP_T01 conditions. Since this steel is usually used in high-temperature applications, this strategy seems to be the best among the other strategies tested in this work.

Thus, the main contribution brought by this work can be stated as the development of a new heat-treatment strategy to be applied in welding of grade 91 steels that is less time consuming and more environmentally friendly, providing very close results compared to the solution currently recommended by construction codes and steel manufacturers.

Author Contributions: Conceptualization, A.P.P. and F.J.G.S.; methodology, A.P.P. and F.J.G.S.; validation, F.J.G.S., O.C.P. and A.B.P.; formal analysis, A.P.P.; investigation, A.P.P.; resources, A.P.P.; data curation, A.P.P. and F.J.G.S.; writing—original draft preparation, F.J.G.S.; writing—review and editing, A.P.P., O.C.P. and A.B.P.; supervision, F.J.G.S., O.C.P. and A.B.P. All authors have read and agreed to the published version of the manuscript.

Funding: This research received no external funding.

Acknowledgments: The authors would like to thanks to ARSOPI company, which provided the best conditions to carry out this work.

Conflicts of Interest: The authors declare no conflict of interest.

References

1. Pandey, C.; Mahapatra, M.M.; Kumar, P.; Saini, N. Some studies on P91steel and their weldments. *J. Alloys Compd.* **2018**, *743*, 332–364. [CrossRef]
2. Standard, B. *Seamless Steel Tubes for Pressure Purposes. Technical Delivery Conditions. Non-Alloy and Alloy Steel Tubes with Specified Elevated Temperature Properties*; BS EN 10216-2:2013; British Standards Institution: London, UK, 2013.
3. Marzocca, A.L.; Luppo, M.I.; Zalazar, M. Identification of Precipitates in Weldments Performed in an ASTM A335 Gr P91 Steel by the FCAW Process. *Procedia Mater. Sci.* **2015**, *8*, 894–903. [CrossRef]
4. Vidyarthy, R.S.; Dwivedi, D.K. Microstructural and mechanical properties assessment of the P91A-TIG weld joints. *J. Manuf. Process.* **2018**, *31*, 523–535. [CrossRef]
5. Dhandha, K.H.; Badheka, V.J. Effect of activating fluxes on weld bead morphology of P91 steel bead-on-plate welds by flux assisted tungsten inert gas welding process. *J. Manuf. Process.* **2015**, *17*, 48–57. [CrossRef]
6. Krishnan, S.; Dulkarni, D.V.; De, A. Pulsed current gas metal arc welding of P91 steels using metal cored wires. *J. Mater. Process. Technol.* **2016**, *229*, 826–833. [CrossRef]
7. Kapal, P.; Surjya, K.P. Effect of pulse parameters on weld quality in pulsedgas metal arc welding: A review. *J. Mater. Eng. Perform.* **2011**, *20*, 918–931.
8. Krishnan, S.; Dulkarni, D.V.; De, A. Multipass pulsed current gas metal arc welding of P91 steel. *Sci. Technol. Weld. Join.* **2016**, *21*, 171–177. [CrossRef]
9. Sireesha, M.; Albert, S.K.; Sundaresan, S. Importance of filler materialchemistry for optimising weld metal mechanical properties in modified 9Cr–1Mo steel. *Sci. Technol. Weld. Join.* **2001**, *6*, 247–254. [CrossRef]
10. Arivazhagan, B.; Sundaresan, S.; Kamaraj, M. A study on influence ofshielding gas composition on toughness of flux-cored arc weld of modified9Cr–1Mo (P91) steel. *J. Mater. Process. Technol.* **2009**, *209*, 5245–5253. [CrossRef]
11. Arivazhagan, B.; Kamaraj, M. Metal-cored arc welding process for joining of modified 9Cr-1Mo (P91) steel. *J. Manuf. Process.* **2013**, *15*, 542–548. [CrossRef]
12. Shanmugarajan, B.; Padmanabham, G.; Kumar, H.; Albert, S.K.; Bhaduri, A.K. Autogenous laser welding investigations on modified 9Cr–1Mo (P91) steel. *Sci. Technol. Weld. Join.* **2011**, *16*, 528–534. [CrossRef]
13. Kundu, A.; Bouchard, P.J.; Kumar, S.; Venkata, K.A.; Francis, J.A.; Paradowska, A.; Dey, G.K.; Truman, C.E. Residual stresses in P91 steel electron beam welds. *Sci. Technol. Weld. Join.* **2013**, *18*, 70–75. [CrossRef]
14. Pai, A.; Sogaladc, I.; Albert, S.K.; Kumar, P.; Mitra, T.K.; Basavarajappa, S. Comparison of microstructure & properties of modified 9Cr-1Mo welds produced by narrow gap hot wire & cold wire gas tungsten arc welding processes. *Procedia Mater. Sci.* **2014**, *5*, 1482–1491.
15. El-Dosoky, O.E.; Abd El-Azim, M.E.; ElKossy, M.R. Analysis of creep behavior of welded joints of P91 steel at 600 °C. *Int. J. Press. Vessel. Pip.* **2019**, *171*, 145–152. [CrossRef]
16. Hyde, T.H.; Saber, M.; Suna, W. Creep crack growth data and prediction for a P91 weld at 650 °C. *Int. J. Press. Vessel. Pip.* **2010**, *87*, 721–729. [CrossRef]
17. Standard, A.S. Standard test method for measurement of creep crack growth rates in metals. In *Annual Book of ASTM Standards*; ASTM: Philadelphia, PA, USA, 2001; Volume 3.
18. Kumar, Y.; Venugopal, S.; Sasikala, G.; Albert, S.K.; Bhaduri, A.K. Studyofcreepcrackgrowthinamodified 9Cr–1Mo steelweldmetal and heataffectedzone. *Mater. Sci. Eng. A* **2016**, *655*, 300–309. [CrossRef]

19. Venugopal, S.; Sasikala, G.; Kumar, Y. Creep Crack Growth Behavior of a P91 Steel Weld. *Procedia Eng.* **2014**, *86*, 662–668. [CrossRef]

20. Baral, J.; Swaminathan, J.; Chakrabarti, D.; Ghosh, R.N. Effect of welding on creep damage evolution in P91B steel. *J. Nucl. Mater.* **2017**, *490*, 333–343. [CrossRef]

21. Divya, M.; Das, C.R.; Albert, S.K.; Goyal, S.; Ganesh, P.; Kaul, R.; Swaminathan, J.; Murty, B.S.; Kukreja, L.M.; Bhaduri, A.K. Influence of welding process on Type IV cracking behavior of P91 steel. *Mater. Sci. Eng. A* **2014**, *613*, 148–158. [CrossRef]

22. Wang, Y.; Li, L.; Kannan, R. Transition from Type IV to Type I Cracking in Heat-Treated Grade 91 Steel Weldments. *Mater. Sci. Eng. A* **2018**, *714*, 1–13. [CrossRef]

23. Sharma, A.; Verma, D.K.; Kumaran, S. Effect of post weld heat treatment on microstructure and mechanical properties of Hot Wire GTA welded joints of SA213 T91 steel. *Mater. Today Proc.* **2018**, *5*, 8049–8056. [CrossRef]

24. Venkata, K.A.; Kumar, S.; Dey, H.C.; Smith, D.J.; Bouchard, P.J.; Truman, C.E. Study on the Effect of Post Weld Heat Treatment Parameters on the Relaxation of Welding Residual Stresses in Electron Beam Welded P91 Steel Plates. *Procedia Eng.* **2014**, *86*, 223–233. [CrossRef]

25. Paddea, S.; Francis, J.A.; Paradowskac, A.M.; Boucharda, P.J.; Shibli, I.A. Residual stress distributions in a P91 steel-pipe girth weld before and after post weld heat treatment. *Mater. Sci. Eng. A* **2012**, *534*, 663–672. [CrossRef]

26. Pandey, C.; Mahapatra, M.M.; Kumar, P.; Kumar, S. Effect of post weld heat treatments on microstructure evolution and type IV cracking behavior of the P91 steel welds joint. *J. Mater. Process. Technol.* **2019**, *166*, 140–154. [CrossRef]

27. Pandey, C.; Mahapatra, M.M.; Kumar, P.; Sirohi, S. Fracture behaviour of crept P91 welded sample for different post weld heat treatments condition. *Eng. Fail. Anal.* **2019**, *95*, 18–29. [CrossRef]

28. Pandey, C.; Mahapatra, M.M.; Kumar, P.; Saini, N.; Srivastava, A. Microstructure and mechanical property relationship for differentheat treatment and hydrogen level in multi-pass welded P91 steeljoint. *J. Manuf. Process.* **2017**, *28*, 220–234. [CrossRef]

29. Pandey, C.; Mahapatra, M.M.; Kumar, P.; Saini, N.; Thakre, J.G.; Vidyarthy, R.S.; Narang, H.K. A brief study on d-ferrite evolution in dissimilar P91 and P92 steel weld joint and their effect on mechanical properties. *Arch. Civ. Mech. Eng.* **2018**, *18*, 713–722. [CrossRef]

30. Widak, V.; Dafferner, B.; Heger, S.; Rieth, M. Investigations of dissimilar welds of the high temperature steels P91and PM2000. *Fusion Eng. Des.* **2013**, *88*, 2539–2542. [CrossRef]

31. Akram, J.; Kalvala, P.R.; Misra, M.; Charit, I. Creep behavior of dissimilar metal weld joints between P91 and AISI 304. *Mater. Sci. Eng. A* **2017**, *688*, 396–406. [CrossRef]

32. Vidyarthy, R.S.; Kulkarni, A.; Dwivedi, D.K. Study of microstructure and mechanical property relationships of A-TIG welded P91-316L dissimilar steel joint. *Mater. Sci. Eng. A* **2017**, *695*, 249–257. [CrossRef]

33. *Welding—Fusion-Welded Joints in Steel, Nickel, Titanium and Their Alloys (Beam Welding Excluded)—Quality Levels for Imperfections*; ISO 5817; International Organization for Standardization: Geneva, Switzerland, 2014.

34. *Metallic Materials—Conversion of Hardness Values*; ISO 18265; International Organization for Standardization: Geneva, Switzerland, 2013.

35. *Non-Destructive Testing of Welds—Visual Testing of Fusion-Welded Joints*; ISO 17637; International Organization for Standardization: Geneva, Switzerland, 2003.

36. *Non-Destructive Testing—Penetrant Testing—Part 1: General Principles*; ISO 3452-1; International Organization for Standardization: Geneva, Switzerland, 2013.

37. *Process Piping*; ASME B31.3; The American Society of Mechanical Engineers: New York, NY, USA, 2010.

38. Newell, W.F. *Guideline for Welding P(T) 91*; Euroweld, Ltd.: Mooresville, NC, USA, 2001; Available online: http://cfs9.blog.daum.net/upload_control/download.blog?fhandle=MEZMejdAZnM5LmJsb2cuZGF1bS5uZXQ6L0lNQUdFLzAvMjQuUERG&filename=24.PDF&filename=Guideline+for+Welding+P91.PDF (accessed on 16 December 2019).

39. *Metallic Materials—Vickers Hardness Test—Part. 1: Test. Method*; ISO 6507-1; International Organization for Standardization: Geneva, Switzerland, 2018.

40. *Destructive Test on Welds in Metallic Materials—Macroscopic and Microscopic Examination of Welds*; BS EN 1321; British Standards Institution: London, UK, 1997.

41. ASME. *ASME Boiler and Pressure Vessel Code An International Code—Qualification Standard for Welding, Brazing, and Fusing Procedures; Welders; Brazers; and Welding, Brazing, and Fusing Operators*; IX: 2015, QW 150; The American Society of Mechanical Engineers: New York, NY, USA, 2015.
42. *Metallic Materials—Tensile Testing—Part 2: Method of Test at Elevated Temperature*; ISO 6892-2; International Organization for Standardization: Geneva, Switzerland, 2018.
43. ASME. *Boiler and Pressure Vessel Code An. International Code—Qualification Standard for Welding, Brazing, and Fusing Procedures; Welders; Brazers; and Welding, Brazing, and Fusing Operators*; IX: 2015, QW 163; The American Society of Mechanical Engineers: New York, NY, USA, 2015.

Article

Structural Aspects of Execution and Thermal Treatment of Welded Joints of Hardox Extreme Steel

Łukasz Konat

Department of Materials Science, Welding and Strength of Materials, Wrocław University of Science and Technology, 50-370 Wrocław, Poland; lukasz.konat@pwr.edu.pl

Received: 17 July 2019; Accepted: 18 August 2019; Published: 21 August 2019

Abstract: The paper presents structure and mechanical properties of welded joints of the high-strength, abrasive-wear resistant steel Hardox Extreme. It was shown that, as a result of welding this steel, structures conducive to lowering its abrasion-wear resistance are created in the heat-affected zone. Width of the zone exceeds 60 mm, which results in accelerated wear in the planned applications. On the grounds of the carried-out examinations of structures and selected mechanical properties, a welding technology followed by heat treatment of heat-affected zones was suggested, leading to reconstruction of *HAZ* structures that is morphologically close to the base material structure. In spite of high carbon equivalent (*CEV*) of Hardox Extreme, the executed welding processes and heat treatment did not result in the appearance, in laboratory conditions, of welding imperfections in the welded joints.

Keywords: wear-resistant martensitic steel; submerged arc welding (SAW); heat treatment; structures; hardness changes; mechanical properties; Hardox Extreme steel

1. Introduction

During recent years, metallic materials classified by their manufacturers as low-alloyed abrasive-wear resistant martensitic steels, have been more and more widely applied. Regardless of their declared abrasion resistance, the common features of all these materials are very high mechanical parameters, which are maintained even for very thick steel sheet material. This feature is obtained with strictly selected chemical compositions depending on sheet thickness (in particular, the microaddition of 0.002% to 0.005% of boron), a reduced content of harmful admixtures of phosphorus and sulfur, and also by thermo-mechanical treatment. Results from generally available advertising information [1] and the author's own experience suggest that the currently available commercial steels of the considered material group reach a tensile strength exceeding 2000 MPa, with maintained satisfactory plasticity and impact strength. It is also worth stressing that these indices are obtained for carbon content up to 0.50 wt%, which is of crucial importance from the viewpoint of welding techniques. Tables 1 and 2 show properties of selected abrasive-wear resistant steels, which are declared by their manufacturers, and chemical compositions of these steels. In Table 2, in addition to the most commonly used *CEV* carbon equivalent (determining the metallurgical weldability of steel according to the International Institute of Welding), the carbon equivalent *CET* is also given. This indicator, according to the SS-EN 1011-2 standard, determines the preheating temperature to avoid hydrogen cracking of welded joints of fine-grained, non-alloyed, and low-alloy steels.

Information from manufacturers [1–5], literature data [6–9] and our own results [10–15] concerning the steels Hardox 400 and Hardox 500 confirm their good weldability and relatively high mechanical properties of welded joints. However, in each of the considered cases, thermal processes occurring during welding caused adverse structural changes in heat-affected zones of the welded materials, resulting in a significant reduction of their abrasive-wear resistance. Results of some research works [16] suggest that changes of structure and hardness distribution occurring in welded joints of low-alloyed martensitic steels, adverse from the viewpoint of abrasive-wear resistance, can be eliminated by the application of additional heat-treatment operations. With regard to this, the author believes that it is worth complementing knowledge about execution and optimization of properties of welded joints of Hardox Extreme steels. In the previously published examination results of Hardox 600 [17], the authors indicated that it was possible to obtain static tensile strength of welded joints over 1500 MPa by proper selection of welding conditions and parameters, followed by suitable heat treatment.

Table 1. Declared mechanical properties of selected abrasive-wear resistant steels [1–5].

Grade of Steel	$R_{p0.2}$ [MPa]	R_m [MPa]	A_5 [%]	KCV_{-40} [J/cm^2]	HBW
Hardox 400	1100	1250	10	56	370–430
Hardox 450	1200	1400	10	50	425–475
Hardox 500	1400	1550	10	46	470–530
Hardox 600	1650	2000	7	25	570–640
Hardox Extreme	NA	NA	NA	NA	650–700
Brinar 400	900	1200	12	25 (−20 °C)	340–440
Brinar 500	1350	1500	8	25 (−20 °C)	480
XAR 600	1700	2000	8	25 (−20 °C)	>550

$R_{p0.2}$—yield strength, R_m—ultimate tensile strength, A_5—percentage elongation after fracture for proportional specimens with the original gauge length L_0 equal to 5 times diameter, KCV_{-40}—Charpy V-notch toughness at −40 °C, HBW—Brinell hardness, NA—no available data.

Table 2. Chemical compositions and declared carbon equivalents of selected abrasive-wear resistant steels [1–5].

Element	Hardox 400	Hardox 450	Hardox 500	Hardox 600	Hardox Extreme	Brinar 400	Brinar 500	XAR 600
				Selected Element [wt%]				
C	0.32	0.26	0.30	0.45	0.47	0.18	0.28	0.40
Mn	1.60	1.60	1.60	1.40	1.40	2.00	1.50	1.50
Si	0.70	0.70	0.70	0.70	0.50	0.50	0.80	0.80
P	0.025	0.025	0.020	0.015	0.015	0.015	0.020	0.025
S	0.010	0.010	0.010	0.010	0.010	0.005	0.005	0.010
Cr	2.40	1.40	1.50	1.20	1.20	1.55	1.50	1.50
Ni	1.50	1.50	1.50	2.50	2.50	NA	NA	1.50
Mo	0.60	0.60	0.60	0.70	0.80	0.60	0.40	0.50
B	0.004	0.005	0.005	0.005	0.005	0.005	NA	0.005
# [mm]	8–20	10–19.9	4–13	6–35	8–19	≤80	≤60	15
CEV^T	0.44	0.48	0.51	0.66	0.66	NA	NA	0.79
CET^T	0.28	0.36	0.37	0.55	0.55	NA	NA	0.53

CEV^T—typical carbon equivalent according to International Institute of Welding, CET^T—typical carbon equivalent according to SS-EN 1011-2, #—sheet thickness for the given chemical properties, NA—no available data.

Development of the subject matter concerning welding of high-strength, abrasive-wear resistant martensitic steels seems to also be well-grounded in the context of numerous negative opinions of users of these steels, undertaking technological activities similar to these discussed here. These opinions predominantly result from big discrepancies between the data published in material data sheets and the results of our own examinations. The carbon equivalent value given by manufacturers is most often a typical value that is significantly different from that calculated on the grounds of real chemical composition of the given grade and sheet thickness. For this reason, activities undertaken by the users on the grounds of inaccurate data often do not provide positive results. This is why, in most cases, the use of the considered grades are given up in favor to their equivalents with much lower mechanical properties, but which are characterized by better carbon equivalents.

The performed examinations concerning the chemical and structural properties of low-alloyed abrasive-wear resistant steels [16,18–26] make it possible to formulate a general statement with regard to the good weldability of steels Hardox 400 and 450, as well as the satisfactory weldability of Hardox 500. This standpoint is also confirmed by the location of chemical compositions of these steels in the diagram *C-CEV* (Figure 1) close to the zones of low (I) or conditions-dependent (II) susceptibility to cracking. However, in the case of Hardox Extreme, this statement does not seem to be well-grounded (Table 3 and Figure 1) because of big discrepancies between the manufacturer's data (designation "P" in Figure 1) and own results (designation "O" in Figure 1). The discrepancy also concerns Hardox 600.

Table 3. Real chemical compositions and carbon equivalent values calculated on their grounds for sheet metal 8–15 mm thick of selected low-alloyed steels [18,24].

Element	Hardox Steels					Brinar Steels		XAR
	400	450	500	600	Extreme	400	500	600
	Selected Element [wt%]							
C	0.17	0.17	0.28	0.45	0.48	0.17	0.28	0.37
Mn	1.00	1.00	0.69	0.51	0.52	1.14	0.95	0.85
Si	0.37	0.32	0.26	0.16	0.16	0.22	0.66	0.19
P	0.010	0.010	0.011	0.012	0.010	0.008	0.012	0.014
S	0.002	0.000	0.001	0.002	0.001	0.000	0.000	0.001
Cr	0.22	0.45	0.66	0.33	0.89	0.60	0.84	0.83
Ni	0.05	0.05	0.08	1.98	1.96	0.38	0.01	1.21
Mo	0.01	0.08	0.03	0.14	0.13	0.30	0.20	0.15
V	0.004	0.005	0.010	0.009	0.008	0.039	0.006	0.002
Cu	0.006	0.018	0.016	0.016	0.021	0.010	0.008	0.030
Al	0.035	0.032	0.050	0.031	0.034	0.073	0.039	0.097
Ti	0.020	0.016	0.005	0.006	0.006	0.009	0.012	0.003
Nb	0.010	0.000	0.000	0.005	0.001	0.043	0.023	0.009
Co	0.010	0.016	0.017	0.026	0.022	0.001	0.005	0.005
B	0.0016	0.0014	0.0016	0.0026	0.0025	0.0023	0.0008	0.0021
# [mm]	8	10	10	10	10	12	12	15
CEV	0.38	0.44	0.54	0.76	0.90	0.58	0.65	0.79
CET	0.28	0.30	0.39	0.58	0.64	0.36	0.43	0.54

CEV [%] = C + Mn/6 + (Cr + Mo + V)/5 + (Cu + Ni)/15; CET [%] = C + (Mn + Mo)/10 + (Cr + Cu)/20 + Ni/40, CEV—carbon equivalent according to International Institute of Welding, CET—carbon equivalent according to SS-EN 1011-2, #—sheet thickness for the given chemical properties.

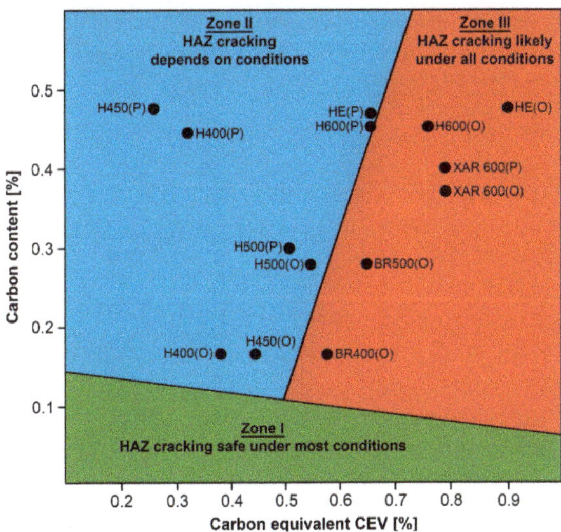

Figure 1. Susceptibility to cracking in function of carbon content and *CEV* of selected abrasive-wear resistant steels. H—Hardox, BR—Brinar, P—manufacturer's data, O—own results. Development based on data from Table 3 and [27].

The above-described discrepancies decidedly provide evidence against these steels with regard to their metallurgical weldability, shifting *CEV* values from the zone (II) to the zone (III) of high susceptibility to cracking in any welding conditions. According to users and manufacturers of abrasive-wear resistant steels, the most often indicated problems of weldability of Hardox Extreme (and also Hardox 600) are related to the susceptibility of the made welds to the brittle cracking (including also delayed cracking) and very wide zones with lower hardness in comparison to the base material. In this connection, the purpose of the presented research work is the identification of macro- and microscopic structures of Hardox Extreme welded joints in the as-welded condition (after welding) and determination of the area of structural changes within the entire welded joint, as well as provoking, through heat treatment, structural changes in order to eliminate or minimize the previously existing adverse structures. It is worth mentioninging that similar examinations of Hardox 600 were also carried out. However, with regard to the very comprehensive set of results, they will be elaborated on separately. It should be also mentioned that the author is currently conducting examinations of Hardox welded joints with regard to their abrasive wear in real conditions of soil abrasive mass.

2. Material and Methodology

Examinations were carried-out on Hardox Extreme steel sheets in an as-delivered condition, which was 1000 mm long and 10 mm thick. Welded joints were made by submerged arc welding *SAW* (121), while considering the welding materials dedicated for low-alloyed high-strength steels. Selected properties of the used welding materials are given in Table 4.

Table 4. Selected properties of welding materials used for Hardox Extreme joints [28].

Weld Metal	C	Mn	Si	Cr	Ni	Mo	$R_{p0.2}$	R_m	A_4	KCV_{-40}
			Chemical Composition [%]				[MPa]		[%]	[J/cm^2]
OK Autrod 13.43 + OK Flux 10.62	0.11	1.50	0.25	0.60	2.20	0.50	700	800	21	94

$R_{p0.2}$—yield strength, R_m—ultimate tensile strength, A_4—percentage elongation after fracture for proportional specimens with the original gauge length L_0 equal to 4 times diameter, KCV_{-40}—Charpy V-notch toughness at −40 °C.

Joints were made using an automatic welding machine ESAB A2 Mini Trac with the power source ESAB LAE 800. The Hardox sheets were joined with a both-sides weld (Figure 2) using the following parameters guaranteeing correct penetration:

- weld type: BW (butt weld),
- welding position: PA (flat),
- electrode diameter: 3.0 mm,
- arc voltage (weld layer: 1, 2): 34/35 V,
- amperage (weld layer: 1, 2): 520/640 A,
- polarity: DC(+),
- welding rate (weld layer: 1, 2): 61/63 cm/min,
- electrode wire: OK Autrod 13.43 (S3Ni2.5CrMo acc. to EN ISO 26304),
- flux: OK Flux 10.62,
- preheating: no,
- interpass temperature: ≤100 °C,
- preparation of sheet edges (chamfering): no.

After welding, test specimens were cut-out in form of cuboids by means of a high-energy abrasive water stream (general specimen geometry) and through electroerosion (V-notch geometry). Next, a part of the specimens was subjected to heat treatment in laboratory conditions by being quenched in oil and tempering. It should be stressed that, with regard to the posed cognitive goals and available technical measures, the heat treatment operations were carried-out on whole specimens, i.e., for both heat-affected zone and base material. Before quenching, the specimens were additionally subjected to normalizing. All the thermal operations were carried-out in gas-tight chamber furnaces Czylok FCF 12SHM/R in a protective atmosphere of 99.95% argon. Quenching was carried-out in quenching oil Durixol W72 with a kinematic viscosity of 21 mm^2/s, heated-up to 50 ± 5 °C. Detailed characteristics of the specimens and heat treatment parameters are given in Table 5.

Chemical analyses were carried-out spectrally by means of a glow discharge spectrometer Leco GDS-500A, using the following parameters: $U = 1250$ V, $I = 45$ mA, 99.999% argon. The results were given as averages of at least five measurements.

Observations of macro- and micro-structures were performed using a multifunctional stereoscopic microscope Nikon AZ100 and a light microscope Nikon Eclipse MA200 coupled with a digital camera Nikon DS-Fi2. Images were recorded and analyzed using software NIS Elements.

Rockwell hardness (HRC/HRA) measurements were taken with a universal hardness tester Zwick/Roell ZHU 187.5 at 1500/600 kgf according to EN ISO 6508-1:2016-10. Measurements were taken on the specimens after microstructure examinations, within the base material (Hardox Extreme sheets) and in the zones subjected to structural analysis (lines A and B marked in Figure 2a).

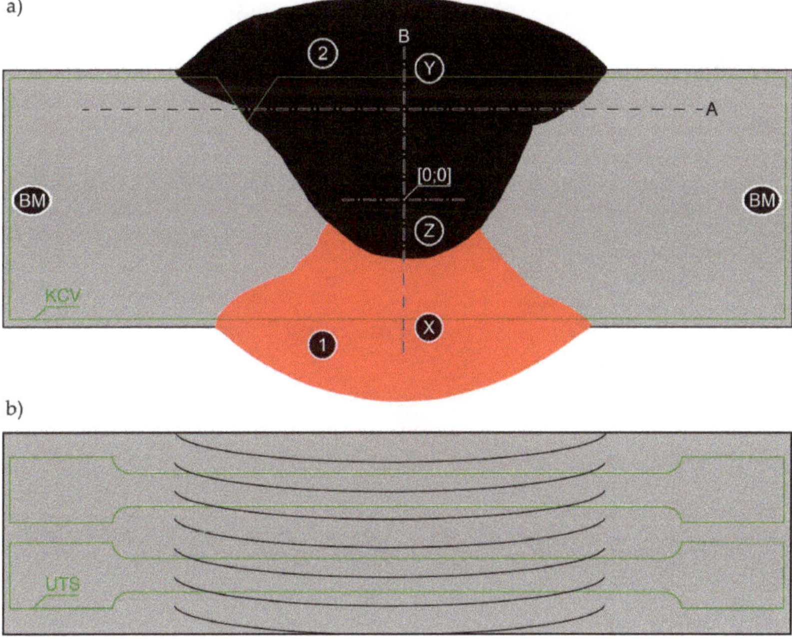

Figure 2. General layout of a Hardox Extreme welded joint: (**a**) cross-section view; (**b**) view from the face of the weld. 1,2—individual welds acc. to their execution order; A,B—lines of hardness measurements; BM—base material; X,Y,Z—places of chemical analyses; KCV—way of cutting out specimens for impact tests—dimensions of the specimen after V-notching: $8.5 \times 10 \times 55$ mm; UTS—way of cutting out specimens for ultimate tensile strength—dimensions of the cuboidal specimens with preset gauge length of $L_0 = 25$ mm: 10×10 mm.

Mechanical tests were carried-out at ambient temperature according to EN ISO 6892-1:2016-09, using a testing machine Instron 5982, on cuboidal specimens with preset gauge length of $L_0 = 25$ mm (Figure 2b). The tensile tests were carried-out under controlled force to ensure a uniform strain rate for the specimens until their failure. Next, tensile strength (R_m) and reduction of area at failure (Z) were determined.

Impact tests of the welded joints were carried-out using a pendulum Charpy tester Zwick/Roell RPK300 with initial energy 300 J, according to EN ISO 148-1:2017-02. The V-notch specimens used in the tests were cut from the entire butt joints (according to EN ISO 9016:2011) and included fusion zones of the analyzed welded joints in the conditions directly after welding and after heat treatment operations (Figure 2a). The tests were carried-out at +20 °C and −40 °C. Observations of fracture surfaces were performed with a stereoscopic microscope and a scanning electron microscope JEOL JSM-6610A. The SEM observations were performed at an accelerating voltage of 20 kV, in material contrast to what occurs when using SE detectors.

3. Results

Basic parameters of heat treatment operations carried-out on the examined welded joints are given in Table 5 and Figures 3 and 4, together with results of mechanical and impact tests of the joints in the conditions directly after welding and after heat treatment operations.

The performed heat treatment operations were aimed at obtaining a microstructure and mechanical properties in the entire welded joints that are similar to those of the base material. As such, the welded joints were volumetrically quenched in oil bath and tempered (stress relieved). The austenitizing temperature before quenching was established on the grounds of real chemical compositions of both the base material (Table 3) and weld metal, while considering complete cross-section of the welded joint (Table 6). Establishing the tempering temperature at 100 °C resulted from the fact that exposure of Hardox Extreme to temperatures over 125 °C causes decomposition of martensitic structure and significant decrease of hardness. This results also in lower indices from static tensile test and can lead to lower abrasive-wear resistance.

Table 5. Heat treatment parameters and selected mechanical properties of Hardox Extreme welded joint:

Specimen	Heat Treatment Parameters	R_m [MPa]		Z [%]		KCV_{+20}		KCV_{-40} [J/cm^2]	
UTS-1		1329		26.0					
UTS-2		1263	1278	16.3	17.4			—	
UTS-3		1242		10.1					
KCV-1						18.7			
KCV-2	No treatment					16.4	17	—	
KCV-3				—		16.6			
KCV-4								14.3	
KCV-5						—		12.7	18
KCV-6								26.1	
UTS-25	Normalization: 800 °C/1h/Air + Quenching: 850 °C/20′/Oil + Tempering: 100 °C/20h/Air	1831		30.6					
UTS-26		1823	1831	31.6	29.1			—	
UTS-28		1839		25.2					
KCV-7						27.1			
KCV-8						27.9	27	—	
KCV-9				—		25.9			
KCV-10								21.6	
KCV-11						—		17.1	19
KCV-12								19.0	

R_m—ultimate tensile strength, Z—percentage reduction of area, KCV_{+20}—charpy V-notch toughness at room temperature, KCV_{-40}—charpy V-notch toughness at −40 °C.

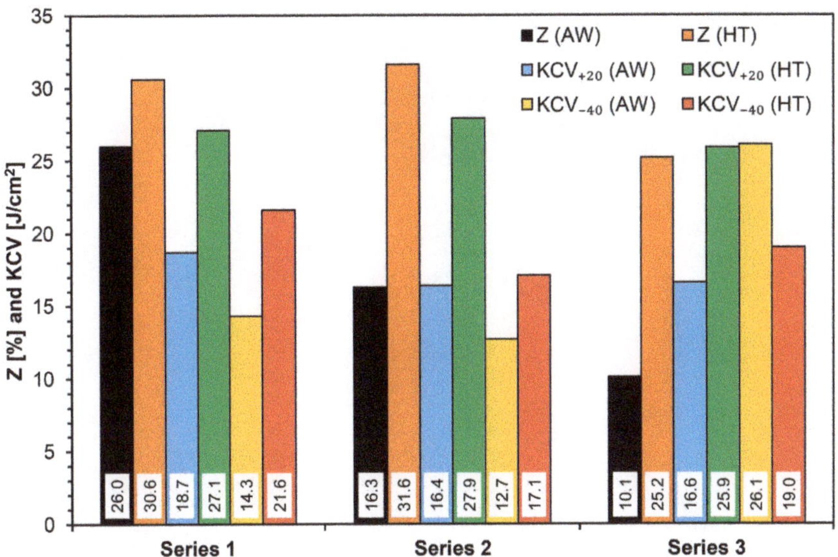

Figure 3. Percentage reduction of area *Z* and toughness *KCV* of tested samples based on data from Table 5. AW—condition after welding, HT—condition after heat treatment.

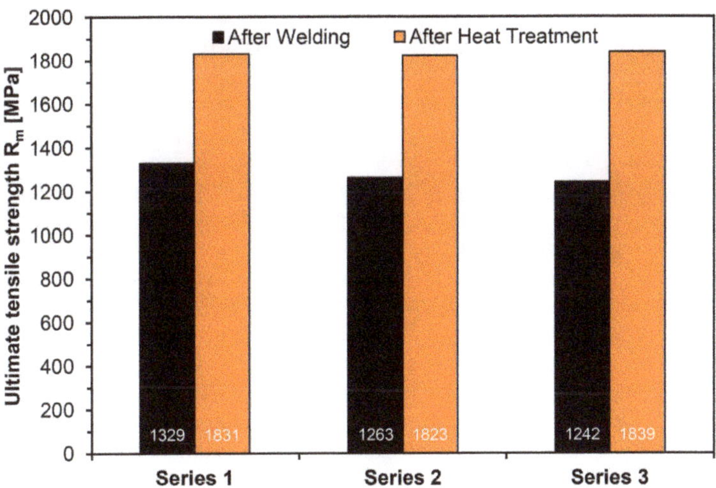

Figure 4. Ultimate tensile strength R_{m} of tested samples based on data from Table 5.

Table 6. Chemical composition in cross-section of the welded joint.

Element	X	Y	Z	OK 13.43 + OK Flux 10.62	Hardox Extreme
	Chemical Composition [wt%]				
C	0.31	0.32	0.31	0.10	0.48
Mn	0.81	0.82	0.81	1.22	0.52
Si	0.23	0.23	0.23	0.30	0.16
P	0.013	0.013	0.014	0.020	0.010
S	0.002	0.002	0.002	0.002	0.001

Table 6. Cont.

Element	X	Y	Z	OK 13.43 + OK Flux 10.62	Hardox Extreme
	Chemical Composition [wt%]				
Cr	0.81	0.82	0.81	0.50	0.89
Ni	2.06	2.08	2.07	1.65	1.96
Mo	0.23	0.23	0.22	0.31	0.13
V	0.009	0.009	0.010	0.009	0.008
Cu	0.040	0.038	0.041	0.105	0.021
Al	0.020	0.021	0.020	0.013	0.034
Ti	0.004	0.004	0.005	0.003	0.006
Nb	0.000	0.000	0.000	0.000	0.001
Co	0.017	0.016	0.018	0.006	0.022
B	0.0021	0.0021	0.0021	0.0014	0.0025
CEV	0.79	0.80	0.80	0.59	0.90
CET	0.51	0.51	0.51	0.33	0.64

CEV [%] = C + Mn/6 + (Cr + Mo + V)/5 + (Cu + Ni)/15; *CET* [%] = C + (Mn + Mo)/10 + (Cr + Cu)/20 + Ni/40.3.1.
X,Y,Z—places of chemical analyses, marked in Figure 2, *CEV*—carbon equivalent according to International. Institute of Welding, *CET*—carbon equivalent according to SS-EN 1011-2.

3.1. Mechanical Properties

Average tensile strength of Hardox Extreme welded joints in the condition after welding reached R_m = 1278 MPa (Table 5 and Figure 4), with maintained moderate plastic properties defined by relative area reduction Z = 17.4% (Table 5 and Figure 3). Even if the obtained strength is very high in comparison to that of constructional steels, it only provides ca. 64% of the value for the base material (assuming minimum tensile strength of Hardox Extreme equal to that of Hardox 600, i.e., 2000 MPa). Here, a significant scatter of the obtained values should be indicated. Even if the test joints were made on an automated welding station with significant lengths of run-off welds of over 150 mm, the results were characterized by rather low repeatability. In the author's opinion, such a behavior of these materials after welding is caused by uncontrolled structural changes occurring in the heat-affected zone, which is the probable cause of the unpredictable behavior of these steels in industrial conditions. This observation can be related mostly to plastic properties of the weld, which, to some extent, is also confirmed by relatively low, widely scattered impact strength values (especially at negative testing temperatures). Average impact strength at both testing temperatures of 17–18 J/cm^2 clearly shows the necessity to classify the obtained welded joints as being susceptible to brittle cracking.

As a result of the applied thermal treatments, all the considered mechanical properties of Hardox Extreme welded joints increased significantly (Table 5). Average tensile strength amounted to 1831 MPa, which makes nearly 90% of the assumed strength of the base material. In addition to very high strength indices, the plastic properties increased. The average value of relative area reduction at failure was Z = 29.1%, and impact energy occurred at an ambient temperature amounting to 27 J/cm^2, allowing the supposition that ductile fracture provides a significant part of the entire fracture area.

3.2. Results of Microscopic Examinations and Hardness Measurements

Figures 5–23 show macroscopic images, hardness measurements and an overview of characteristic microstructures in the entire area of the welded joint. It should be stated on the grounds of the performed examinations that the welding process provoked in the considered steel diverse structural changes that resulted in wide zones with decreased hardness, generally designated as heat-affected zones (Figures 5 and 7). Width of this zone was mainly decided by the used technology, parameters and conditions of welding, the value of delivered heat input, chemical composition of steel, and also by the structure of the base material before welding (Figure 9). High influence of the latter parameter on properties of the welded joint can be explained by tempering the processes of post-martensitic structures occurring mainly in the heat-affected zone. In the case of Hardox Extreme steel, the entire zone affected by the temperature exceeding 125 °C is characterized by wide areas of reduced hardness (Figure 7). Experiences of the author indicate that, in many cases, the width of this zone exceeds 60–100 mm. In turn, application of high-energy welding methods to a steel with chemical composition close to that of Hardox Extreme very often results in cold cracking and reduced brittleness threshold determined in impact tests.

Conclusions similar to these above cannot be formulated in relation to the specimens after heat treatment. In the macroscopic image of the joint in this condition (Figure 6) no heat-affected zone is observed, which is additionally confirmed by hardness measurements (Figures 7 and 8). It should be also stressed that, from the viewpoint of real chemical composition of Hardox Extreme ($CEV = 0.90$; Table 3 and Figure 1), application of generally accepted weldability criteria to low-alloyed steels can give rise to serious doubts. In this connection, the author believes that, with regard to very high mechanical properties of the base material, operations of post-weld heat treatment should be obligatorily considered at welding processes of the steel Hardox Extreme. Such an approach makes it possible to restore microstructure and mechanical properties of the entire area of a welded joint.

Analysis of structural changes of the examined welded joints showed, in the zone of base material after welding (BM in Figure 5), microstructure characteristic for toughening processes, i.e., tempered sorbite (Figure 9b). This structure shows the tempering processes occurring in the welded material, resulting in significant loss of hardness in comparison to the condition before welding (Figure 7). The carried-out thermal operations made it possible to obtain, in the analogous zone, structures very well corresponding with the Hardox base material in an as-delivered condition. After heat treatment, the base material zone (BM in Figure 6) showed microstructure of fine-lath hardening martensite, almost identical to that of Hardox Extreme in as-delivered condition (Figure 9a,c). The above finding is confirmed by hardness measurements taken in the considered zones. After welding, level of hardness in the BM zone reached 320–350 HV (Figure 7). Heat treatment resulted in an increase of hardness to over 650 HV (Figure 7), which only slightly declines from the value declared by the manufacturer of Hardox Extreme in an as-delivered condition, i.e., ca. 700 HV (min. 650 HBW, see Table 1).

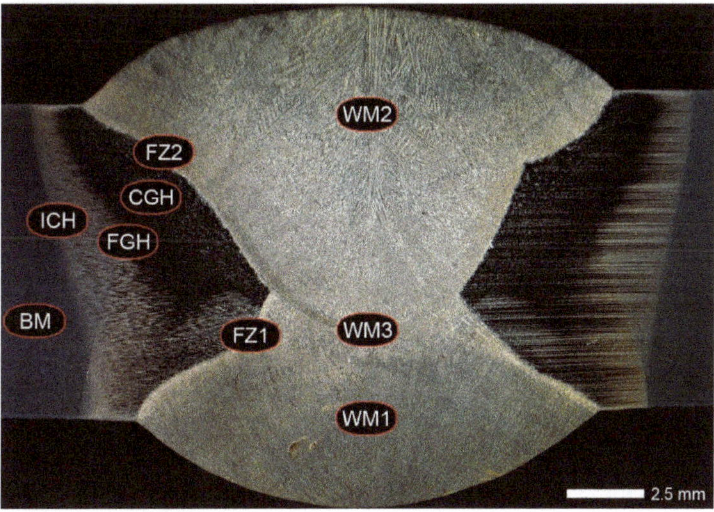

Figure 5. Macroscopic image of cross-section of a Hardox Extreme welded joint after welding. WM—weld metal zone, FZ—fusion zone, BM—base material zone, CGH(AZ)—coarse-grained heat-affected zone (overheating), FGH(AZ)—fine-grained heat-affected zone (normalization and recrystallisation), ICH(AZ)—intercritical heat-affected zone (incomplete normalization). Stereoscopic microscopy, etched with 3% HNO_3 and Adler's etchant.

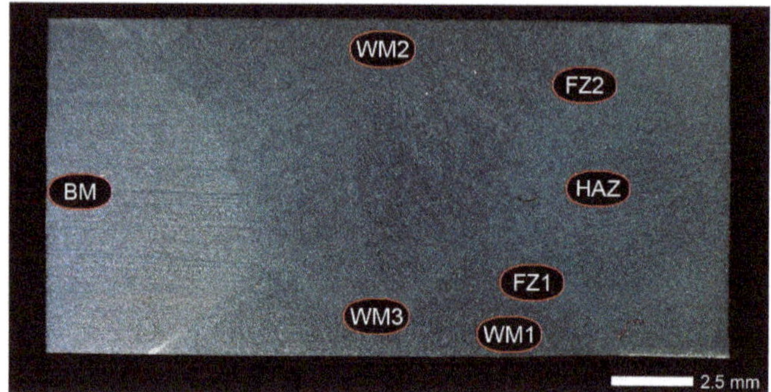

Figure 6. Macroscopic image of cross-section of a Hardox Extreme welded joint after heat treatment. WM—weld metal zone, FZ—fusion zone, BM—base material zone, HAZ—heat-affected zone. Stereoscopic microscopy, etched with 3% HNO_3.

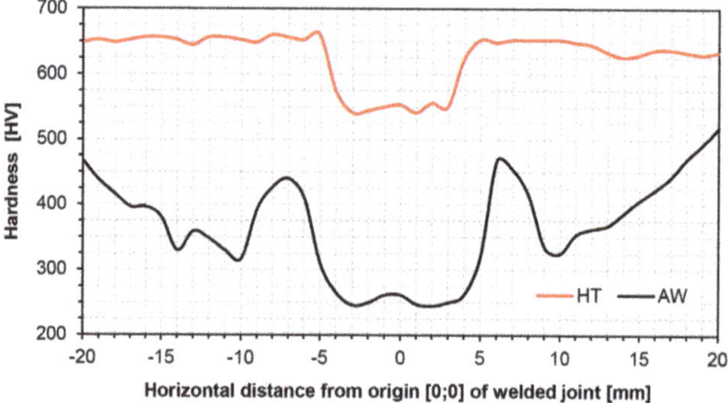

Figure 7. Hardness distribution of a Hardox Extreme welded joint along the line A shown in Figure 2. AW—condition after welding, HT—condition after heat treatment.

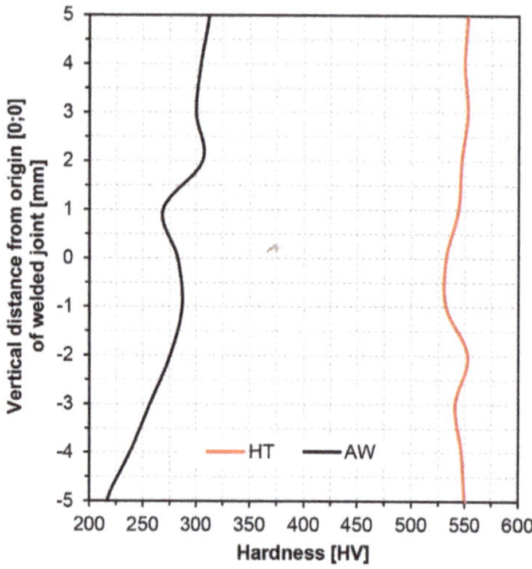

Figure 8. Hardness distribution of a Hardox Extreme welded joint along the line B shown in Figure 2. AW—condition after welding, HT—condition after heat treatment.

Hardness level in the weld metal zone (WM1, WM2 and WM3 in Figure 6) after heat treatment was ca. 550 HV, slightly lower than that of the base material (Figure 7). A drop of hardness by ca. 100 HV can be explained by slightly lower percentage of carbon in the weld axis, which decides saturation degree of ferrite and thus hardening capacity of the material. However, it is worth mentioninging that the obtained hardness on the entire thickness of the heat-treated welded joint was nearly linear at the level of 550 HV (Figure 8). This feature clearly distinguishes this material condition from that after welding, where hardness along the weld thickness changes within 220 to 310 HV, which also determines individual usable properties of the joint depending on sequence of applying the weld layers. Structural examinations indicate that the above conclusions concerning hardness should be related to microstructure changes in representative areas of the joint. In the zones designated WM1, WM2 and WM3 in Figure 5, microstructure after welding was typical for diverse temperatures and cooling rates. Generally, the microstructure of the weld metal is of a dendritic nature and composes of bands of martensite and small quantity of bainite on the background of non-equilibrium ferrite grains with features of a Widmanstätten structure in the WM1 zone (Figure 10).

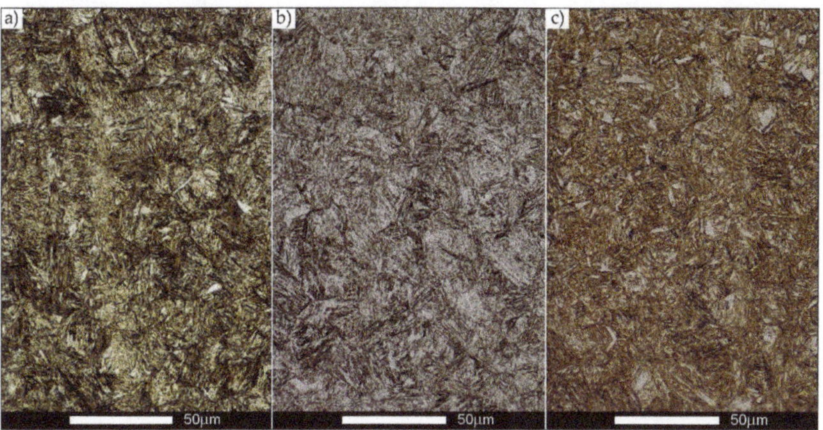

Figure 9. Microstructure of base material Hardox Extreme: (**a**) condition before welding—structure of tempered martensite; (**b**) condition after welding—BM zone in Figure 5, characteristic structure of tempered sorbite; (**c**) condition after heat treatment—BM zone in Figure 6, structure of hardening martensite. Light microscopy, etched with 2% HNO_3.

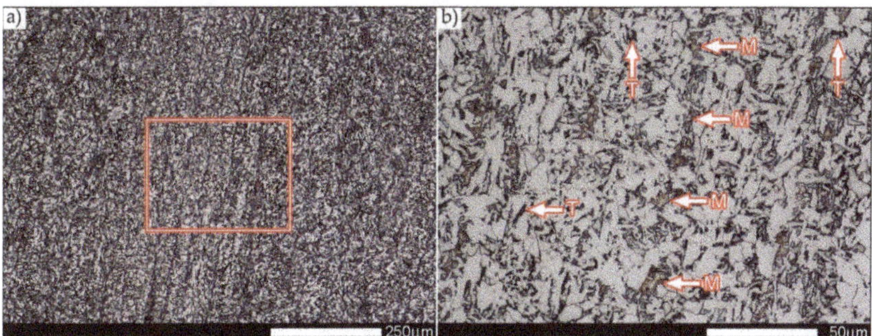

Figure 10. Microstructure of Hardox Extreme welded joint after welding: (**a**) in the area marked WM1 in Figure 5; (**b**) an enlarged image marked with frame in Figure 10a. Morphologically diverse structure composed of bands of martensite (M) with colonies of hardening troostite (T) on the background of non-equilibrium grains of ferrite. Light microscopy, etched with 2% HNO_3.

Microstructure in the zone WM2 is composed of lath hardening martensite on the background of tempered sorbite, small quantity of bainite and colonies of troostite (Figure 11). Microstructure in the transition zone between the weld layers 1 and 2 (WM3 in Figure 5) is composed mostly of hardening sorbite with band-like martensite, hardening bainite and colonies of troostite (Figure 12). It is also worth to indicate the significant change of structure in the zone WM3, which can determine general impact strength of the entire welded joint. Thus, from the viewpoint of fracture mechanics, the microstructure of the transition zone between the weld layers 1 and 2 (Figure 2) should be taken into account during the selection of technology and welding parameters for this type of metallic materials.

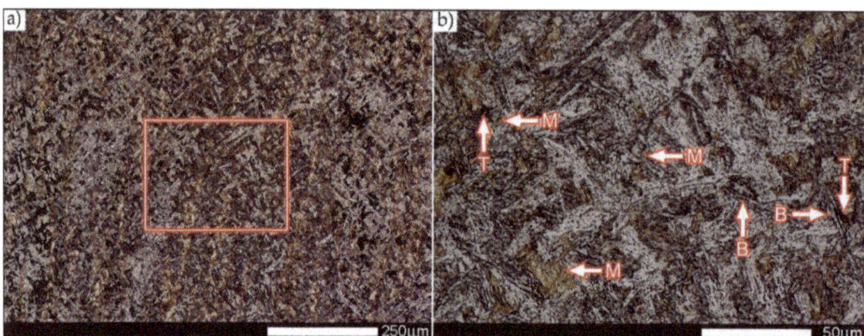

Figure 11. Microstructure of Hardox Extreme welded joint after welding: (**a**) in the area marked WM2 in Figure 5; (**b**) an enlarged image marked with frame in Figure 11a. Strongly diversified structure with dendritic nature, composed of bands of martensite (M) and bainite (B) and few colonies of troostite (T) on the grounds of hardening sorbite. Light microscopy, etched with 2% HNO_3.

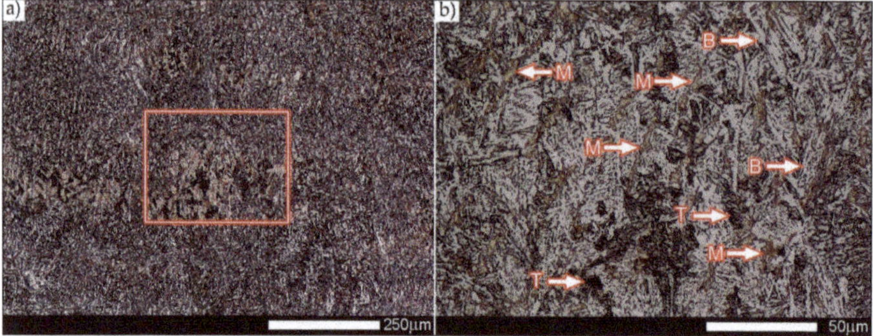

Figure 12. Microstructure of Hardox Extreme welded joint after welding: (**a**) in the area marked WM3 in Figure 5; (**b**) an enlarged image marked with frame in Figure 12a. Clearly visible, morphologically diversified fusion lines 1 and 2 composed of bands of martensite (M) and bainite (B) and colonies of troostite (T) on the background of hardening sorbite. Light microscopy, etched with 2% HNO_3.

After heat treatment, in all zones of the additional material (WM1, WM2 and WM3 in Figure 6), very similar microstructures (corresponding with that of as-delivered base material) were obtained, composed of fine-lath hardening martensite (Figures 13–15). With regard to the applied technological operations and their parameters, a remainder of band-like structure after welding can be observed in the recorded microstructures. Nevertheless, in the context of industrial applications, its complete removal would not be technologically and economically justified. However, it is worth mentioning that the applied heat treatment practically eliminated the observed structure changes in the zone marked WM3 in Figure 15.

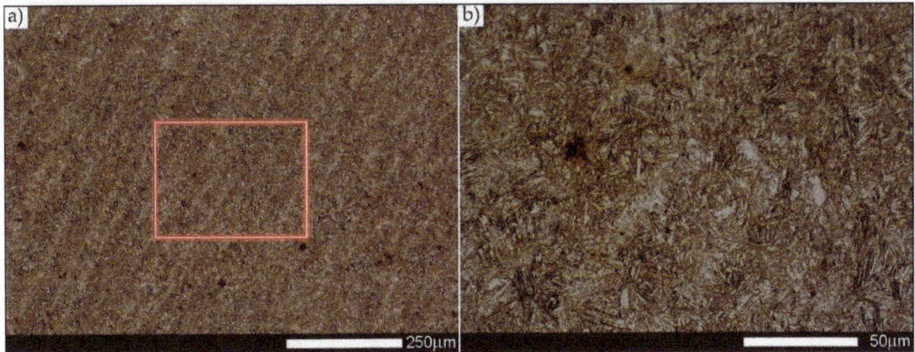

Figure 13. Microstructure of Hardox Extreme welded joint after heat treatment: (**a**) in the area marked WM1 in Figure 6; (**b**) an enlarged image marked with frame in Figure 13a. Structure of fine-lath hardening martensite with clear banding features resulting from former dendritic structure. Light microscopy, etched with 2% HNO_3.

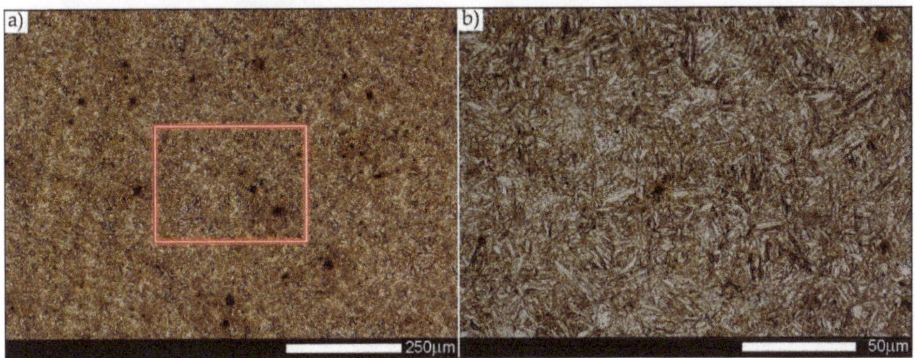

Figure 14. Microstructure of Hardox Extreme welded joint after heat treatment: (**a**) in the area marked WM2 in Figure 6; (**b**) an enlarged image marked with frame in Figure 14a. Structure of fine-lath hardening martensite. Light microscopy, etched with 2% HNO_3.

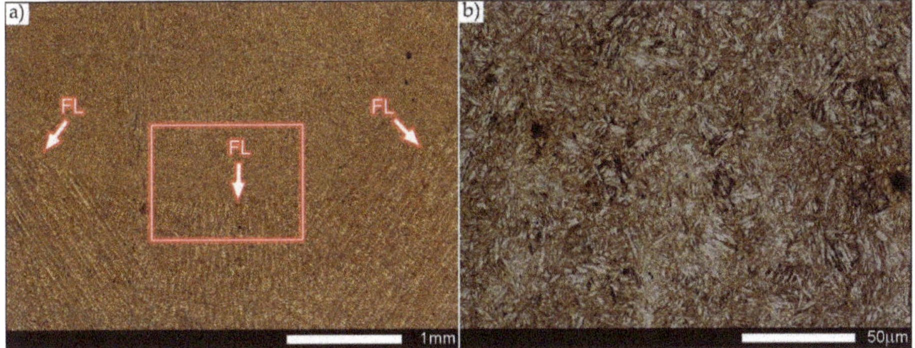

Figure 15. Microstructure of Hardox Extreme welded joint after heat treatment: (**a**) in the area marked WM3 in Figure 6; (**b**) an enlarged image marked with frame in Figure 15a. Structure of fine-lath hardening martensite with weak banding features resulting from former dendritic structure. Faintly outlined fusion line between weld layers 1 and 2 (FL) is marked with arrows. Light microscopy, etched with 2% HNO_3.

Observations of microstructures in the fusion zones indicate a very wide range of their kinds and morphologies. It can be generally found that the zones marked FZ1 and FZ2 in Figure 5 are characterized by a very clearly outlined fusion line (FL in Figures 16 and 17) composed of coarse-lath hardening martensite, areas of upper bainite and non-equilibrium grains of ferrite, and also hardening sorbite and troostite (Figures 16 and 17). In addition, the recorded microstructures are characterized by strongly heterogeneous morphology, even within the same type of structure. It should be stressed that creation of a complete characteristic of structures in the considered Hardox Extreme welded joint requires examinations with the use of transmission electron microscopy (TEM) that is currently being carried-out.

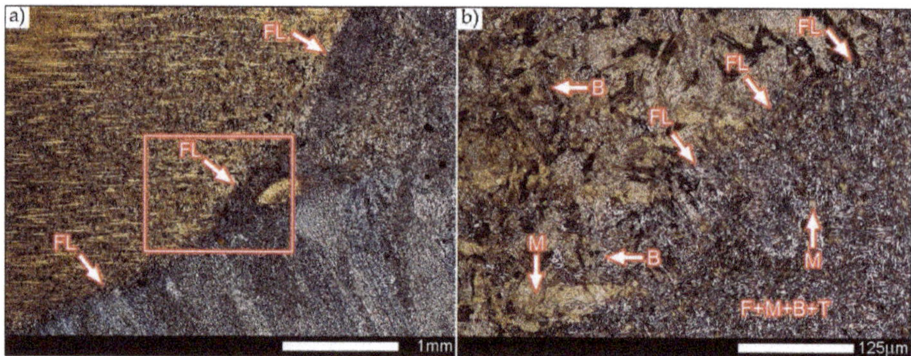

Figure 16. Microstructure of Hardox Extreme welded joint after welding: (**a**) in the area marked FZ1 in Figure 5; (**b**) an enlarged image marked with frame in Figure 16a. Very clearly outlined fusion line (FL) with strongly diversified microstructure. Visible islands of hardening martensite (M) and bainite (B), troostite areas (T) and acicular ferrite (F). Light microscopy, etched with 2% HNO_3.

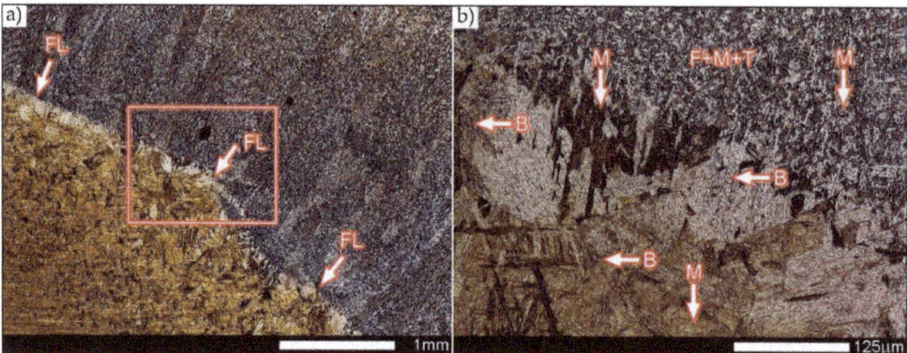

Figure 17. Microstructure of Hardox Extreme welded joint after welding: (**a**) in the area marked FZ2 in Figure 5; (**b**) an enlarged image marked with frame in Figure 17a. Very clearly outlined fusion line (FL) with strongly diversified microstructure. Visible islands of hardening martensite (M) and bainite (B), small number of troostite colonies (T) and acicular ferrite (F). Light microscopy, etched with 2% HNO_3.

The performed heat treatment operations led to homogeneous structure in the fusion zones (FZ1 and FZ2 in Figure 6), with respect to both type and fineness. In both considered fusion zones, structures of fine-lath hardening martensite were observed on the very weakly outlined fusion line (Figures 18 and 19). The above findings are also confirmed by hardness measurements (Figure 7) that do not show clear differences in comparison to the joint after welding.

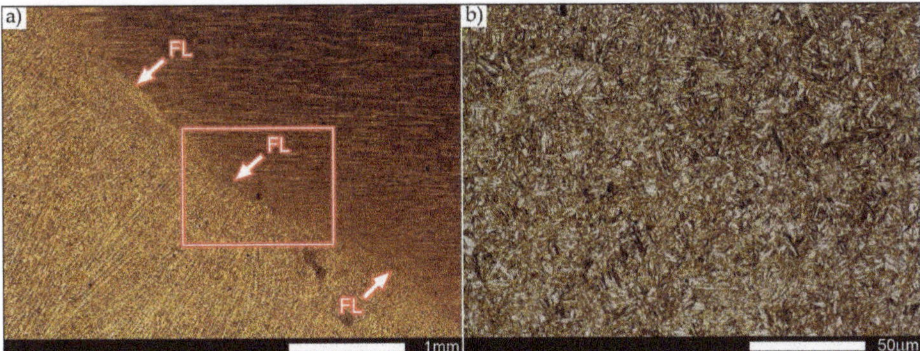

Figure 18. Microstructure of Hardox Extreme welded joint after heat treatment: (**a**) in the area marked FZ1 in Figure 6; (**b**) an enlarged image marked with frame in Figure 18a. Structure of fine-lath hardening martensite. The arrows (FL) indicate the very weakly outlined fusion line. Light microscopy, etched with 2% HNO$_3$.

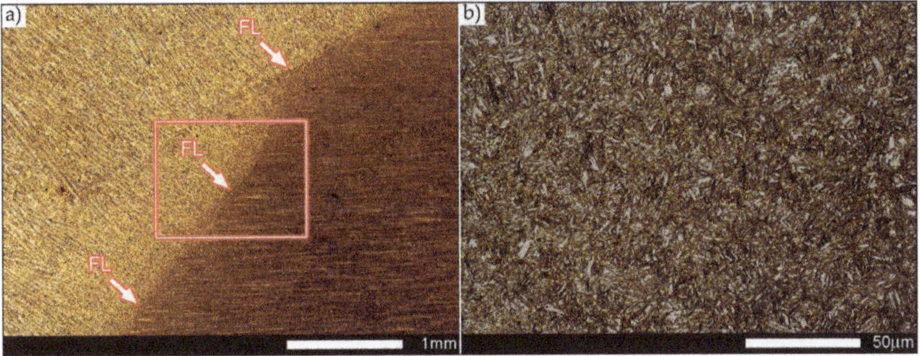

Figure 19. Microstructure of Hardox Extreme welded joint after heat treatment: (**a**) in the area marked FZ2 in Figure 6; (**b**) an enlarged image marked with frame in Figure 19a. Structure of fine-lath hardening martensite. The arrows (FL) indicate the very weakly outlined fusion line. Light microscopy, etched with 2% HNO$_3$.

Analysis of the other characteristic areas of heat-affected zone in the condition after welding, i.e., CGHAZ, FGHAZ and ICHAZ (marked in Figure 5), indicates much diversified structures in each case, resulting in significantly varying hardness levels. In the coarse-grained heat-affected zone (CGHAZ in Figure 20), fine- and coarse-lath hardening martensite is observed, with bainite separated on grain boundaries of former austenite, and also small quantity of hardening sorbite. In the fine-grained heat-affected zone (FGHAZ in Figure 21), microstructure includes mostly fine-lath hardening martensite with colonies of troostite and small number of bainitic areas. The intercritical heat-affected zone (ICHAZ in Figure 22) is composed of very fine, band-like arranged martensite with tempered sorbite. The morphology of this zone is clearly different than that of the base material, composed of fine-lath martensite (left side in Figure 22a,b).

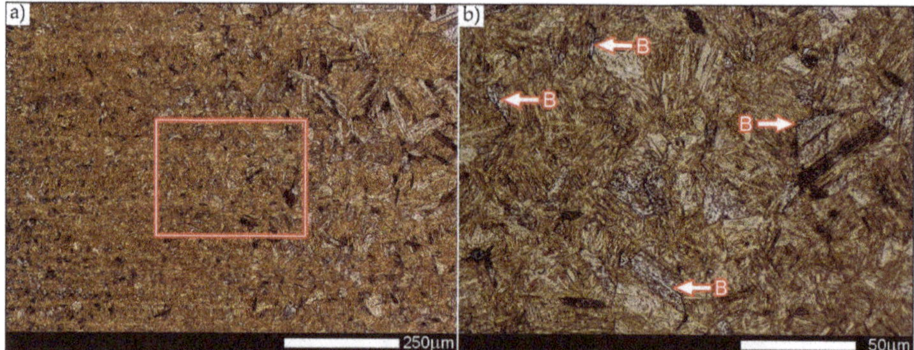

Figure 20. Microstructure of Hardox Extreme welded joint after welding: (**a**) in the area marked CGHAZ in Figure 5; (**b**) an enlarged image marked with frame in Figure 20a. Lath-like and locally acicular martensitic structure with bainite (B) on grain boundaries of former austenite. Light microscopy, etched with 2% HNO$_3$.

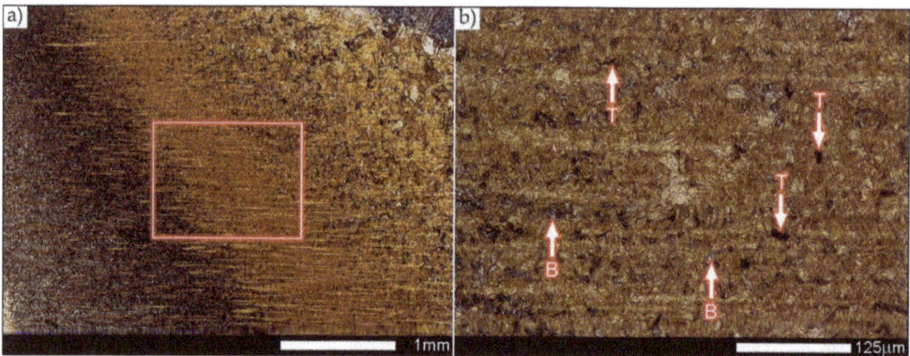

Figure 21. Microstructure of Hardox Extreme welded joint after welding: (**a**) in the area marked FGHAZ in Figure 5; (**b**) an enlarged image marked with frame in Figure 21a. Structure of fine-lath hardening martensite with inclusions of bainite (B) and troostite (T). Light microscopy, etched with 2% HNO$_3$.

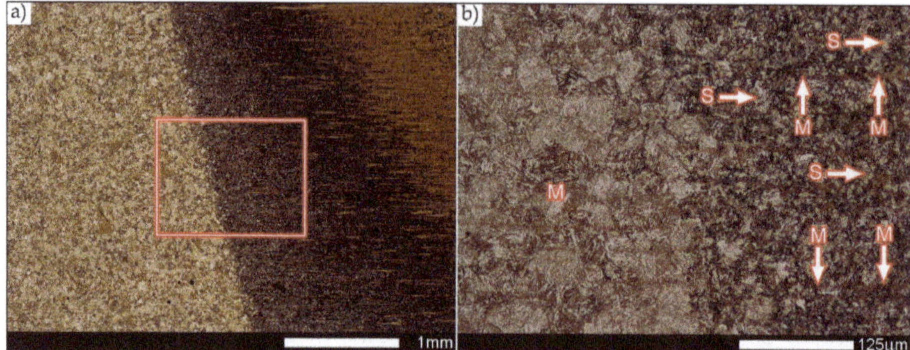

Figure 22. Microstructure of Hardox Extreme welded joint after welding: (**a**) in the area marked ICHAZ in Figure 5; (**b**) an enlarged image marked with frame in Figure 22a. Structure of fine-lath band-like arranged hardening martensite (M) with hardening sorbite (S). Light microscopy, etched with 2% HNO$_3$.

The performed heat treatment operations of the joint resulted in uniform microstructure in the entire heat-affected zone. The microstructure of this zone (HAZ in Figure 6) is composed of fine-lath hardening martensite with maintained banding features resulting from the process of thermo-mechanical rolling of Hardox sheet metal (Figure 23). As such, the hardness course in this zone is nearly linear at the average level of 650 HV (Figure 7). It can be stated on this ground that heat treatment of the joint brought the structure of heat-affected zone to the structure of the base material in the as-delivered condition.

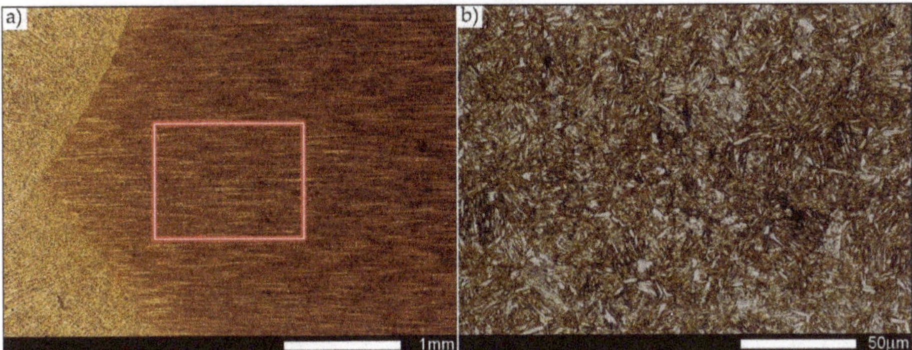

Figure 23. Microstructure of Hardox Extreme welded joint after heat treatment: (**a**) in the area marked HAZ in Figure 6; (**b**) an enlarged image marked with frame in Figure 23a. Structure of fine-lath hardening martensite. Light microscopy, etched with 2% HNO_3.

3.3. Results of Fractographic Analysis

Figures 24–28 show representative images of fracture surfaces of Hardox Extreme welded joints in conditions both after welding and after heat treatment operations. Fractographic analyses were made at the temperatures of impact testing, i.e., +20 °C and −40 °C. In each of the analyzed cases, fractures after impact tests do not show a significant share of plastic side zones (Figure 24), which proves relatively small energy expenditure during their creation.

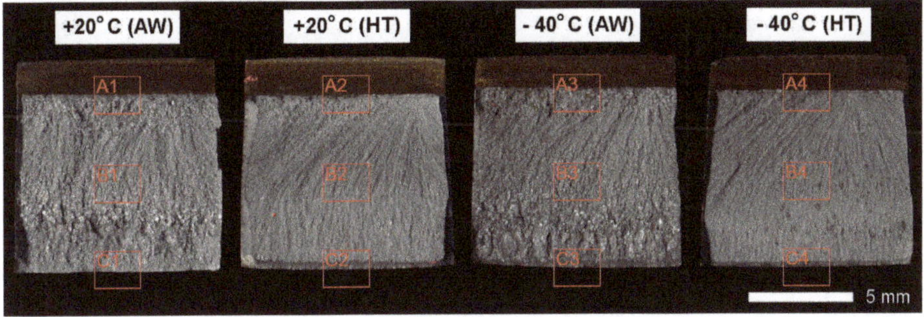

Figure 24. Macroscopic images of representative specimens of Hardox Extreme welded joints after impact testing. AW—condition after welding, HT—condition after heat treatment. Marked with frames: A—under-notch zone, B—central zone, C—final fracture zone. Stereoscopic microscopy, unetched.

This statement especially refers to fractures of the joint after welding, which were subjected to impact testing at the negative temperature. In addition, macroscopic analysis showed that fractures of the specimens after welding are characterized by highly diversified surface topography, resulting from the presence of coarse-grained structure on the fusion line, partially under the notch (frames A1 and A3 in Figure 24), in the central zone (frames B1 and B3 in Figure 24) and in major part of the final fracture zone (frames C1 and C2 in Figure 24). In turn, fracture surfaces of the heat treated joint can be found to be uniform and rough on their entire area, showing a characteristic run according to the crystallization direction after welding (Figure 24). In order to reveal a detailed structure of individual zones, all fracture surfaces were subjected to further examinations by using scanning electron microscopy.

Transcrystalline fractures of Hardox Extreme welded joints in the condition after welding, subjected to examination at both ambient and reduced temperatures, are fractures of mixed nature, with irregularities on the separation surface (steps) and with a clearly visible "river" system (Figures 25 and 26).

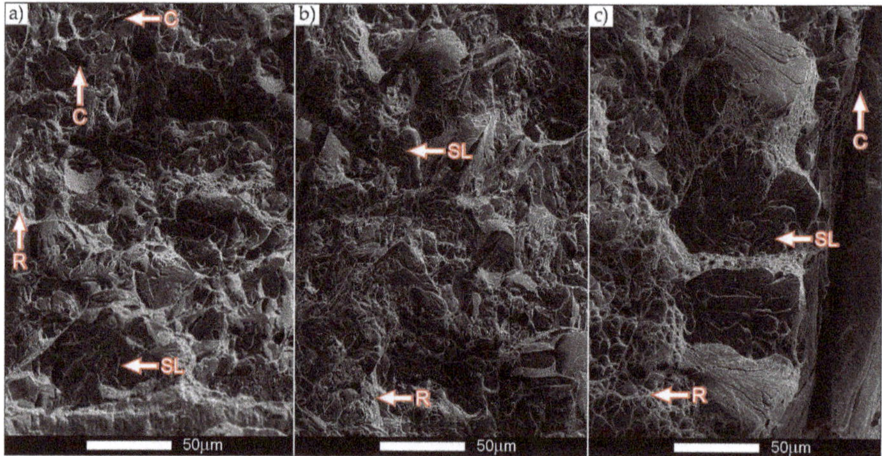

Figure 25. SEM images of fracture surfaces of Hardox Extreme welded joints after welding, shown in Figure 24, after impact testing at +20 °C. (**a**) Area marked with the frame A1; (**b**) area marked with the frame B1; (**c**) area marked with the frame C1. R—"river" system; SL—slides; C—microcracks with fine steps. Scanning microscopy, unetched.

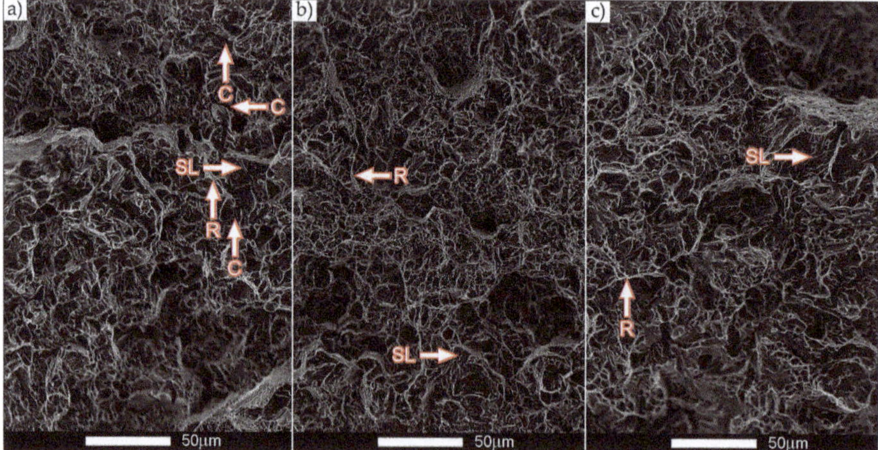

Figure 26. SEM images of fracture surfaces of Hardox Extreme welded joints after welding, shown in Figure 24, after impact testing at −40 °C. (**a**) Area marked with the frame A3; (**b**) area marked with the frame B3; (**c**) area marked with the frame C3. R—"river" system; SL—slides; C—microcracks with fine steps. Scanning microscopy, unetched.

In both cases, structures with micro-voids of different sizes can be also distinguished on fracture surfaces, where no inclusions of phases derived from alloying microadditives are observed. A characteristic feature of fracture surfaces of the specimens after welding is presence of numerous transverse microcracks with agglomerations of fine steps. Such a state is observed mostly in the zone under the notch (Figures 25a and 26a) and locally in the final fracture zone (Figure 25c). Moreover, almost in all areas of the fractures not subjected to heat treatment, slides are visible, which is characteristic for cleavage fractures.

Qualitative differences in fracture structures are clearly demonstrated on the specimens subjected to heat treatment after welding. On the surfaces, areas of micro-voids are observed, separated by plastic areas with band-like arrangement of "scaly" steps (Figures 27b,c and 28c).

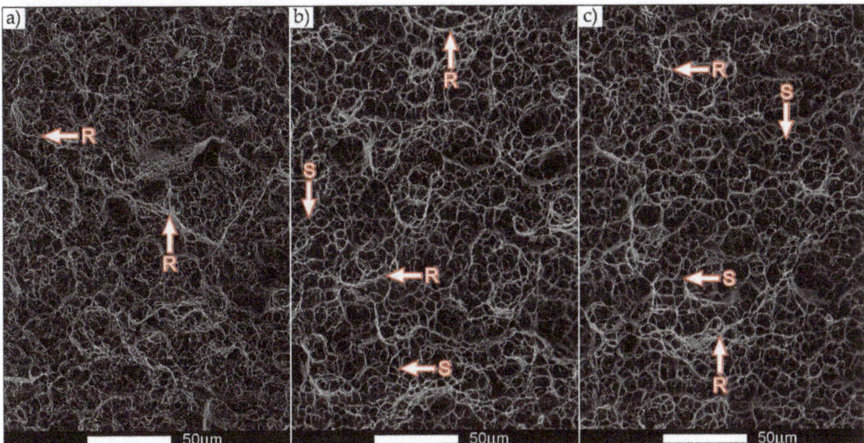

Figure 27. SEM images of fracture surfaces of Hardox Extreme welded joints after heat treatment, shown in Figure 24, after impact testing at +20 °C. (**a**) Area marked with the frame A2; (**b**) area marked with the frame B2; (**c**) area marked with the frame C2. R—"river" system; C—microcracks with fine steps; S—"scaly" steps. Scanning microscopy, unetched.

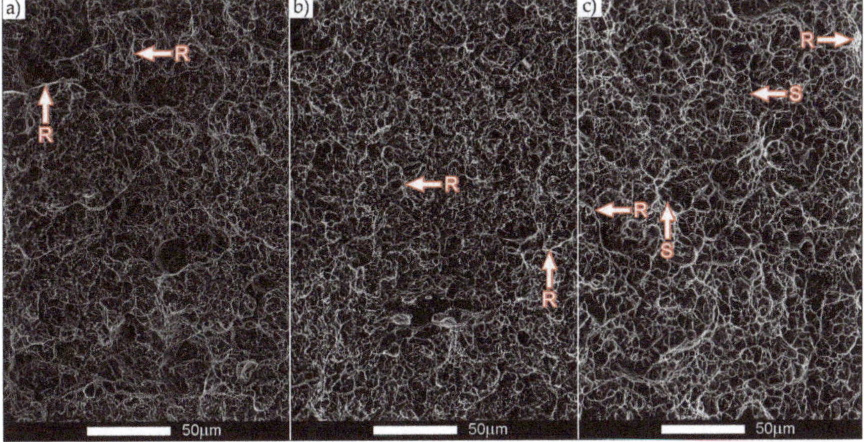

Figure 28. SEM images of fracture surfaces of Hardox Extreme welded joints after heat treatment, shown in Figure 24, after impact testing at −40 °C. (**a**) Area marked with the frame A4; (**b**) area marked with the frame B4; (**c**) area marked with the frame C4. R—"river" system; C—microcracks with fine steps; S—"scaly" steps. Scanning microscopy, unetched.

The described topography of a fracture surface is created by slips and decohesion, which results in appearance of microcracks in the planes {100} [29] and creation, after their separating walls are merged, of scales overlapping in a characteristic way. Parabolic contours of micro-voids indicate that the fracture is initiated by plastic deformation—slip—provoked by tangent forces in the process of fracture creation. It is worth mentioning that the cracking itself proceeds along the specific crystallographic planes.

In addition, it can be stated that identification of these planes is practically impossible because of the characteristic "river" relief occurring on fracture surfaces of the heat-treated specimens. This is caused by the fact that the meandering "river" system creates on a large area a system of micro-void coalescence characteristic for a plastic fracture. Mixed nature of the fractures contributes to the creation, during cracking, of steps increasing the amount of absorbed energy and thus decreasing the brittleness threshold.

4. Summary

It can be said on the grounds of the carried-out examinations that, in spite of limited weldability declared in the manufacturer's information materials, elements of the martensitic abrasive-wear resistant steel Hardox Extreme can be joined by welding techniques. It was shown in this elaboration that, through a proper selection of technology and welding parameters, it is possible to obtain an imperfection-free welded joint that is also characterized by very favorable strength indices. The above-mentioned assumptions were realized by proper selection of technology and welding parameters, as well as by use of additional heat treatment after welding, composed of normalization followed by volumetric quenching in oil and low tempering (stress-relieving). In the considered case, the applied thermal operations made it possible to restore, in the entire area of the welded joint, a structure and—to a large degree—a hardness level similar to those of the base material. The presented statement was positively verified in course of the carried-out strength and impact testing. The problems of execution and heat treatment of welded joints made of Hardox Extreme sheet metal can be characterized in the following ways:

- After welding, a highly morphologically diversified microstructure occurs in the entire area of the welded joint that shows, in comparison to the base material, lower hardness levels, tensile strength and impact energy. The obtained and relatively low mechanical properties of the welded joint make it possible to state that welding operations result in lowering the abrasive-wear resistance of the steel Hardox Extreme. The expected drop can occur in both the weld material and the area of base material directly adjacent to the very wide heat-affected zone.
- The additional thermal treatments carried-out after welding make it possible to favorably modify the structure in the entire welded joint and wide heat-affected zone, to obtain the structures similar to those of Hardox Extreme steel in the as-delivered condition from the manufacturer's plant.
- The obtained average hardness level of the welded joint after welding, amounting to only 17 J/cm^2 at +20 °C, clearly indicates susceptibility of this steel to brittle cracking. From practical point of view, it excludes application of welding techniques for joining Hardox Extreme sheets (irrespective of the relatively very high average tensile strength R_m = 1278 MPa obtained) with no additional heat treatment operations. The above statement is additionally confirmed by the performed fractographic analysis.
- Examination results of heat-treated welded joints of Hardox Extreme steel indicate a possibility of restoring structural, mechanical, and impact properties "degraded" as a result of welding to the level corresponding to the base material. In the case of tensile strength, the obtained result R_m = 1831 MPa makes a good reason for undertaking the problems of welding and heat treatment of the considered steel. An additional justification of this question is also obtained through heat treatment of other mechanical properties that are much better than those existing in the as-welded condition. In spite of a significant increase of the R_m value (and also of the yield point, not cited in the reference), a nearly 12% increase percentage reduction of area reaching Z = 29.1% (Table 5) was noted, as well as an increase of impact strength at ambient temperature to KCV = 27 J/cm^2.

It is worth noting that the brittleness threshold of constructional materials is accepted as impact strength of 35 J/cm^2 [30], which results from maintaining at least 50% share of ductile fracture. Therefore, the obtained impact strength value after heat treatment and results of fractographic analysis make it possible to conclude that there was an occurrence of a favorable "shift" of plastic properties of the welded joint beyond the accepted brittleness threshold.

Irrespective of the conclusions formulated above, it should be also mentioned that the examination results of the steel Hardox Extreme, presented in this paper, constitute a fragment of the cycle of research works concerning low-alloyed martensitic, abrasive-wear resistant steels with an addition of boron, realized by the author for several years. Thus, it can be stated that problems of weldability, announced before, are not limited to the steel Hardox Extreme only, but concern almost the entire considered group of materials. The materials already considered by the author include the steels: Hardox 400, Hardox 500, Hardox 600, HTK700H, HTK900H, AR400, XAR 600, Creusabro 4800, Creusabro 8000, TBL Plus, B27, Brinar 400, Brinar 500, and others. In almost all of the considered cases, problems with welding of these steels were observed, although the manufacturers declared their weldability. Therefore, attempts to weld high-strength abrasive-wear resistant steels with subsequent post-weld heat treatment seem to be very well grounded. In addition, it is also worth considering the application of advanced methods of hybrid welding and dedicated induction heating stations for welding and heat treatment of the above-mentioned materials. Optimum usage of this type infrastructure would make it possible to apply welding techniques for joining high-strength abrasive-wear resistant martensitic steels, while maintaining their very profitable mechanical properties and usable features.

Funding: This research received no external funding.

Acknowledgments: The author would like to thank Józef Ptak from Stal-Hurt S.C. company for providing sheets of Hardox steels and for Zbigniew Konat and Eugeniusz Szymanowicz from the Wroclaw Shipyard for assistance in the implementation of welded joints.

Conflicts of Interest: The author declares no conflict of interest.

References

1. SSAB Data Sheet Hardox. Available online: https://www.ssab.com/products/brands/hardox (accessed on 11 August 2019).
2. Ilsenburger Grobblech GmbH Data Sheet Brinar 400. Available online: https://www.ilsenburger-grobblech.de/fileadmin/mediadb/ilg/infocenter/downloads/werkstoffblaetter/eng/abrasion_resistant_brinar400.pdf (accessed on 11 August 2019).
3. Ilsenburger Grobblech GmbH Data Sheet Brinar 500. Available online: https://www.ilsenburger-grobblech.de/fileadmin/mediadb/ilg/infocenter/downloads/werkstoffblaetter/eng/abrasion_resistant_brinar500.pdf (accessed on 11 August 2019).
4. Stal-Hurt Data Sheet Hardox. Available online: https://www.stal-hurt.com/ (accessed on 5 April 2019).
5. ThyssenKrupp Steel Europe Data sheet XAR 600. Available online: https://www.flinkenberg.fi/wp-content/uploads/DATASHEET-XAR600.pdf (accessed on 5 April 2019).
6. Bugłacki, H.; Smajdor, M. Mechanical Properties of Abrasion-Resistant Hardox 400 Steel and Their Welded Joints. *Adv. Mater. Sci.* **2003**, *4*, 5–8.
7. Magdalena, M.; Robert, U.; Otakar, B. The impact of welding wire on the mechanical properties of welded joints. *Mater. Eng. Mater. Inžinierstvo* **2014**, *21*, 122–128.
8. Eva, B.; Harold, M.; Pavol, R. Welding of High Strength Materials Used in the Manufacture of Special Equipment. *Univ. Rev.* **2014**, *8*, 51–61.
9. Uzunali, U.Y.; Cuvalci, H. The effects of post weld heat treatment on the mechanical properties of tempered martensite and high strength steel welded joints. In Proceedings of the 2015 World Congress on Advances in Structural Engineering and Mechanics (ASEM15), Incheon, Korea, 25–29 August 2015.
10. Pękalski, G.; Haimann, K.; Konat, Ł.; Koniarek, K.; Krugła, M.; Mroczkowski, L.; Orłowski, J.; Oskwarek, M.; Ptak, T.; Szymczak, H. *Material Testing of Hardox 400 and Hardox 500*; Wrocław Univertsity of Science and Technology Report SPR: Wrocław, Poland, 2005. (In Polish)

11. Konat, Ł. Structures and Properties of Hardox Steels and their Application Possibilities in Conditions of Abrasive Wear and Dynamic Loads. Ph.D. Thesis, Wroclaw University of Technology, Wroclaw, Poland, 2007. (In Polish).

12. Dudziński, W.; Konat, Ł.; Pękalski, G. Modern constructional steels. In *Maintenance Strategy of Surface Mining Machines and Facilities with High Degree of Technical Degradation*; Dudek, D., Ed.; Publishing House of Wroclaw University of Technology: Wroclaw, Poland, 2013; pp. 346–366. (In Polish)

13. Dudziński, W.; Konat, Ł.; Pękalski, G. Structural and strength characteristics of wear-resistant martensitic steels. *Arch. Foundry Eng.* **2008**, *8*, 21–26.

14. Cegiel, L.; Konat, Ł.; Pawłowski, T.; Pękalski, G. Hardox Steels New generations of construction materials for surface mining machinery. *Brown Coal* **2006**, *3*, 24–29. (In Polish)

15. Pękalski, G.; Konat, Ł.; Oskwarek, M. Macro- and microstructural properties of welded joints of Hardox 400 and Hardox 500 steels. In Proceedings of the XIX Scientific Conference on Development Problems of Working Machines, Zakopane, Poland, 23–26 January 2006.

16. Frydman, S.; Konat, Ł.; Pękalski, G. Structure and hardness changes in welded joints of Hardox steels. *Arch. Civ. Mech. Eng.* **2008**, *8*, 15–27. [CrossRef]

17. Konat, Ł.; Białobrzeska, B.; Białek, P. Effect of Welding Process on Microstructural and Mechanical Characteristics of Hardox 600 Steel. *Metals* **2017**, *7*, 349. [CrossRef]

18. Pawlak, K.; Białobrzeska, B.; Konat, Ł. The influence of austenitizing temperature on prior austenite grain size and resistance to abrasion wear of selected low-alloy boron steel. *Arch. Civ. Mech. Eng.* **2016**, *16*, 913–926. [CrossRef]

19. Łętkowska, B. Influence of Heat Treatment on Structure and Selected Properties of B27 and 28MCB5 Steels. Ph.D. Thesis, Wroclaw University of Technology, Wroclaw, Poland, 2013. (In Polish).

20. Konat, Ł.; Pękalski, G. Structures and selected properties of Hardox steels in the context of their use in surface mining machinery construction. In Proceedings of the XV International Symposium on Mine Planning Equipment Selection (MPES 2006), Torino, Italy, 20–22 September 2006.

21. Frydman, S.; Konat, Ł.; Łętkowska, B.; Pękalski, G. Impact resistance and fractography of low-alloy martensitic steels. *Arch. Foundry Eng.* **2008**, *8*, 89–94.

22. Dudziński, W.; Konat, Ł.; Pękalska, L.; Pękalski, G. Structures and properties of Hardox 400 and Hardox 500 steels. *Mater. Eng.* **2006**, *3*, 139–142. (In Polish)

23. Dudziński, W.; Konat, Ł.; Białobrzeska, B. Fractographic analysis of selected boron steels subjected to impact testing. *Arch. Met. Mater.* **2015**, *60*, 2373–2378. [CrossRef]

24. Dudziński, W.; Białobrzeska, B.; Konat, Ł. Comparative analysis of structural and mechanical properties of selected low-alloy boron-containing abrasive-wear resistant steels. In *Polish Metallurgy in the Years 2011–2014*; Świątkowski, K., Ed.; Akapit: Kraków, Poland, 2014; pp. 871–888. (In Polish)

25. Białobrzeska, B.; Konat, Ł.; Jasiński, R. The influence of austenite grain size on the mechanical properties of low-alloy steel with boron. *Metals* **2017**, *7*, 26. [CrossRef]

26. Białobrzeska, B.; Konat, Ł.; Jasiński, R. Fractographic Analysis of Brinar 400 and Brinar 500 Steels in Impact Testing. *Scanning* **2018**, *2018*, 17. [CrossRef] [PubMed]

27. Graville, B.A. Cold Cracking in welds in HSLA steels, welding of HSLA (Microalloyed) structural steels. In Proceedings of the AIM/ASM Conference, Rome, Italy, 9–12 November 1976.

28. ESAB. *Welding Consumables*; ESAB: Gothenburg, Swedish, 2012; p. H17.

29. Maciejny, A. *Brittleness of Metals*; Publishing House Śląsk: Katowice, Poland, 1973. (In Polish)

30. Wyrzykowski, J.W.; Pleszakow, E.; Sieniawski, J. *Deformation and Cracking of Metals*; WNT: Warszawa, Poland, 1999. (In Polish)

 *metals* 

Article

Effects of Cerium on Weld Solidification Crack Sensitivity of 441 Ferritic Stainless Steel

Shuangchun Zhu [1,2] **and Biao Yan** [1,*]

[1] School of Materials Science and Engineering, Tongji University, Shanghai 201804, China; zhushuangchun@baosteel.com
[2] Central Research Institute, Baoshan Iron & Steel Co., Ltd., Shanghai 201900, China
[*] Correspondence: yan_biao@tongji.edu.cn; Tel.: +86-21-6958-2007

Received: 26 February 2019; Accepted: 13 March 2019; Published: 22 March 2019

Abstract: The addition of rare earth element Ce in ferritic stainless steel can improve the high temperature performance to meet the service requirements of automobile exhaust systems at high temperatures. Automobile exhaust systems are generally applied as welded pipes, so it is necessary to study the effect of Ce on the weldability of ferritic stainless steel. In this study, the Trans-varestraint test method was used to test the solidification crack sensitivities of 441 and 441Ce ferritic stainless steel. The 441Ce steel, which has added Ce, showed poor resistance to weld solidification cracking. Using Thermo-Calc software, Ce was observed to expand the solidification temperature range of 441 ferritic stainless steel, increase the time for solid–liquid coexistence during solidification, and increase the sensitivity of solidification cracking. Further, from scanning electron microscopy and energy dispersive spectrometer analysis, the addition of Ce was found to reduce high temperature precipitation (Ti,Nb)(C,N), reduce or even eliminate the "pinning" effect during solidification, and increase solidification crack sensitivity of 441 ferritic stainless steel.

Keywords: ferritic stainless steel; cerium; solidification crack; Trans-varestraint test

---

1. Introduction

As an important type of stainless steel, ferritic stainless steel has been a popular focus of research and application in recent years, not only because of its lack of nickel content but also because it is economical. Its cost is not affected by fluctuations in international nickel prices, and compared with austenitic stainless steel, ferritic stainless steel too has good thermal conductivity, low thermal expansion coefficient, good high temperature oxidation resistance, and good high temperature thermal fatigue resistance; thus, it has a wide range of applications in automotive exhaust systems [1–4]. However, with the constantly increasing requirements of automobile exhaust emission standards, automobile engine technology is continuously improving, and the exhaust gas temperature is continuously increasing. Better high temperature performance requirements are therefore imposed on the high temperature end material of the exhaust system [5–7].

Rare earth elements have a unique electronic layer structure and active chemical properties. In stainless steel, they can clean the steel, metamorphose inclusions, control solidification structure, and refine the grain [8–12], so they have received much attention from scholars these days. Some scholars have shown that the addition of rare earth elements in ferritic stainless steel can improve its corrosion resistance, mechanical properties, high temperature performance, and other service properties [13–16]. In order to study and improve the high temperature oxidation resistance, previously reported studies [17,18] show that adding Ce can increase densification of the oxidation film of ferritic stainless steel and improve its high temperature oxidation resistance. However, research data on the influence of rare earth elements on weld solidification crack sensitivity is still lacking. Automobile exhaust systems are generally applied as welded pipes, so the alloying effect on the weldability of ferritic stainless steel needs to be investigated. Only with excellent weldability can the alloyed stainless steel be utilized commercially. In this study, the 441 type ferritic stainless steel is the research object. The effect of Ce on the weld solidification crack sensitivity of 441 ferritic stainless steel and its mechanism were studied using the Trans-varestraint test, and the theoretical support for solidification crack control of Ce-containing ferritic stainless steel is provided.

2. Materials and Methods

The test materials used were two types of hot-rolled ferritic stainless steel. The chemical compositions of the test materials are shown in Table 1. The Trans-varestraint test method was used to evaluate the solidification crack sensitivities of these materials. The test equipment utilized the MTV2500 testing system produced by D. L. Wright Inc. Samples of dimensions 120 mm × 30 mm × 6 mm without the weld groove were machined for the Trans-varestraint tests. The weld was completed by arc remelting, and the remelting position and experimental principle are shown schematically in Figure 1 [19]. The principle of this method is to apply different strain values to study the crack generation. An autogenous gas tungsten arc welding method was used to melt the weld bead. The welding current and arc voltage were 150 A and 17 V, respectively. The welding speed was 100 mm/min. The welding shielding gas used was 99.999% argon gas, and the gas flow rate was 12 l/min. Nine series of strains (i.e. 0%, 0.25%, 0.50%, 1%, 2%, 3%, 4%, 5% and 6%) were used in the tests. The Trans-varestraint tests were performed in increasing order of strain from small to large values.

The surface cracks of the welds of the specimens that have completed the Trans-varestraint test are observed by scanning electron microscopy (SEM, ZEISS EVO MA 25, ZEISS Group, Cambridge, UK) to note the occurrence of cracks and to measure the lengths of these cracks. Metallographic specimens of the weld cross section were cut from the tested specimens under 0% strain. The specimens were machined, polished, corroded (corrosive solution, $FeCl_3$:HCl:H_2O = 5 g:50 mL:100 mL), cleaned with anhydrous ethanol, and dried. The microstructures of the specimens were observed by SEM (FEI QUANTA 600F, FEI Company, Eindhoven, Netherlands) and equipped with an energy dispersive spectrometer (EDS, OXFORD INCAx-act, Oxford Instruments, Oxfordshire, UK). The Scheil solidification simulation module of the thermodynamic calculation software Thermo-Calc was used along with the TCFE9 database to simulate the stainless steel welding solidification process.

Table 1. Chemical compositions of test materials (wt%).

Steel	C	Si	Mn	P	S	Cr	Nb	N	Ti	Ce
441	0.005	0.380	0.340	0.008	0.002	18.630	0.430	0.008	0.140	-
441Ce	0.008	0.360	0.310	0.008	0.001	18.440	0.460	0.008	0.190	0.024

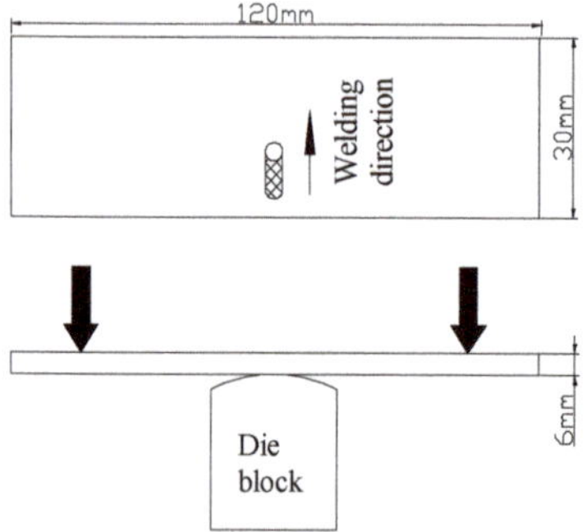

Figure 1. Trans-varestraint test.

3. Results and Discussion

3.1. Threshold Strain and Maximum Crack Distance

There are many quantitative indicators for evaluating the thermal crack sensitivities of materials. In this study, the threshold strain and maximum crack distance (MCD) are used as evaluation indexes of weld solidification cracks. Threshold strain refers to the strain at crack initiation during a Trans-varestraint test for a series of small to large strains. Saturation strain refers to the strain that exceeds the threshold value where the maximum crack length no longer shows a significant change; the MCD refers to the maximum crack length (MCL) measured for each value of the test strain above the saturation strain. Figure 2 shows the MCLs of the test materials for the transverse variable restraint welding solidification crack test under various strains. It can be seen from the graph that the threshold strain of the 441 ferritic stainless steel is between 0.5% and 1%, whereas the threshold strain of the 441Ce ferritic stainless steel is between 0% and 0.25%. The MCD values for 441 and 441Ce ferritic stainless steels are 239.5 ± 0.1 µm and 290.0 ± 0.1 µm, respectively. Thus, it can be concluded that the addition of Ce increases the weld solidification crack sensitivity of 441 ferritic stainless steel.

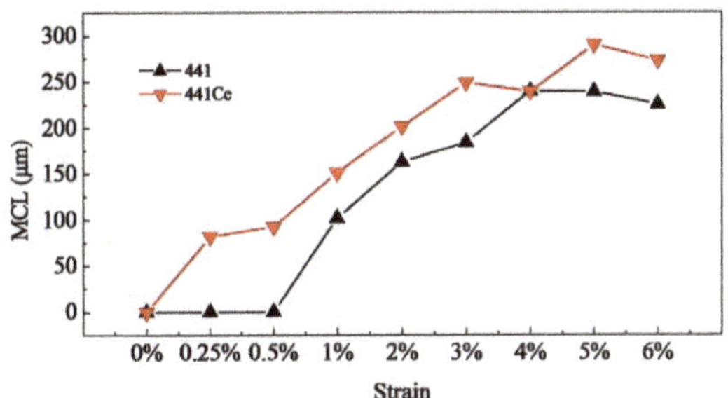

Figure 2. MCLs of the test materials under various strains.

3.2. Thermodynamic Calculation of Solidification Process

Table 2 shows the solidification temperature range obtained from simulation calculation. It can be seen that the solidification temperature range of the 441Ce material with the rare earth element Ce reached 199 °C, which is approximately 14 °C higher than that of the 441 material, indicating that Ce expands the solidification temperature range. Thus, Ce addition increases the area and time of solid–liquid coexistence during solidification, greatly increases the risk of solidification cracking, and reduces the ability of stainless steel to resist solidification cracking.

Table 2. The solidification temperature range obtained by simulation calculation.

Steel	Solidification Start Temperature/°C	Solidification End Temperature/°C	Solidification Temperature Range/°C
441	1504	1319	185
441Ce	1502	1303	199

3.3. Solidification Crack Distribution and Morphology

Figure 3 depicts the crack distribution and morphology of the test specimen under Trans-varestraint test for the 5% strain condition. The direction of the columnar crystal of the weld is mostly perpendicular to the fusion line during solidification crystallization. This is because the liquid metal solidifies and crystallizes along the direction of the temperature gradient. When the columnar crystal grows, there is a liquid phase film between the front columnar crystals of the solidification interface. At this time, a transverse strain is applied to the welding metal, and the liquid phase film that has not solidified at the grain boundary is insufficient to fill the gap generated by the strain, thus forming cracks at the grain boundary. As can be seen from Figure 3, the cracks are primarily situated at the grain boundary position, verifying that the solidification cracks occur between the columnar crystals.

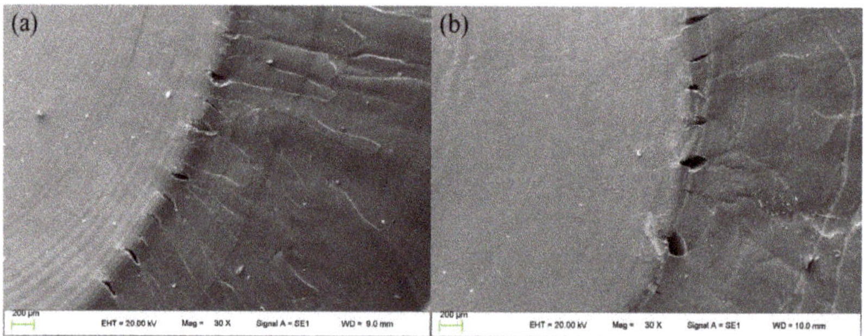

Figure 3. Crack distributions and morphologies under 5% strain for (**a**) 441 and (**b**) 441Ce type materials.

Figure 4 shows the typical shape of a crack formed in the Trans-varestraint test, which applies instantaneous strain to the specimen. Under the application of a sudden strain, owing to the temperature difference of the instantaneous cracking zone, the solidification crack morphology produced by the Trans-varestraint test has the characteristics of both high and low temperature zones [20]. The high temperature zone is close to the crystal front and shows the characteristics of liquid phase fracture; the liquid film traces at the time of fracture are visible between the crystals. Peng et al. [21] believe that the crystallization process of the columnar crystal at the moment of cracking in the high temperature region is complete or nearly complete, and the connection, contact, and void formation of adjacent columnar crystals are also complete; this means that at this moment, the adjacent columnar crystals have only the connections of the residual low-melting of the liquid phase film and the liquid bridge forms. The low temperature zone refers to the zone of solidification cracking that is away from the crystallization front. In this zone, it is observed that the columnar crystal plane with complete crystal morphology has annular or striped ridges, and the arrangement direction of each annular ridge and the main direction of the striped ridges are roughly parallel to the direction of the columnar crystals, similar to a waterfall; the low temperature zone is close to final solidification (i.e., in a liquid–solid zone with poor fluidity), which is pulled apart under a large tensile strain and forms the characteristic shapes of the annular and striped ridges.

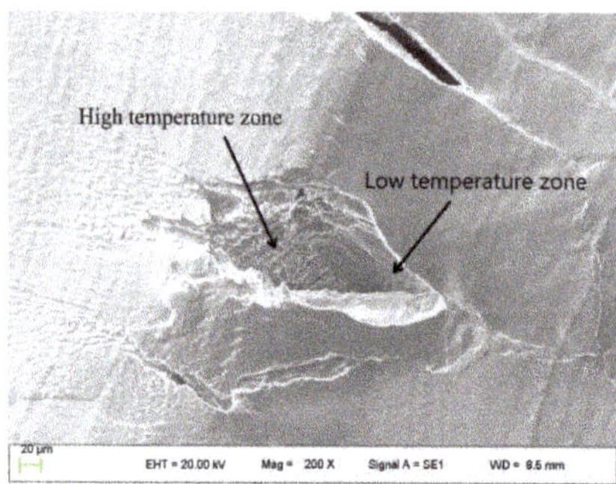

Figure 4. Typical morphology of a crack. Test material: 441 steel, test strain: 5%.

3.4. Precipitation

Figure 5 shows the metallographic structure of the weld cross-section after application of the Trans-varestraint test for the 0% strain condition, as observed under scanning electron microscopy (SEM). It can be seen from the figure that a large number of precipitates appear in the intragranular and grain boundaries of the 441 material. The precipitates of 441Ce are relatively small. This indicates that the addition of the alloying element Ce reduces the generation of the precipitated phase and has a purifying effect on the grain boundary as well as the crystal inside, thus greatly influencing the hot crack sensitivity and strength of the material.

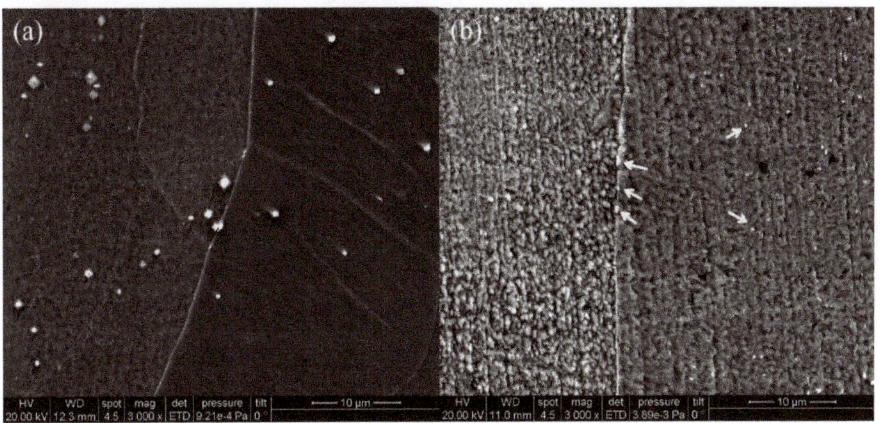

Figure 5. Metallographic structures of the weld cross-sections of (**a**) 441 and (**b**) 441Ce type steel.

The grain boundary of the metallographic structure of each weld cross-section was further magnified under SEM, and energy dispersive spectrometer (EDS) analysis was performed on the precipitation on the grain boundary. The results are shown in Figure 6. The precipitation of the 441 weld grain boundary is granular, and the size is approximately 1 μm. The EDS result shows that it is a composite precipitation rich in Fe, Cr, Ti, and Nb. The precipitation of the 441Ce weld grain boundary is relatively small, and the spectrum of the precipitation is similar to that of the 441 weld grain boundary. The thermodynamic calculation results in Table Table 2 indicate that Ce expands the solidification temperature range, increases the degree of undercooling during the solidification process, promotes heterogeneous nucleation and growth of the liquid metal during solidification, increases the grain boundary, and refines the grain, so that the precipitated phase is dispersed. In addition, rare earth elements can reduce the activity of C, increase the solubility of C, and reduce the precipitation of carbides of Ti and Nb during solidification [22]. Ce is easily segregated near the grain boundary because in the cooling process following the solidification of the weld, the radius of the Ce atom is large, which tends to segregate Ce near the grain boundary, thus causing lattice expansion near the grain boundary and increasing its energy. It is easy to nucleate the carbide [23] so that the precipitation of the grain boundary is fine and exhibits a dispersed distribution.

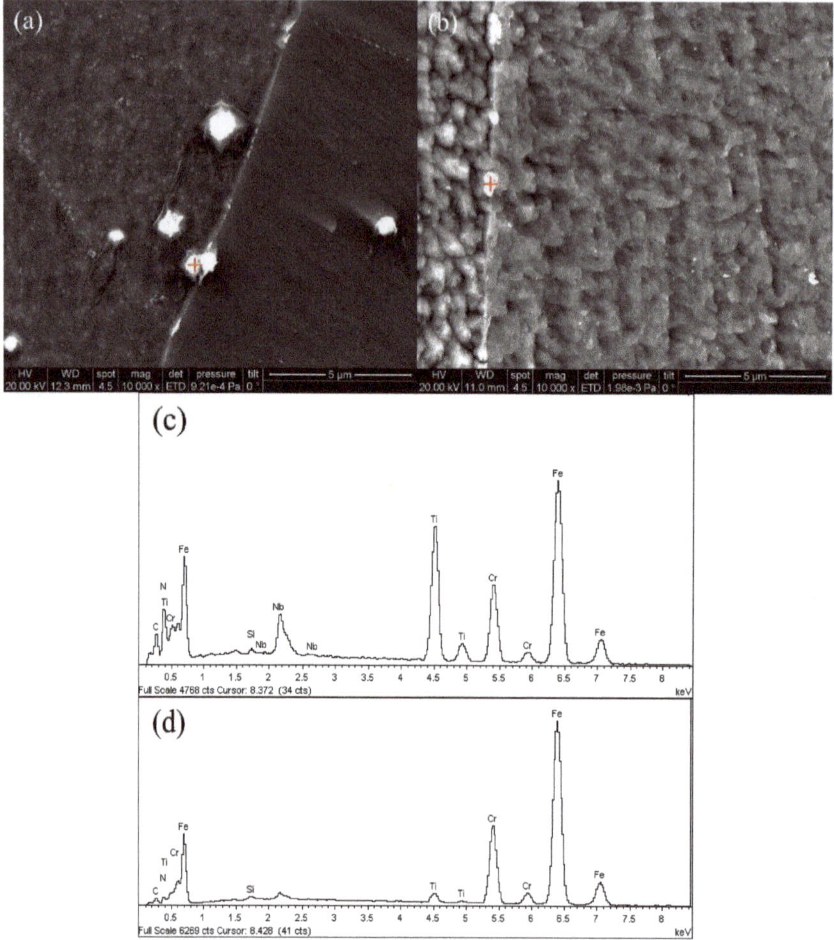

Figure 6. SEM images and EDS analyses of precipitations on the grain boundaries of (**a**,**c**) 441 and (**b**,**d**) 441Ce type materials, respectively.

Figure 7 depicts the phase-temperature equilibrium diagrams of 441 and 441Ce steel using Thermo-Calc software. When 441 ferritic stainless steel solidifies, it produces high melting point precipitates (Ti,Nb)(C,N). Some of these precipitations occur on the solidification grain boundary, destroying the liquid film that has not yet solidified at the boundary, improving the bonding strength of the boundary, and forming a "pinning" effect, as shown in Figure 8. This effect reduces the sensitivity of the weld solidification crack. Compared to the 441 stainless steel without added Ce, 441Ce increases the solidification temperature range, prolongs the total time of the weld in the solid–liquid mixing stage during the welding cooling process, and greatly increases the risk of solidification cracking. In addition, the Ce atoms are segregated on the grain boundary during solidification [23], which easily forms a low-melting liquid film. Because Ce promotes a solid solution of C [23], it is not conducive to precipitation of carbides of Ti and Nb, thus reducing such precipitations. Further, it reduces the pinning effect of precipitation, so the addition of Ce greatly improves the weld solidification crack sensitivity of 441 ferritic stainless steel.

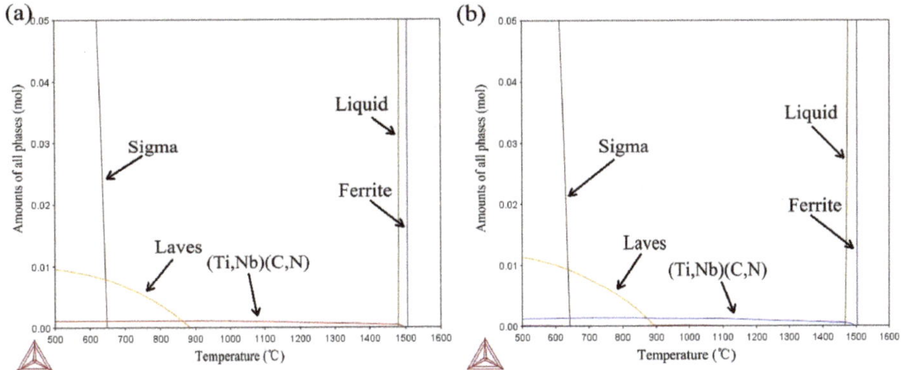

Figure 7. Phase-temperature equilibrium diagrams for (**a**) 441 and (**b**) 441Ce type ferritic stainless steel materials.

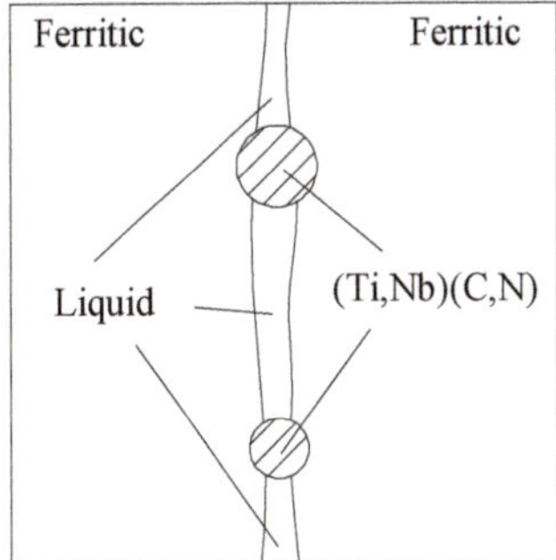

Figure 8. Pinning effect of precipitation in the solidification grain boundary.

4. Conclusions

The sensitivity of the solidification crack of 441 ferritic stainless steel can be improved by adding the rare earth element Ce; the observations and conclusions of such an addition are as follows:

(1) Ce widens the solidification temperature range of 441 ferritic stainless steel, so in the actual welding process, it increases the duration of the weld metal in the solid–liquid mixing stage, thus improving the solidification crack sensitivity of stainless steel.

(2) The high temperature precipitates produced during solidification of 441 ferritic stainless steel weld metal, especially at the solid–liquid interface, can pin the interface, improve the bonding strength of the interface, and prevent the solidification cracks. After adding Ce to 441 ferritic stainless steel, the precipitates produced in the weld can be reduced and the weld structure can be purified. It is precisely because of this purifying effect that the pinning due to high temperature precipitates in the solidification process of the weld metal is weakened and solidification crack sensitivity is improved.

(3) In a future work, the influence of Ce on solidification temperature range and the precipitates will be studied in depth, and the improvement of the anti-solidification crack properties of Ce-containing ferritic stainless steel will be explored.

Author Contributions: Conceptualization: S.Z. and B.Y.; investigation: S.Z.; original draft preparation: S.Z.; review and editing: S.Z. and B.Y.

Funding: This research received no external funding.

Acknowledgments: This work was supported by the Baoshan Iron & Steel Co., Ltd.

Conflicts of Interest: The authors declare no conflict of interest.

References

1. Bi, H.Y.; Wang, Z.Y.; Li, X.; Wang, B.S. Stainless steel for automotive exhaust system and pipes welding. *World Iron Steel* **2011**, *5*, 1–9. [CrossRef]
2. Kim, J.K.; Kim, Y.H.; Uhm, S.H.; Lee, J.S.; Kim, K.Y. Intergranular corrosion of Ti-stabilized 11 wt% Cr ferritic stainless steel for automotive exhaust systems. *Corros. Sci.* **2009**, *51*, 2716–2723. [CrossRef]
3. Ali-Löytty, H.; Jussila, P.; Valden, M. Optimization of the electrical properties of Ti–Nb stabilized ferritic stainless steel SOFC interconnect alloy upon high-temperature oxidation: The role of excess Nb on the interfacial oxidation at the oxide–metal interface. *Int. J. Hydrogen Energy* **2013**, *38*, 1039–1051. [CrossRef]
4. Inoue, Y.; Kikuchi, M. Present and future trends of stainless steel for automotive exhaust system. *Nippon Steel Tech. Rep.* **2003**, *88*, 62–69.
5. Sato, E.; Tanoue, T. Present and future trends of materials for automotive exhaust system. *Nippon Steel Tech. Rep.* **1995**, *64*, 13–19.
6. Miyazaki, A.; Hirasawa, J.; Satoh, S. Advanced stainless steels for stricter regulations of automotive exhaust gas. *Kawasaki Steel Tech. Rep.* **2000**, *43*, 21–28.
7. Li, X.; Bi, H.Y.; Zhang, Z.X. Development of ferritic stainless steels at Baosteel for the hot end of the exhaust system. *Baosteel Technol.* **2013**, *5*, 15–21. [CrossRef]
8. Keming, F.; Ruiming, N. Research on determination of the rare earth content in metal phases of steel. *Metall. Trans. A* **1986**, *17*, 315–323. [CrossRef]
9. Du, T. The effect and mechanism of rare earth elements in metals. *Chin. J. Nonferrous Met.* **1996**, *6*, 12–18. [CrossRef]
10. Chen, Z.P.; Xu, Y.T.; Gu, L.M. Effect of Rare Earth on the Inclusions of Ultra Clean Ferritic Stainless Steel. In Proceedings of the 2012 National Conference on Metallurgical Physical Chemistry, Kunming, China, 15–17 August 2012; pp. 652–656.
11. Wang, L.M. Application prospects and behavior of RE in new generation high strength steels with superior toughness. *J. Chin. Rare Earth Soc.* **2004**, *22*, 48–54.
12. Zhang, S.H.; Yu, Y.C.; Li, H.; Ren, X.; Wang, S.B. Effect of cerium on solidification structure of SUS434 ferritic stainless steel. *Foundry Technol.* **2016**, *37*, 2641–2644. [CrossRef]
13. Dong, F.; Zhao, X.H.; Yang, L. Effect of Ce on intergranular corrosion of 00Cr12 ferrite stainless steel. *Chin. Rare Earths* **2014**, *35*, 86–91. [CrossRef]
14. Zheng, F.; Cheng, T.Y.; Zhang, Q.Y. Application and development of cerium in stainless steel. *Nonferrous Met.* **2011**, *63*, 78–80. [CrossRef]
15. Sun, S.Y.; Yuan, S.Q.; Zhou, G.S.; Chai, D.L. Influence of rare earth element on the mechanical properties and microstructure of ferritic stainless steel. *Trans. Mater. Heat Treat.* **2007**, *28*, 22–24. [CrossRef]
16. Zhu, J.X.; Chen, D.X.; Qi, G.P.; Ge, Q.; Chen, X.; Cai, Z.; Wang, L.M. Modifications on Resistance to High-Temperature-Oxidation of AISI430 Ferrite Stainless Steel by Addition of Rare Earth Elements. In Proceedings of the 2006 National Conference on Metallurgical Physical Chemistry, Jinan, China, 17–20 October 2006; pp. 490–496.
17. Wei, L.L.; Zheng, J.H.; Chen, L.Q.; Misra, R.D.K. High temperature oxidation behavior of ferritic stainless steel containing W and Ce. *Corros. Sci.* **2018**, *142*, 79–92. [CrossRef]
18. Wei, L.L.; Chen, L.Q.; Ma, M.Y.; Liu, H.L.; Misra, R.D.K. Oxidation behavior of ferritic stainless steels in simulated automotive exhaust gas containing 5 vol.% water vapor. *Mater. Chem. Phys.* **2018**, *205*, 508–517. [CrossRef]
19. Lippold, J.C. *Welding Metallurgy and Weldability*; John Wiley & Sons, Inc.: Hoboken, NJ, USA, 2015; pp. 345–346.
20. Lu, W.X.; Meng, Q.S.; Yang, S.J. The crystal crack form and the fracture character of austenitic weld under the Trans-varestraint hot crack test. *J. Taiyuan Univ. Technol.* **1986**, *1*, 21–27. [CrossRef]
21. Peng, R.H.; Li, M.Z.; Sun, L.M. Investigation on the dendrites in weld metals of austenitic heat-resistant cast steels and on the fracture appearance of solidification cracks. *J. Iron Steel Res.* **1985**, *5*, 307–311. [CrossRef]

22. Zhao, L.P.; Zhang, H.M.; Sun, X.S.; Cui, C.Y. Effect of adding methods and addition amount of Ce on microstructure of 2Cr13 stainless steel. *Trans. Mater. Heat Treat.* **2012**, *33*, 91–94. [CrossRef]
23. Li, Y.B.; Wang, F.M.; Li, C.R. Effect of cerium on grain and carbide in low chromium ferritic stainless steels. *J. Chin. Rare Earth Soc.* **2009**, *27*, 123–127.

 *metals*

Article

Interface Characterization of Ultrasonic Spot-Welded Mg Alloy Interlayered with Cu Coating

Amir Badamian [1],*, Chihiro Iwamoto [2], Shigeo Sato [2] and Suguru Tashiro [2]

[1] Graduate school of science and engineering, Ibaraki University, Hitachi, Ibaraki 316-8511, Japan
[2] Department of Materials Science and Engineering, Ibaraki University, Hitachi, Ibaraki 316-8511, Japan; chihiro.iwamoto.77@vc.ibaraki.ac.jp (C.I.); shigeo.sato.ar@vc.ibaraki.ac.jp (S.S.); suguru.tashiro.5045@vc.ibaraki.ac.jp (S.T.)
* Correspondence: amir_badamian@yahoo.co.jp

Received: 8 April 2019; Accepted: 5 May 2019; Published: 8 May 2019

Abstract: The effect of Cu coating metallic interlayer on the weldability, joint strength, and interfacial microstructure during high-power ultrasonic spot welding (HP-USW) of AZ31B Mg alloy has been studied. Interlayered samples exhibited good weldability and they resulted in strong sound joints with nearly the same strength as joints without interlayer, with the distinction of lower energy being required. The Cu interlayer affected the thermal and vibrational properties of the interface, as the maximum interface temperature decreased and approached better uniformity across the weld nugget. The base metal grain structure changed to equiaxed larger grains after ultrasonic welding and a chain of parent metal small grains were observed around the interface. A binary intermetallic compound product of Mg-Cu, which was rich in Mg, has been found around the interface that was diffused toward base metal. According to the electron probe micro-analyzer (EPMA) results, alongside temperature measurements and hardness data, the formation of Mg_2Cu is suggested in this region. At the interface centerline, a narrow region was identified that was composed of Mg, Cu, and Al. Complementary transmission electron microscopy analysis estimated that Al-containing reaction product is a ternary alloy of the $MgCu_xAl_y$ type. The dispersion of fine grain intermetallic compounds as discrete particles inside Mg substrate in both interfacial regions formed a composite like structure that could participate in joint strengthening.

Keywords: high-power ultrasonic welding; interface; magnesium; cu interlayer; intermetallic compound

1. Introduction

Mg alloys are increasing in demand due to their high specific strength (strength to density ratio) and they have attractive applications in automotive industries, aerospace, medical implants, electronic appliances, and sport equipment [1,2]. Different methods are available for the joining of magnesium alloys. The primary methods include tungsten arc welding (TIG), laser beam welding (LBW), resistance spot welding (RSW), electromagnetic welding (EMW), electron beam welding (EBW), diffusion bonding, soldering, brazing, adhesive bonding, and mechanical fastening. Fusion based welding methods, like TIG, LBW, RSW, etc., have major problems that are mostly related to low boiling temperature of magnesium alloys that leads to porosities and hot cracking in the weld line and heat affected zone (HAZ). The other main limitations include high energy consumption and the need of consumables [3,4]. Therefore, the development of solid-state processes, like friction stir welding (FSW), friction stir spot welding (FSSW), and ultrasonic spot welding (USW) helped to avoid problems that are related to melting and provide opportunities for dissimilar material welding [5–7]. However, FSW and FSSW have the limitations of making "keyhole" and the need to fixture and a long process time [5]. USW is considered to be a solid-state welding process, with low energy consumption and time saving process and no need for consumables or atmosphere control. It is using relatively high frequency ultrasonic vibrations (20–40 KHz) that induce the workpieces to be held together with simultaneous pressure, which leads to frictional shear forces parallel to the interface. Under the effect of frictional shear forces, surface oxide layers break down and two joining parts approach bare contact and the temperature increases due to friction, and then local welding occurs between the asperities of two faying surfaces. Finally, the spreading of local micro-welds and coalescence results in the creation of a metallurgical bond in the weld spot [5,7–11].

High-power ultrasonic spot welding (HP-USW) is an effective way of achieving sound strong joints among different USW methods. This method uses higher power and lower energy and time when compared to other spot-welding processes [12]. The use of high energy levels (or equally longer welding times) may cause extra-thinning of the joint and induce cracks in the weld nugget that limits access to maximum joint strength [9]. Furthermore, the use of interlayer in USW mostly related to weldability improvement, especially in dissimilar joining. Typically, use of metallic foil interlayer is more convenient and cost effective than other methods, like sputtering, plating, etc. [8]. The first trials of current research also performed using Cu foil in different thicknesses and USW energies. However, those experiments resulted in relatively strong joints, but the use of Cu coating was more effective in achieving better weldability and mechanical properties.

According to authors knowledge and a review on previous research studies, the use of Cu as an interlayer in the AZ31B Mg alloy "similar USW" has not been recorded. Previous efforts mostly were related to processes other than USW, the same as TLP (transient liquid phase), eutectic, or diffusion bonding. HP-USW of AZ31 Mg alloy classifies similar joining, but, when Cu used as an interlayer in contact with Mg, it forms a dissimilar welding combination at the interface. Hence, it is beneficial to review the Mg-Cu bonding processes and diffusion behavior, as well as AZ31 that is interlayered with Cu through other processes to find out the possible interactions in this system.

Diffusion kinetics and the identification of interfacial layers between Mg and Cu (pure metals) were studied by Nonaka et al. [13] and they reported the formation of Mg_2Cu as major interfacial layer near Mg and a thin layer of $MgCu_2$ adjacent to Cu (410–475 °C). Additionally, they suggested that the diffusion of Cu in Mg is faster than the inverse. However, in a nearly same recent study by Dai et al. [14], there was agreement about formation of two intermetallic layers between Mg and Cu, but they reported a different interdiffusion priority based on measured diffusion coefficients and activation energy data. These are basic studies regarding the Mg-Cu reaction diffusion couple, but, in the field of ultrasonic welding, Macwan and Chen [15] applied HP-USW for joining AZ31 to Cu (2000 W, 1000–2500 J) and observed the formation of a diffusion layer around the interface containing $Mg + Mg_2Cu$ that spread just towards the Mg side. They proposed a mechanism that is mainly based on the preferable diffusion of Cu in Mg to form Mg_2Cu intermetallic compound besides liquid phase formation through the eutectic reaction. This behavior was mostly attributed to the differences in the physical properties between Cu and Mg, such as melting point, thermal conductivity, heat capacity, and atomic radius.

One of trials to use the Cu interlayer was related to the field of eutectic bonding, where Elthalabawy et al. [16,17] studied joining AZ31 to 316L steel using a Cu insert (at 530 °C), and found that the eutectic reaction between Mg and Cu after 1 min. resulted to formation of 4 distinct interfacial layers in Mg part. Middle interfacial layers included Mg_2Cu and a layer of ternary $Mg + Cu + Al$ composition wherein Al content depleted from AZ31 base metal to participate in the Al-containing product.

The participation of AZ31 alloying elements in interfacial phenomena makes it more complicated, which can lead to non- stoichiometric compositions. Anyway, the formation of intermetallic compound (IMC) reaction products in dissimilar joining is prevalent and, in most cases, it has negative effect on mechanical properties of the joint, because they can embrittle the interface, especially when a continuous layer formed. Panteli et al. [18–20], during HP-USW of AZ31 to Al-6111, observed a 5 μm thickness continuous IMC layer along the interface, which resulted in low tensile shear strength with interfacial fracture. Additionally, they found that, under dynamic high strain rates of USW, the rate of IMC product formation was twice when compared to normal static conditions [18–20]. Patel et al. [21,22] also used HP-USW to join AZ31-Al 5754 and AZ31-HSLA steel combinations while applying a 50 μm thickness Sn interlayer and, finally, the Sn interlayer improved the weld strength in lower energies. The improvement was attributed to the formation of a composite-like Sn + Mg_2Sn eutectic structure instead of forming a brittle IMC product without an interlayer.

The objective of the current research is to study the effects of using a Cu coating interlayer in HP-USW of AZ31B Mg alloy on the weldability, mechanical strength, and process optimization from the point of view of energy, time, and interface characterization.

2. Materials and Methods

Commercial AZ31B rolled plates of magnesium alloy (3%Al, 1%Zn in wt%) with dimensions of 50 mm in length, 20 mm in width, and thickness of 0.8 mm were used as the initial material. The plates faying surfaces were polished until #2000 by emery paper, and finally cleaned while using acetone prior to USW. Final polishing resulted in a decrease of plate thickness, hence the applicable plate thickness was 0.78 mm. Joint welding carried out using a 2500 W, 20 KHz single wedge-reed USW machine MH2026/CLF2500 that was equipped with weld control monitoring system TDM-10AC (TMEIC/SONOMAC JAPAN Inc. Yokohama, Japan). Figure 1 presents a schematic of this machine. Additionally, a real photo of the sonotrode tip, as shown in Figure 1b (up), which has a circular shape with special design of two surfaces, including: "flat region" and "slope region". The flat region of tip head surface (5 mm in diameter) serrated with a knurling pattern that consisted of 14 × 14 rows of indentations (see a magnified view in Figure 1b; down), which can help to better the friction and plastic deformation on the weldment.

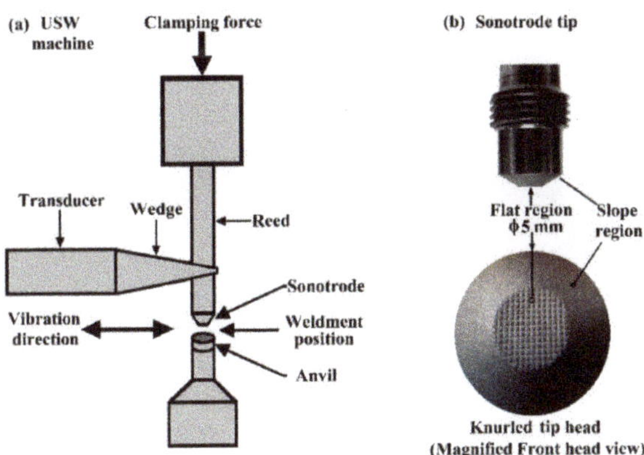

Figure 1. (a) Schematic of single wedge-reed type ultrasonic spot welding (USW) machine used in current research; (b) sonotrode tip head (up: side view; down: Magnified normal view of head surface).

An overlap of 20 mm × 20 mm between two plates was considered in order to perform lap shear tensile test on welded samples and, in order to prevent rotation of plates under effect of ultrasonic vibrations, the plates were kept together prior to welding using slight pressing on two points near the tip affecting area. A schematic representation of the welded sample for lap shear tensile testing and/or microstructure analysis is shown in Figure 2a. Additionally, Figure 2b shows a typical macrography of the welded joint intersection including Cu interlayer.

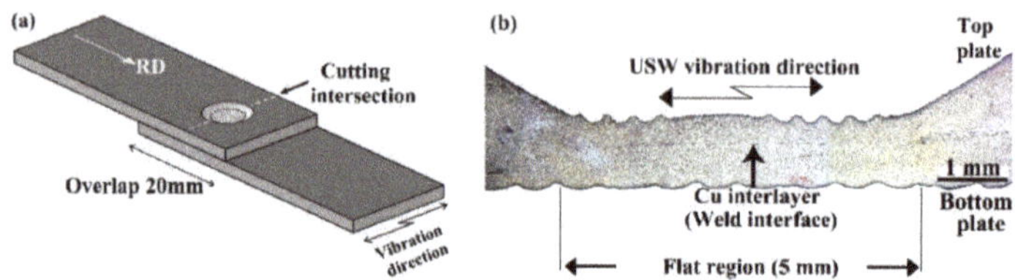

Figure 2. (**a**) Schematic of welded sample; (**b**) typical welded joint intersection (RD: Rolling direction).

Lap shear tensile tests on the welded specimens were performed using 0.5 ton-f, DS-3100 (Showa inc., Tokyo, Japan) desk type tensile machine under 1mm/min. cross head speed. The tensile axis was along the length of plates (i.e., perpendicular to vibration direction). For hardness evaluations, an MVK-H1 microhardness testing machine (AKASHI inc., Yokohama, Japan) used according to ASTM E-384-2011. A hardness measurement for welded samples has been conducted inside the flat region.

From the point of view of using or not using interlayer, two types of USW joints were produced, including:

- Joint type A: Direct welding of AZ31B plates without using interlayer; and,
- Joint type B: Cu coated AZ31B top plate welded to uncoated AZ31B bottom plate.

The high-power of 2450W was selected as the maximum capability of USW machine that was allowed by maker under safe conditions. USW parameters (same as impedance and clamping pressure) were optimized to access the maximum mechanical strength for all series of joints. Since that impedance is a variable that changes under different conditions of weldments, like coatings, oxide layers, or machine related design, it should be optimized under the maker's instructions before experiments. Subsequently, pressure optimization has been done under optimized impedance and a desired medium constant energy (e.g., 1500 J energy for joint type A and 1000 J for joint type B). Finally, the main experiments were performed under the optimized obtained pressure of 0.56 MPa for both joint types. The range of applied energy was selected between 800–1600 J, which covers low to high levels of energy and it has been monitored test by test according to lap shear tensile test results. The welding time and tip penetration depth were dependent variables that changed with change of energy. Ultrasonic vibration direction was along the plate width, perpendicular to rolling direction (RD) (see Figure 2a,b). The welded samples were cut from the width intersection (parallel to the vibration direction) for microstructure evaluations, including light microscopy, scanning electron microscopy (SEM) and electron backscatter diffraction analysis (EBSD) by SU5000 (Hitachi high-Technologies Ltd., Hitachi, Japan), electron probe micro-analyzer (EPMA) JXA8200T (JEOL Ltd., Tokyo, Japan), and transmission electron microscopy (TEM) analysis by JEM-2010 (JEOL Ltd., Tokyo, Japan). TEM sample cutting was performed using low speed diamond saw blades and polishing was continued until 0.5 μm (60000 grit) with abrasive polishing paper and finally dimpled with GATAN 656 and PIPS-GATAN 691 (Gatan Inc., Pleasanton, CA, USA). Oil-based diamond slurry was used for the final polishing of light microscopy and SEM-EBSD samples, to prevent oxidative corrosion. For a revealing microstructure with light microscopy, acetic-picral etchant was used, which is a common etchant that is capable of revealing Mg grain boundaries. Mean grain size measurement has been performed with using the line intercept method. Final polishing was continued with ion milling by HITACHI IM-4000 series (Hitachi high-Technologies Ltd., Hitachi, Japan) for EBSD and EPMA analysis.

Cu interlayer was deposited through the vacuum vapor deposition method. Deposition was applied just on the overlapping contact area (20 mm × 20 mm) of the AZ31B top plate.

A notch was created at the center part of bottom plate overlap surface for temperature measurements, as shown in Figure 3, and a 0.5 mm K-type thermocouple was inserted inside the notch to near the center of the weld nugget surface.

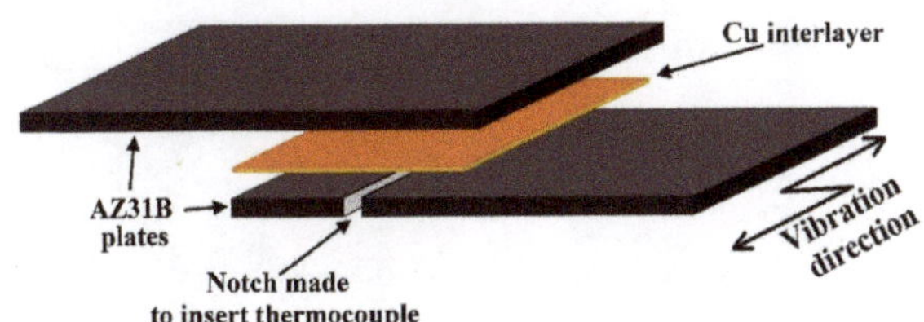

Figure 3. Schematic configuration for temperature measurement.

3. Results and Discussion

3.1. Lap Shear Tensile Test Results

Figure 4 shows a comparative graph of maximum lap shear tensile force for two types of samples in different energy levels. All of the tests were optimized to use 0.56 MPa clamping pressure. Each data in the curve is the average of three samples.

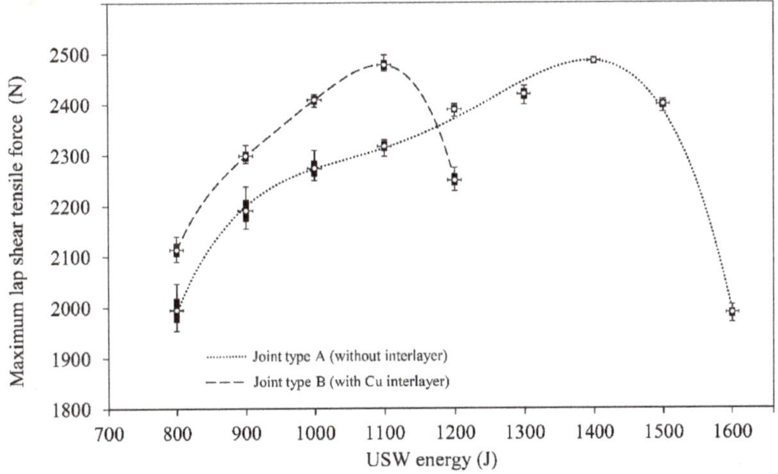

Figure 4. Comparative lap shear tensile force data vs energy, in high-power ultrasonic spot welding (HP-USW) of AZ31B, with and without interlayer (A: without interlayer; B: using Cu coating as interlayer).

The average maximum force that was achieved with joint type A, was 2485 N, while, in the case of joint type B, it was 2477 N, which is comparable with type A, but the used energy to achieve peak point for interlayered joint of type B was 20% lower, being equal to 1100 J. There were some cases that the maximum force data of joint type B was higher than that for joint type A, or the failure occurred in the base metal (out of weld spot).

The presented curves in Figure 4 imply that after a certain amount of applied USW energy, in both series of samples, with increasing USW energy, the curves reached a maximum and higher levels of used energy resulted in a relatively sharp drop in the maximum lap shear tensile force. For a better understanding of this effect, a macrography of failed samples after lap shear tensile testing is shown in Figure 5, which correlates change in failure behavior with USW energy. Figure 5a–d are related to the joint type B samples in energies of 900, 1000, 1100, and 1200 J respectively. Figure 5e that is related to the strongest sample of joint type A (1400 J) is presented to compare with 5c (strongest sample among joint type B).

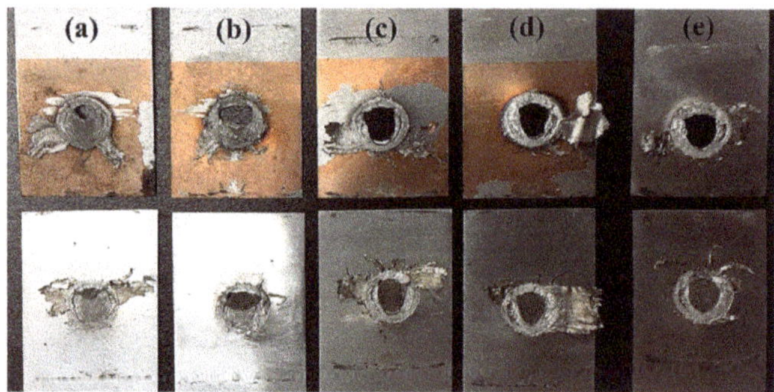

Figure 5. Failed samples after lap shear tensile testing for joint type B at (**a**) 900 J; (**b**) 1000 J; (**c**) 1100 J; (**d**); 1200 J; and, (**e**) joint type A at 1400 J USW energy.

The full nugget pull-out in samples of joint type B occurred when higher USW energies were applied (≥1100 J), while in lower USW energies (<1100 J), weld button was partly pulled out like a mix of pull out and the interfacial shear fracture. Hence, partial pull-out gradually changed to a complete pull-out of weld button, which is correspondent with the data attained in the related graph in Figure 4. With a comparison between the strongest samples of both joint types (Figure 5c,e), it can be found that the Cu-interlayered sample and the sample without interlayer mostly showed the same failure behavior.

The relationship between different parameters of USW is not simple and the main parameters are energy, power, and clamping force (pressure), and other parameters can be assumed as the dependent parameters like welding time or tip penetration depth that change with changing of energy. There is a famous relationship between power, energy, and time, as that energy equals to the multiplication of power and time. However, according to current research results and many other HP-USW studies, it can be found that this relationship is not completely responsible in HP-USW. For example, in the welding conditions of 2450 W and 800 J energy, the recorded welding time was 0.7 s, while, according to the energy-power relationship, it should be around 0.33 s. The other point is the role of clamping force, which is normally used to stablish a continuous intimate contact between two plates during USW. Logically, it can be considered that with an increasing of clamping force, a preventive force is working opposite to frictional force, which can result in the prohibition of easy movement between two joining parts. However, the application of HP-USW results in the need of increased higher clamping forces to keep the weldments together. Liu et al. [23], who studied the vibration amplitude effects on ultrasonic welding of Cu-Al joints, also noticed that interactions between different USW parameters have great influence on the mechanical performance of the joints, however a moderate clamping force is indispensable in generating intimate contact between the welded samples, and high power levels are required for extensive deformation that is associated with high clamping force.

Energy is one of the more attentional parameters in USW. Generally, USW energy can influence final strength in different ways comprising an increase in sonotrode tip penetration depth (or decreasing the joint thickness as a result), expanding the weld nugget surface area, and raising the temperature or heat delivered to the weldment. In lower energies, the produced frictional heat is not enough to bring the interface temperature up to levels that are necessary for completely occurring interfacial phenomena, like diffusion, eutectic reaction, etc. In other words, in lower energy levels, if IMC's formed at the interface during USW, they cannot diffuse and distribute correctly in the base metal and may play a role of imperfection at the interface that attenuates joint strength, because of less energy and inadequate heat input.

On the other side, when the applied energy is more than that needed to achieve a perfect joint (such as energies > 1100 J in joint type B), it can lead to an increase in the maximum interface temperature and extra melting possibly occurs at the interface layer, which encounters the highest rate of temperature rise. Thermochemistry data available for Mg-Cu intermetallic system implies that the formation of both IMC products in this system (Mg_2Cu and $MgCu_2$) is exothermic due to the negative enthalpy of formation [24]. It means that generated heat in these exothermic reactions may assist in temperature rising, however in small amounts. Consequently, if significant amounts of liquid phase are present at the interface, the rapid cooling after USW may result in brittle solidified reaction products, especially near the nugget edge walls. Several researchers reported the occurrence of melting during the use of interlayer, especially at the nugget edge walls. In recent research by Peng et al. [25] for dissimilar USW of Mg-ZEK100 to Al-6022 while using a 50 µm Cu interlayer, they reported locally melting at the nugget edges that were attributed to the high stress concentration that caused increase in temperature and lead to molten eutectic reaction products of Mg + Mg_2Cu, which were squeezed in nugget edges that resulted in interfacial fracture. The same evidence was observed with Macwan for HP-USW of Mg to Cu, and in high levels of USW energy (1500 J) the diffusion eutectic layer was squeezed out of the nugget edge [15,26]. Some melting at the nugget edge walls were rather visible, as it can be found in a higher energy sample of Figure 5d. The pulled-out weld button has a rather smaller area when compared to 5c, which is evidence of the high energy effect.

In the other hand, higher energy effects on joint thickness and it results in the thinning of the joint, thereupon introducing more stress concentration, leading to crack initiation in the weld nugget. Other researchers have frequently reported this behavior. For example, Peng et al. [27], during HP-USW of Al-6022, achieved the maximum mechanical strength at the intermediate energies of 1200–1400 J and the final strength was decreased by increasing the energy to higher levels, hence they attributed this to crack growth and stress concentration that were caused by deeper tip penetration. The measured average joint thickness for A and B joint types showed that total weldment initial thickness of 1.56 mm before USW (2 mm × 0.78 mm), was decreased to about 0.81 mm for joint A (1400 J energy) after HP-USW, while, in the case of type B, which used the Cu interlayer and made in lower energy of 1100 J, the final joint thickness was 0.95 mm. This can be found from Figure 6a,b in a critical region at the end of interface flat part that continues with the slope region.

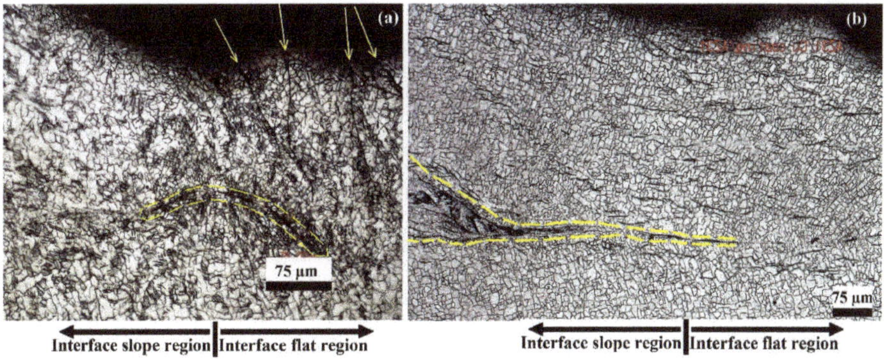

Figure 6. Comparison between macrographs of joint type A and B in transition region from flat part to slope part of the interface; (**a**) Joint type A (1400 J); (**b**) joint type B (1100 J).

In Figure 6a, which is related to joint type A, the arrow markers are notifying some cracks near the slope region that are introduced due to high energy and deep penetration of the sonotrode tip, resulting in extra thinning of joint thickness, especially in the top plate. Moreover, the region in Figure 6a that is confined within dashed lines shows a transitional region between the flat and slope part, with extreme plastic deformation and flow, which led to discontinuity in nugget edges under the effect of higher energy levels. Opposite to Figure 6b, when using the Cu coating interlayer, the joint thickness improved under lower energy USW, followed with an appearance of low melting point IMC products in a transitional region from flat to the slope part of interface (marked within a dashed line area). In comparison with Figure 6a, cracks cannot be seen in similar area. However, in this case, severe difference in hardness of the IMC product and the Mg base metal can cause brittleness in weld nugget edge (end of flat region). IMC formation and its effect will be discussed in more detail in next sections, which refer to temperature measurements and hardness evaluations, etc.

3.2. Interface Temperature Measurements

It is important be informed regarding rising temperature level during USW. The main factor affecting maximum process temperature (T_{max}) is friction between the contacting surfaces. The friction is a result of shear force that is induced by transversely ultrasonic vibrations. As shown in Figure 3, temperature measurements have been conducted through the insertion of an accurate 0.5 mm diameter K-type thermocouple inside a notch at the centerline of joint in bottom plate. Temperature profiles that were obtained for different types of weldment combinations, as shown in Figure 7a and variation of T_{max} against distance from weld nugget center presented in the graphs of Figure 7b. Measurements were just performed on the strongest joint for each joint type. It means HP-USW under 1400 J for joint type A and 1100 J for joint type B.

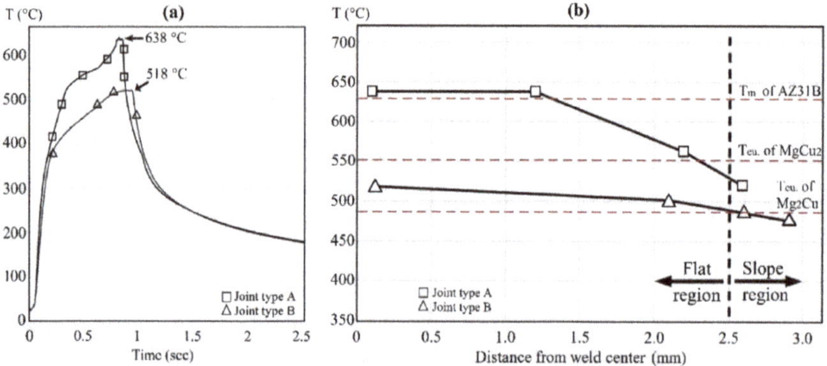

Figure 7. Interface temperature measurements for joint type A (1400 J) and joint type B (1100 J): (a) Comparative temperature profile for HP-USW of joint types A and B; (b) Maximum temperature of interface in different distances from weld center.

With looking at the temperature profiles of Figure 7a, it can be found that temperature firstly was increased from ambient temperature to a high level around 400 °C for joint B and 520 °C for joint A, sharply, just in a short duration of 0.2–0.3 s. Then, continuing of ultrasonic vibrations, led to completion of the bonding process, and temperature was increased to near the peak of T_{max} were observed. There is a clear difference between joint A and B at the maximum point, except the difference in T_{max} level. With using thin interlayer of Cu for joint type B, the peak temperature cycle in the curve flattened and T_{max} has longer persistence when compared to the joint type A. This can be predicated to the role of Cu that serves as a heat sink during the USW process. Without using the Cu interlayer (as in joint type A), the interface temperature increased and reached to T_{max} as a sharp peak with a duration of only 30 ms at the maximum point, and, while using the Cu interlayer besides lower energy, the peak shape somewhat flattened and was maintained near the T_{max} for a duration of 450 ms. After the T_{max} welding process is completed and ultrasonic vibrations stopped, weldment temperature decreased to near 300 °C very fast and then cooled down to ambient temperature, which continued for long time and nearly the same rate between two joint types.

Melting temperature (T_m) for pure Mg is 650 °C, while T_m of AZ31B is 630 °C [28]. Knowing this, the T_{max} for joint type A (Power: 2450 W; Energy: 1400 J) was recorded at 638 °C, which is around the melting point of the AZ31B Mg alloy. However, referring to the microstructure observations that will be shown in Section 3.3 it is difficult to summarize that melting took place and it seems that the process still occurs in the solid state. One reason might be cited to the short time of maximum temperature persistence with an appearance of a sharp peak in temperature profile. Additionally, looking at the temperature distribution in graphs of Figure 7b for joint type A represents T_{max} as accessible near the weld center and it quickly drops more than 120 °C in the half way of flat region. In Figure 7b, the melting point (T_m) of AZ31B pointed out with a dashed line as a useful guide. The calculation of the surface area of joint type A where the temperature remains near the melting point of AZ31B proves that just 25% of weld nugget area can reach the maximum temperature (i.e., 4.6 mm^2 from nugget surface area exposed to T_{max}, against 19.6 mm^2 total nugget surface area in flat region). While, 75% of weld nugget area in joint type B fairly reach to maximum temperature and do not show a quick drop. Haddadi et al. [29] also evaluated the temperature drop from the nugget center to the weld edge and found that the peak temperature dropped by 50–130 °C at the weld edge for HP-USW of 0.93 mm thickness Al-6111.

In the other side, for joint type B (Power: 2450 W; Energy: 1100 J), the recorded T_{max} was 518 °C, which is apparent in Figure 7a, and the temperature distribution for this joint is presented in Figure 7b. In order to obtain a better understanding, two dashed lines that are related to critical temperatures in Mg-Cu binary system, as drawn in Figure 7b to show the situation of interfacial region. It can be found that most of interface flat region lies in the temperature values over Mg_2Cu IMC formation eutectic temperature (T_{eu}), which is 487 °C, but it cannot exceed T_{eu} of $MgCu_2$ IMC product, which is 552 °C. Therefore, it is predictable to achieve a kind of interfacial area that contains Mg_2Cu as the main composition.

It is inferable that the Cu interlayer affects temperature distribution through improving the heat conductivity of interface. Copper has good electrical and heat conductivity and its thermal conductivity is four times larger than AZ31B Mg alloy (385 W/m-K for Cu; 96 W/m-K for Mg). Referring to this, the Cu effects on the interface to have a better heat distribution during the short time of process. Accordingly, with the matching graphs of Figure 7a,b for joint type B, it is clear that all parts of interface flat region have a chance of staying near maximum temperature for longer duration.

It is noteworthy to say that the Cu coating thin layer not only affected the thermal characteristics, but also the vibration behavior of interface. When using a Cu interlayer, USW impedance (a characteristic of vibration behavior) needs to set in lower levels to achieve proper resonance at the contact area of the two plates. In a research by Haddadi and Fadi [30], for dissimilar ultrasonic welding between steel and Al, they found that using a soft zinc coating on the steel surface altered the vibration behavior of the plate and the long sharp amplitude spike in the amplitude-time curve obviously decreased. They concluded that the surface condition of the steel sheet, which was confirmed by measurement of the net weld energy for each combination, affected the power delivery (when a fixed weld time was set). Furthermore, in that research, a eutectic reaction between Zn and Al led to a low melting point IMC reaction product that is molten under temperature rise that is caused by friction. The presence of this molten phase affected the amplitude of vibrations [30,31]. In the case of Mg-Cu, as it was explained, there is a eutectic point lower than process temperature that has the potential of making a molten phase and affecting the interface vibrational behavior. The effect of interlayer on impedance is also experimentally monitored through lap shear tensile force data before optimizing the energy effect on mechanical strength. At a constant USW energy and pressure, a change in impedance affected the process time and tip penetration depth and, accordingly, the optimized parameters that were used in experiments related to the effect of energy on maximum lap shear force.

3.3. Interface Characterization

3.3.1. Light Microscopy Observation

Initial microstructure of base metal is depicted in Figure 8a, where it is related to the plate thickness intersection. Figure 8b depicts a schematic of raw material plate and observation surface for a better understanding. The microstructure of the as-rolled Mg base metal shown in Figure 8a included large elongated grains inside the small equiaxed ones, with a mean grain size of 3.5 µm. The deposited Cu interlayer had a random thickness instead of a uniform layer, as the average thickness of deposited Cu layer was measured to be 2.5 ±1.0 µm.

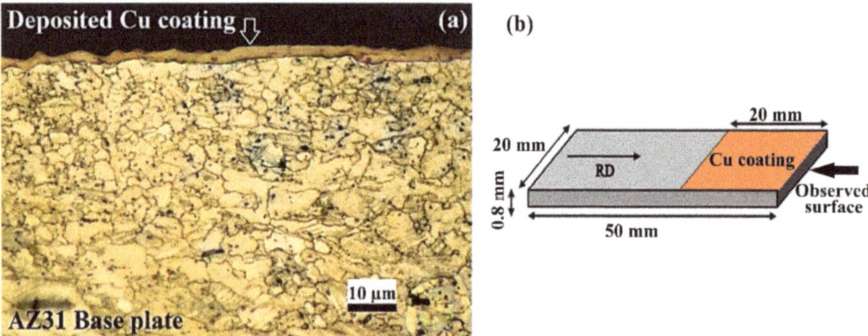

Figure 8. Light microscopy observation of AZ31 base metal with Cu coating, before USW (**a**) microstructure with Cu coating; (**b**) schematic of observed surface; (RD: Rolling direction).

Light microscopy evaluation on USW samples has been completed for two types of samples (A and B), but it was limited to strongest sample among all the samples in each type. The microstructure after ultrasonic welding is presented in Figure 9a,b for two joint types A and B, respectively.

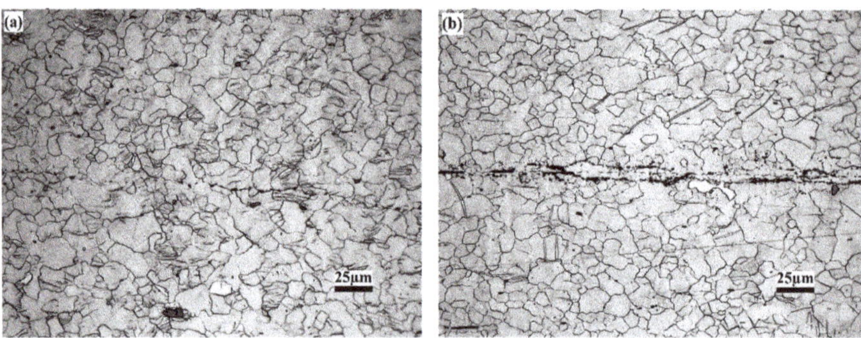

Figure 9. Light microscopy microstructure of interface flat region: (**a**) joint type A (1400 J); (**b**) joint type B (1100 J).

Firstly, through light microscopy microstructure for joint type B (Figure 9b) and comparing it with type A (Figure 9a), it was found that the use of copper coating resulted in a joint without discontinuity along the interface flat region. The observed interface showed a smooth flat interface. Some parts of the interface joined in bare contact between two Mg plates and it seems that the thin layer of Cu coating was scratched and removed due to scrubbing under high frequency ultrasonic vibrations and severe friction. As it was mentioned, the deposited Cu layer was a non-uniform coating layer and some parts were very thin when compared to total Cu layer (look at Figure 8a).

Afterwards, a comparison between microstructures before and after ultrasonic welding revealed amounts of grain growth and changing grain structure morphology. Grain size measurement for welded samples base metal was limited to nugget flat region inside a rectangular area of ±200 µm parallel to the interface centerline (except centerline and IMC layer). Morphology after ultrasonic welding for both joint types changed to larger grains with equiaxed structure, as grains were grown in the scale of about three times larger than initial base metal grain size. Figure 10 graphically summarizes the results.

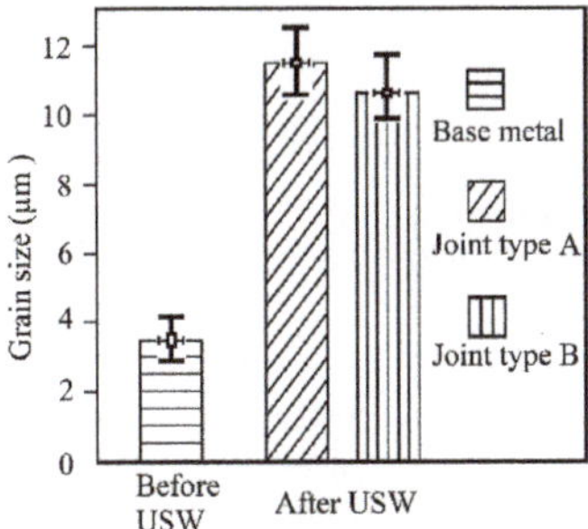

Figure 10. Comparison of base metal mean grain size before and after HP-USW for joint type A (1400 J) and joint type B (1100 J).

In joint type A, the mean grain size around the interface measured to be 11.5 μm ± 0.3 in the average (11.2 μm for the top plate and 11.8 μm for the bottom plate). It means that grain growth for this joint was in the scale of 3.3 times compared to initial base metal. For the joint type B, mean grain size around the interface was measured to be about 10.7 μm (10.2 μm for the top plate and 11.2 μm for the bottom plate). In the average, the mean grain size around the interface was about 7% smaller than the non-interlayered joint. This value also shows grain growth after ultrasonic welding that was about three times larger than initial base metal.

Grain growth after USW, has been frequently reported in previous researches in this field. For example, Macwan et al. [26] used USW for welding a rare-earth containing ZEK100 magnesium alloy in different USW energies 500–2000 J and observed that the average grain size in the NZ (nugget zone) increased from 7.6 to 13.6 μm, while grains that were close to the weld interface were relatively coarser. The main effective parameter on grain growth is the enhanced temperature of joint interface due to friction. Accordingly, grain growth due to temperature rising can lead to soften the material, together with the fact that magnesium and its alloys have poor ductility and formability at room temperature, which can improve in elevated temperatures. Poor ductility was mainly related to their hexagonal closest packed (HCP) crystal structure and a lack of enough slip systems in low temperature due to highly anisotropic dislocation slip behavior. According to critical resolved shear stress (CRSS) analysis, only two independent slip systems (related to basal plane) are active in room temperature, while, based on Taylor/Von Mises criterion, at least five independent slip systems are needed for uniformly deformation without failure. As shown in Figure 7, the interface temperature increased to high levels around the melting point of AZ31B. At elevated temperatures, the formability of magnesium improves, owing to the activation of other non-basal slip systems, like secondary (c+a) pyramidal slip systems and easier movement and dislocation climbing [2,32,33]. Moreover, the process time in USW is very short and mentioning the role of high strain rates governing in USW that has an important effect on phenomena to occur in dynamic conditions is needed. Previously, other researchers discussed the role of dynamic recovery (DRV) and dynamic recrystallization (DRX) occurring in USW [9,27,34–37]. High strain rates in USW were reported to be in the scale of 10^3 s^{-1} [38]. Under high strain rate deformation, DRX occurs that leads to new equiaxed grain structure, followed by DRV under the effect of elevated temperature that affects on grain size to enlarging. Larger grains decrease the hardness of base metal that lead to softening the material. As vibrations periodically produce, this process can occur several times during the short cycle of the process. Material softening facilitates plastic deformation due to softer structure additionally to the activation of more slip systems in elevated temperature, as previously mentioned. Bakavos et al. described this procedure in more detail [9]. To ensure the effect of grain coarsening on softening the material and tracking the effect of IMC formation on interface mechanical properties, microhardness evaluations were done with a focus on the flat region, which presented in the next section.

3.3.2. Microhardness Evaluation

Hardness data after HP-USW obtained from corresponding area of flat region wherein the metallography grain size measurement was performed. Measurements verified the softening of weld base metal after USW, which is attributed to recovery in elevated temperature. The hardness of base metal before USW was 75 HV/10 gf/15 s (average of 10 points), while after the HP-USW of joint type A, it was decreased to 59.9 HV, shows about a 20% reduction in hardness. For joint type B, the average microhardness of base metal that was measured in nearly the same area as joint type A, out of IMC layer, was around 66.7 HV, which shows an 11% reduction in hardness when compared to the as-received material. According to grain size measurements, this evidence is reasonable, as that T_{max} for joint B was 120 °C lower than joint type A, hence the extent of softening is lower. Moreover, according to mean grain size evaluations in the previous section, the grain size of joint type B was 7% smaller than joint type A, which is in agreement with the hardness data achieved.

Hardness evaluation has also been performed on the IMC layer of joint type B, but it cannot assure real data, because the size of indentation effect was larger than the width of the IMC layer. Therefore, the acquired data is an average of the analyzed region, including an IMC layer and base metal. Accordingly, IMC containing region showed a clear increase in the hardness, as its average hardness was 105 HV/10 gf/15 s. This value is 40% more than raw material hardness. The hardness data reported by other researchers on thick IMC layer between Mg and Cu also confirmed an increase in interface hardness. For example, Macwan et.al. reported a sharp increase in hardness at the interface of ultrasonically welded AZ31 to Cu in the scale of 278 HV as related to the Mg_2Cu intermetallic compound [15]. By the way, the high hardness of interfacial region is evidence of a eutectic reaction and the formation of IMC products.

3.3.3. Electron Microscopy Analysis

Magnesium etchant that used in metallography trials cannot simultaneously attack pure Cu, and, if some parts of Cu remained unreacted after USW, it is distinctly recognizable with a red tint in light microscopy photos, as that visible in Figure 8a. Meanwhile, interfacial layer in joint type B has a black and white appearance, which affirms the complete reaction of Cu and the formation of IMC products, alongside microhardness findings. Accordingly, for the better interface characterization of joint type B, more precise analysis methods based on electron microscopy are needed.

SEM Interface Observation

Figure 11a,b present the SEM observation on joint type B. Two white lines were graphically added inside the photos to create better perception regarding the interfacial region. The total thickness of Cu deposited layer before USW was only about 2.5 μm, but the affected interfacial region is much wider than the initial thickness of the Cu coating interlayer and it has an average width of around 9 μm, i.e., near four times larger than the initial thickness of the Cu coating. This is evidence of diffusion during the eutectic reaction.

This region can be seen in more detail at higher magnification, as in Figure 11b. The main reaction products are visible as discrete white particles, instead of a continuous layer. Apart from the composition of these particles, the dispersion of such hard IMC products inside a soft matrix of AZ31B base metal performs a composite like structure at the interface that can participate in the strengthening of the welded joint. Previously, in Section 3.1 it was reported that failure occurred in the pull out mode from the nugget edge walls. It means that crack initiation was not introduced from the flat region interface and it is a sign of strong sound joint.

The occurrence observed in Figure 11b specifies that reaction products were found in farther distances, like diffusion in base metal. This can be stimulated under the effect of enhanced temperature and ultrasonic vibrations. As it was noticed before, the interface temperature rises to over T_{eu} of Mg_2Cu, which predicts the presence of a liquid phase that is able to diffuse through material imperfections, like grain boundaries. The EBSD results in next section will show this matter in more detail.

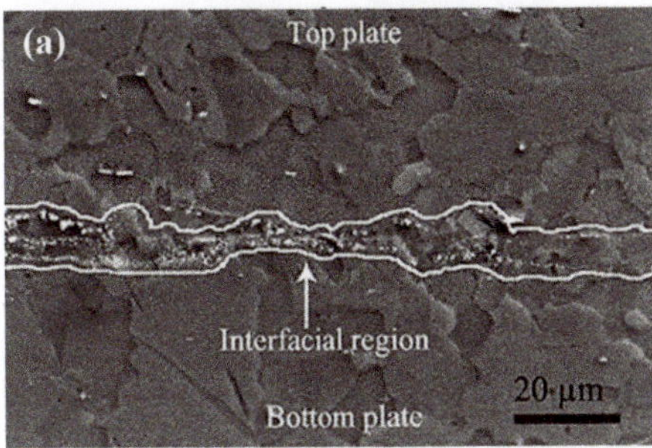

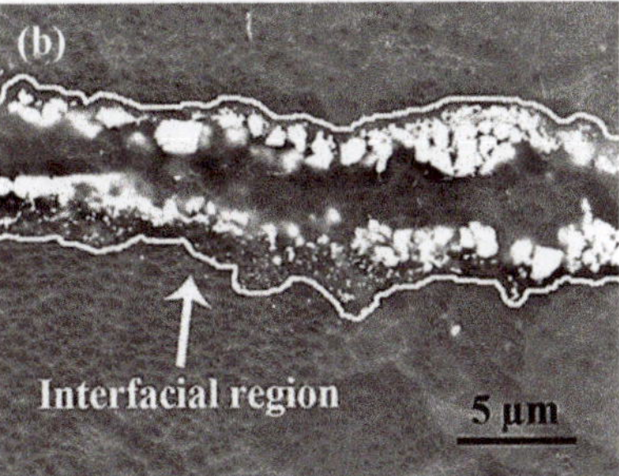

Figure 11. Interface SEM observation of joint type B (1100 J), (**a**) low magnification; (**b**) high magnification.

Interdiffusion of Mg and Cu in steady state conditions were studied before and discussed briefly in the introduction, but the lack of consensus in diffusion data makes diffusion behavior difficult to interpret. For example, activation energy data for the diffusion of Cu in Mg or vice versa are not compatible in different studies. Nonaka et al. [13] studied the reaction diffusion in Mg-Cu diffusion couple between 410–475 °C and observed that Kirkendall markers shifted towards the Cu-rich side, so they suggested that the diffusion of Cu is faster than Mg and they measured the activation energy of Mg_2Cu as 156 kJ/mole. Additionally, the growth rate of Mg_2Cu was much larger than $MgCu_2$. Other research by Dai et al. [14] in nearly the same conditions (400–460 °C) similarly found that, at all examined temperatures, two intermetallic layers formed between Mg and Cu, including the Mg_2Cu layer close to Mg side and $MgCu_2$ in contact with Cu side. The Mg_2Cu layer (with almost no solubility) was much thicker than $MgCu_2$. They also measured the activation energies for both IMCs that resulted in a lower value for Mg_2Cu (139 kJ/mole) when compared to $MgCu_2$ (147 kJ/mole). As it is clear, their data were a little different from reference [13]. The other important information that was obtained in reference [14] was related to activation energies for diffusion of Mg in Cu (139 kJ/mole) and Cu in Mg (164 kJ/mole), which imply different interdiffusion behavior when compared to reference [13].

However it is not the objective of current study to derive the diffusion data and, as was mentioned before, the USW process is more complicated than the original diffusion couples in steady state static conditions, due to high strain rates and vibrations in short time (less than 1 s for USW when compared to 24–72 h in reaction diffusion), but some findings, such as formation priority and layer thickness of Mg_2Cu, are in relative concurrency.

The formation of these two reaction products is based on a simple Mg-Cu system, while, in the presence of major alloying elements of AZ31B, the synthesis of other by-products also is expectable. For example, the Al content in AZ31B Mg alloy can contribute to form non-stoichiometric binary or ternary IMC products, together with Cu.

EBSD Analysis

For better clarification of the interfacial structure, EBSD analysis observations on the strongest joint of type B are done and taken in Figure 12. Figure 12a–c shows the interface area in different magnifications prepared by EBSD phase map (12a) and inverse pole figure map (12b,c).

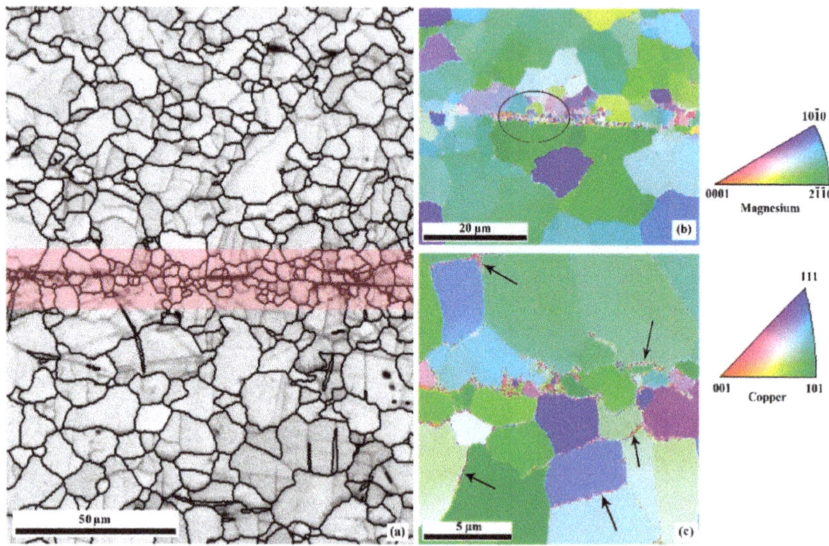

Figure 12. EBSD analysis on joint type B (1100 J), for (**a**) low magnification phase map with grain boundaries, (**b**), and (**c**) IPF colored map of interface in different magnifications.

Figure 12a illustrates a low magnification phase map of the interfacial region. A chain of base metal relatively fine grains (in comparison to large grains of base metal after HP-USW), adjacent to the interface centerline, is distinguishable. For better understanding, this region was graphically covered with a pink mask. Other researchers observed this type of fine-grained structure and they called it a "necklace-like" grain structure. The main effective parameter to produce this chain of smaller grains is occurrence of DRX, because, among interfacial area, this region suffers higher strain rates and plastic deformation that were caused due to friction between two faying surfaces [9,27,29]. Other regions far from the interface, as described in 3.3.1, shows an equiaxed grain structure, where the size of grains is larger than the base metal before welding. The IMC containing layer in Figure 12a has a dark feature at the interface centerline. Since that IMC grains are ultrafine, the grain boundaries covered the face of grains to be darkened. With looking overall at the interface in the metallography and EBSD photos, it can be found that some regions lost the Cu coating because of high frequency vibrations and friction. Moreover, the deposited layer thickness is not uniform in various parts of interface; thereupon, the width of interfacial reaction layer differs part by part. There is a direct bare contact between top plate and bottom plate without the presence of an interlayer in some regions of the interface.

In Figure 12b, the IMC reaction products are visible as fine grains that are distributed along the interface and neighborhood areas. The noticed region inside an oval in Figure 12b is an IMC-containing area at the interface, which is shown in more detail in Figure 12c. Figure 12c shows that a portion of ultrafine particles that are probably molten during HP-USW diffused through the grain boundaries of base metal to farther distances, in addition to the presence at the interface centerline. Some of diffused particles were typically pointed with black arrows to have better perception. With use of EBSD computer software, it was found that the diffused particles are Cu-containing reaction products. More accurate identification of particles was made possible through EPMA and TEM analysis.

Consequently, according to EBSD observations, good evidences were obtained for grain size strengthening based on well-known "Hall-Petch" equation, as supported by fine grains of base metal around the interface and ultrafine grains of hard IMC particles distributed in interfacial region.

EPMA Analysis

EPMA analysis also accomplished for joint type B and is presented in Figure 13. This elemental map analysis on joint type B indicated high levels of Mg at the interface and smaller levels of Cu and Al. The elemental map of Zn (not presented here) included a uniform distribution of it in all areas as a low percentage trace element.

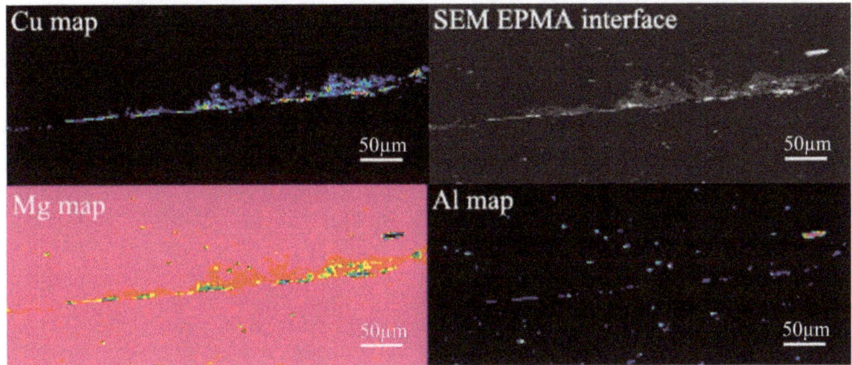

Figure 13. EPMA analysis on joint type B (1100 J) interface: Map of Mg, Cu and Al.

Two main regions are recognizable while looking at three elemental EPMA maps in Figure 13, and comparing them with SEM-EPMA interface photo. The main region is an area around the interface, wherein maps of Mg and Cu exhibit good overlapping together. Mg is prevalent in this region, and Cu is in a lower concentration that predicts a composition rich in Mg with lower content of Cu. As it was explained before, there are two main compositions in the binary phase diagram of Mg-Cu, including Mg_2Cu and $MgCu_2$. Keeping this in mind, it should be again noticed that the maximum interface temperature obtained for this joint was 518 °C (see Figure 7), which is over T_{eu} of Mg_2Cu and lower than that for $MgCu_2$, hence it can be suggested that the main area around the interface is composed of Mg_2Cu intermetallic compound.

Another interfacial region recognized in the centerline of the interface, consisted of a composition containing Al, Mg, and Cu. It seems that the resolution of EPMA analysis is not enough to identify the reaction product in the centerline. Therefore, TEM analysis also done on this region.

TEM Interface Analysis

The Al-containing region at the interface centerline that explored with EPMA analysis has been studied by TEM. Samples for TEM were extracted from the center of interface flat region in a direction similar to SEM-EPMA samples. Ultrafine particles that distributed inside an Mg substrate were found in this region and DP (diffraction pattern) analysis on them are shown in Figure 14a,b.

Based on measurements completed on DP's, it was found that the substrate (Figure 14a) shows crystal structure that precisely matches the HCP structure of magnesium. The DP of particles indicated a material with lattice spacing values of 0.42 nm; 0.44 nm; and, 1.68 nm. According to library data, the closest match is a ternary composition of type $MgCu_xAl_y$. Based on Buhler (1998) and Mel'nik (1981), the composition of $MgCu_{1.1}Al_{0.9}$ 4H (hp24, P63/mmc), with cell parameters of 0.51 nm; 0.51 nm; and, 1.676 nm, that is a Friauf–Laves phase C36 is the nearest estimation [39,40]. Change in indices x and y were attributed to the non-stoichiometry intermetallic compounds. Other products in this ternary system do not match with our findings.

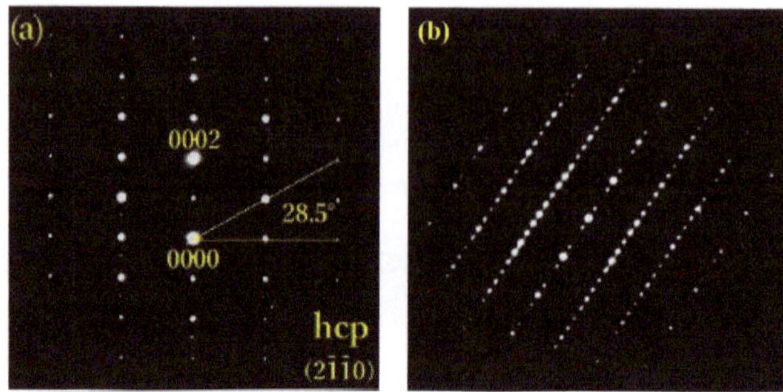

Figure 14. TEM analysis on joint type B (1100 J); diffraction pattern (DP) of (**a**) Mg matrix; (**b**) intermetallic compound (IMC) particles, at the interface centerline region.

4. Conclusions

AZ31B Mg alloy joints welded through HP-USW, with and without the use of Cu coating interlayer, and the following main results concluded:

1. Microstructure observations on interface, together with high mechanical strength and nugget pull-out fracture, revealed good weldability of the Cu interlayered joint.
2. Highest strength for interlayered joints was achieved in 20% lower USW energy, as compared to joints without the interlayer.
3. The microstructure of base metal after USW, converted to enlarged equiaxed grains with grain growth in the scale of three times larger than initial value, which is attributed to enhanced temperature, alongside high strain rates of HP-USW under the effect of DRX and DRV phenomena.
4. Temperature data, alongside EPMA analysis, suggested the formation of Mg_2Cu as the main IMC product around the interface. Additionally, EBSD analysis showed the diffusion and spreading of IMC's to farther the distances from the interface centerline, especially trough grain boundaries.
5. Proper strength of the interlayered joint is cited to the presence of fine grains of base metal adjacent to the interface and the dispersion of ultrafine hard IMC reaction products inside the soft matrix of Mg.
6. TEM analysis declared that the Al-containing reaction product that is found in the interface centerline is a composition of type $MgCu_xAl_y$, which is a Friauf–Laves phase.

Author Contributions: Experiments conceived and designed by A.B. and C.I.; Manuscript wrote by A.B.; Supervision and TEM analysis, C.I.; Consultant and EBSD analysis, S.S.; Consultant and performing coating by vapor deposition, S.T.

Funding: This research received no external funding.

Conflicts of Interest: The authors declare no conflict of interest.

References

1. Avedesian, M.M.; Baker, H. *ASM Specialty Handbook: Magnesium and Magnesium Alloys*; ASM International: Novelty, OH, USA, 1999; ISBN 978-0-87170-657-7.
2. Ropp, R.C. *Encyclopedia of the Alkaline Earth Compounds*; Elsevier: Oxford, UK, 2013; ISBN 978-0-444-59550-8.
3. Harooni, M.; Kovacevic, R. Laser Welding of Magnesium Alloys: Issues and Remedies. In *Magnesium Alloys*; Aliofkhazraei, M., Ed.; IntechOpen: London, UK, 2017; Available online: https://www.intechopen.com/books/magnesium-alloys (accessed on 1 April 2019).
4. Czerwinski, F. Welding and Joining of Magnesium Alloys. In *Magnesium Alloys—Design, Processing and Properties*; InTech: Rijeka, Croatia, 2011; pp. 469–490, ISBN 978-953-307-520-4.

5. Haddadi, F. Ultrasonic Spot Welding. In *Advanced Manufacturing Technologies Modern Machining, Advanced Joining, Sustainable Manufacturing*; Gupta, K., Ed.; Springer International Publishing AG: Cham, Switzerland, 2017; pp. 185–209, ISBN 978-3-319-56098-4.

6. Wang, T.; Shukla, S.; Gwalani, B.; Komarasamya, M.; Reza-Nieto, L.; Mishraa, R.S. Effect of reactive alloy elements on friction stir welded butt joints of metallurgically immiscible magnesium alloys and steel. *J. Manuf. Process.* **2019**, *39*, 138–145. [CrossRef]

7. Edison, K.G. Ultrasonic metal welding. In *New Developments in Advanced Welding*; Ahmed, N., Ed.; Woodhead Publishing Ltd.: Cambridge, UK, 2005; pp. 241–269, ISBN 978-1-85573-970-3.

8. Devine, J. Ultrasonic Welding. In *ASM Metals Handbook*; Lienert, T.J., Babu, S.S., Siewert, T.A., Acoff, V.L., Eds.; ASM International: Materials Park, OH, USA, 2011; Volume 6A, pp. 725–730, ISBN 978-1-61503-133-7.

9. Bakavos, D.; Prangnell, P.B. Mechanisms of joint and microstructure formation in high power ultrasonic spot welding 6111 aluminum automotive sheet. *Mater. Sci. Eng. A* **2010**, *527*, 6320–6334. [CrossRef]

10. Fujii, H.T.; Goto, Y.; Sato, Y.S.; Kokawa, H. Microstructural evolution in dissimilar joint of Al alloy and Cu during ultrasonic welding. *Mater. Sci. Forum* **2014**, *783–786*, 2747–2752. [CrossRef]

11. Sanga, B.; Wattal, R.; Nagesh, D.S. Mechanism of joint formation and characteristics of interface in ultrasonic welding: Literature review. *Period. Eng. Nat. Sci. (PEN)* **2018**, *6*, 107–119. [CrossRef]

12. Jahn, R.; Cooper, R.; Wilkosz, D. The effect of anvil geometry and welding energy on microstructures in ultrasonic spot welds of AA6111-T4. *Metall. Mater. Trans. A* **2007**, *38*, 570–583. [CrossRef]

13. Nonaka, K.; Sakazawa, T.; Nakajima, H. Reaction diffusion in Mg-Cu system. *Mater. Trans. JIM* **1995**, *36*, 1463–1466. [CrossRef]

14. Dai, J.; Jiang, B.; Zhang, J.; Yang, Q.; Jiang, Z.; Dong, H.; Pan, F. Jiang binary system. *J. Phase Equilib. Diffus.* **2015**, *36*, 613–619. [CrossRef]

15. Macwan, A.; Chen, D.L. Microstructure and mechanical properties of ultrasonic spot-welded copper to magnesium alloy joints. *Mater. Des.* **2015**, *84*, 261–269. [CrossRef]

16. Elthalabawy, W.M.; Khan, T.I. Eutectic bonding of austenitic stainless steel 316L to magnesium alloy AZ31 using copper interlayer. *Int. J. Adv. Manuf. Technol.* **2011**, *55*, 235–241. [CrossRef]

17. Elthalabawy, W.; Khan, T.I. Liquid phase bonding of 316L stainless steel to AZ31 magnesium alloy. *J. Mater. Sci. Technol.* **2011**, *27*, 22–28. [CrossRef]

18. Panteli, A.; Robson, J.D.; Chen, Y.C.; Prangnell, P.B. The effectiveness of surface coatings on preventing interfacial reaction during ultrasonic welding of aluminum to magnesium. *Metall. Mater. Trans. A* **2013**, *44*, 5773–5781. [CrossRef]

19. Panteli, A.; Robson, J.D.; Brough, I.; Prangnell, P.B. The effect of high strain rate deformation on intermetallic reaction during ultrasonic welding aluminium to magnesium. *Mater. Sci. Eng. A* **2012**, *556*, 31–42. [CrossRef]

20. Panteli, A.; Chen, Y.C.; Strong, D.; Zhang, X.; Prangnell, P.B. Optimization of aluminium to magnesium ultrasonic spot welding. *JOM* **2012**, *64*, 414–420. [CrossRef]

21. Patel, V.K.; Bhole, S.D.; Chen, D.L. Improving weld strength of magnesium to aluminium dissimilar joints via tin interlayer during ultrasonic spot welding. *Sci. Technol. Weld. Join.* **2012**, *17*, 342–347. [CrossRef]

22. Patel, V.K.; Bhole, S.D.; Chen, D.L. Ultrasonic spot welding of lightweight alloys. In Proceedings of the 13th International Conference on Fracture, Beijing, China, 16–21 June 2013.

23. Liu, J.; Cao, B.; Yang, J. Effects of vibration amplitude on microstructure evolution and mechanical strength of ultrasonic spot welded Cu/Al joints. *Metals* **2017**, *7*, 471. [CrossRef]

24. Mezbahul-Islam, M.; Mostafa, A.M.; Medraj, M. Essential magnesium alloys binary phase diagrams and their thermochemical data. *J. Mater.* **2014**, *2014*, 704283. [CrossRef]

25. Peng, H.; Chen, D.L.; Bai, X.F.; She, X.W.; Li, D.Y.; Jiang, X.Q. Ultrasonic spot welding of magnesium to aluminum alloys with a copper, interlayer: Microstructural evolution and tensile properties. *J. Manuf. Process.* **2019**, *37*, 91–100. [CrossRef]

26. Macwan, A. Ultrasonic Spot Welding of Similar and Dissimilar Alloys for Automotive Applications. Ph.D. Thesis, Ryerson University, Toronto, ON, Canada, 2016.

27. Peng, H.; Chen, D. Microstructure and mechanical properties of an ultrasonic spot welded aluminum alloy: The effect of welding energy. *Materials* **2017**, *10*, 449. [CrossRef] [PubMed]

28. Brandes, E.A.; Brook, G.B. *Smithells Light Metals Handbook*; Butterworth-Heinemann: Oxford, UK, 1998; ISBN 978-0-7506-3625-4.

29. Haddadi, F.; Tsivoulas, D. Grain structure, texture and mechanical property evolution of automotive aluminum sheet during high power ultrasonic welding. *Mater. Charact.* **2016**, *118*, 340–351. [CrossRef]
30. Haddadi, F.; Fadi, A. The effect of interface reaction on vibration evolution and performance of aluminum to steel high power ultrasonic spot joints. *Mater. Des.* **2016**, *89*, 50–57. [CrossRef]
31. Derks, P.L.L.M. Parameters that influence the ultrasonic bond quality. *Electron. Compon. Sci. Technol.* **1983**, *10*, 269–275. [CrossRef]
32. Al-Samman, T.; Gottstein, G. Dynamic recrystallization during high temperature deformation of magnesium. *Mater. Sci. Eng. A* **2008**, *490*, 411–420. [CrossRef]
33. Itakura, M.; Kaburaki, H.; Yamaguchi, M.; Tsuru, T. Novel cross-slip mechanism of pyramidal screw dislocations in magnesium. *Phys. Rev. Lett.* **2016**, *116*, 225501. [CrossRef]
34. Patel, V.K.; Bhole, S.D.; Chen, D.L. Influence of ultrasonic spot welding on microstructure in a magnesium alloy. *Scr. Mater.* **2011**, *65*, 911–914. [CrossRef]
35. Xiea, J.; Zhua, Y.; Bianc, F.; Liud, C. Dynamic recovery and recrystallization mechanisms during ultrasonic spot welding of Al-Cu-Mg alloy. *Mater. Charact.* **2017**, *132*, 145–155. [CrossRef]
36. Shimizu, S.; Fujii, H.T.; Sato, Y.S.; Kokawa, H.; Sriraman, M.R.; Babu, S.S. Mechanism of weld formation during very-high-power ultrasonic additive manufacturing of Al alloy 6061. *Acta Mater.* **2014**, *74*, 234–243. [CrossRef]
37. Fujii, H.T.; Sriraman, M.R.; Babu, S.S. Quantitative evaluation of bulk and Interface microstructures in Al-3003 alloy builds made by very high power ultrasonic additive manufacturing. *Metall. Mater. Trans. A* **2011**, *42*, 4045–4055. [CrossRef]
38. Gunduz, I.E.; Ando, T.; Shattuck, E.; Wong, P.Y.; Doumanidis, C.C. Enhanced diffusion and phase transformations during ultrasonic welding of zinc and aluminum. *Scr. Mater.* **2005**, *52*, 939–943. [CrossRef]
39. Buhler, T.; Fries, S.G.; Spencer, P.J.; Lukas, H.L. A Thermodynamic assessment of the Al-Cu-Mg ternary system. *J. Phase Equilib.* **1998**, *19*, 317–333. [CrossRef]
40. Mel'nik, E.V.; Kinzhibalo, V.V. Investigation of the Mg-Al-Cu and Mg-Ga-Cu systems from 33.3 to 100 at. % Mg. *Russ. Metall.* **1981**, *3*, 154–158.

Article

Internal Material Flow Layers in AA6082-T6 Butt-Joints during Bobbin Friction Stir Welding

Abbas Tamadon [1,*], Dirk J. Pons [1], Don Clucas [1] and Kamil Sued [2]

[1] Department of Mechanical Engineering, University of Canterbury, Christchurch 8140, New Zealand; dirk.pons@canterbury.ac.nz (D.J.P.); don.clucas@canterbury.ac.nz (D.C.)
[2] Fakulti Kejuruteraan Pembuatan, Universiti Teknikal Malaysia Melaka, Durian Tunggal 76100, Malaysia; kamil@utem.edu.my
* Correspondence: abbas.tamadon@pg.canterbury.ac.nz; Tel.: +64-021-0281-2680

Received: 9 August 2019; Accepted: 26 September 2019; Published: 28 September 2019

Abstract: Bobbin friction stir welding with a double-sided tool configuration produces a symmetrical solid-state joint. However, control of the process parameters to achieve defect-free welds is difficult. The internal flow features of the AA6082-T6 butt-joints in bobbin friction stir welding were evaluated using a set of developed reagents and optical microscopy. The key findings are that the dark curved patterns (conventionally called 'flow-arms'), are actually oxidation layers at the advancing side, and at the retreating side are elongated grains with a high-density of accumulation of sub-grain boundaries due to dynamic recrystallization. A model of discontinuous flow within the weld is proposed, based on the microscopic observations. It is inferred that the internal flow is characterized by packets of material ('flow patches') being transported around the pin. At the retreating side they experience high localized shearing at their mutual boundaries, as evidenced in high density of sub-grain boundaries. Flow patches at the advancing side are stacked on each other and exposed to oxidization.

Keywords: bobbin friction stir welding; materials flow; metallography; AA6082-T6; weld defect

1. Introduction

Bobbin friction stir welding (BFSW) is an innovative variant of the friction stir welding (FSW) process [1], whereby a double-sided rotating tool physically ploughs along the interface of two butted plates [2] without needing the backing anvil and the axial force during the process [3]. The dynamic interaction between the workpiece and the non-consuming tool creates a severe friction condition at the contact interface [4]. Consequently, significant heat is generated due friction which can locally soften the workpiece material sufficient for plastic yielding and stirring at the bonding track [5]. The stirring action arises from the rotation speed (ω) and advance speed (V). While the tool advances along the weld line (Figure 1), the mutual interaction of the speeds (ω, V) transports the softened mass from leading edge of the rotating tool to be deposited at the rear or trailing edge of the tool [6].

The side of the weld-seam where the direction of tool rotation is the same as the tool progress is called the advancing side (AS) and the opposite side of the weld-seam is the retreating side (RS) [7]. The region located between the AS and RS borders, named the stirring zone (SZ) [8], experiences a thermomechanical plasticizing and then deposition and consolidation in the weld locus [9]. The adjacent region outside the SZ is called the Thermo-Mechanical Affected Zone (TMAZ). The microstructure of this region is formed by the stress-strain fields and heating flux which are induced by the friction and heat generation effects of the stirring action [10]. The next region between the TMAZ and the Base Metal (BM) is the Heat Affected Zone (HAZ). This region is exposed to thermal fields of the stirring process which alter the microstructure [11].

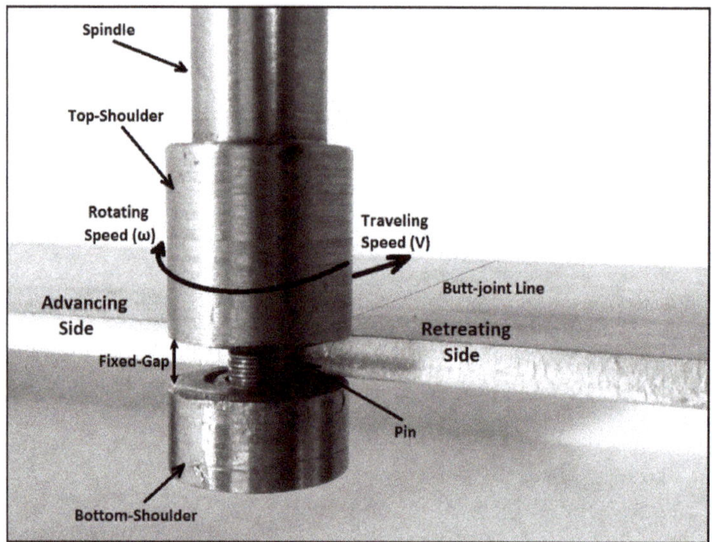

Figure 1. Schematic of the BFSW process for a butt-joint position.

Comparing with the Conventional-FSW (CFSW), in the BFSW the fully-penetrated pin requires more control during the process, as an inconsistency between the process variables can cause more severe failures. The double opposing shoulders system provide a greater contact surface for the frictional heat generation from both sides of the workpiece [9]. The process also replaces the CFSW backing or anvil support plate with the BFSW rotating shoulder at the lower side of the workpiece [12,13]. The CFSW case requires a downward axial load [14] on the tooling, whereas the BFSW requires a compression ratio [15] (the variance between the inner edge-to-inner edge biting gap of the shoulders and the actual thickness of the plate). These differences cause differences in the flow regimes of the two processes [16,17]. In the FSW literature, the onion rings in the conventional-FSW weld structure have been attributed to discontinuous flow during tool-material interaction [18].

As the temperature in FSW processes is lower than in fusion welding, it is categorized as a metal forming process [19]. The internal flow regimes related to the plasticized mass play the main role in the welding mechanism rather than metallurgical transformations of melt-and-solidification. In general, the FSW technique is proposed for materials with a high capability of the dynamic plastic deformation. Aluminium that responds well to large plastic deformation is a good candidate material for FSW. In particular, marine grade AA6082-T6 aluminium alloy with good machinability properties would be attractive to be processed by the BFSW, but shows poor weldability in the conventional fusion welding. One of the obstacles to a better understanding of the actual flow regimes is the need to visualise the details of the flow features for the cross section of the weld [8,20]. This is challenging for AA6082-T6 alloy as the weld region responds poorly to conventional etchant reagents [21]. This problem arises because of the low contrast between grains and grain boundaries for the AA6082-T6 microstructure where the precipitate particles are uniformly dispersed within a supersaturated solid solution treated by artificially ageing per the T6 cycle. Also, severe plastic deformation and grain fragmentation during the BFSW process reduce grain size to ultrafine.

While there is an extensive literature on the microstructural characteristics of the FSW welds [22], the flow mechanism also needs to be identified within the weld structure [23–25]. In this regard, there have been attempts to elucidate the heat flow [26] and material flow [27] mechanisms in BFSW, as it is expected to be different to conventional-FSW [28].

The aim of this work is to identify the causality between flow regimes and physical defects. The approach is to visualise plastic deformation features of AA6082-T6 BFSW welds with new etchants [21], using optical metallography. These reagents show the details and complexity of the plastic flow patterns within the stirring zone, even when there is an element of grain refinement due to the thermomechanical plastic deformation. A benefit of this approach is achieving a detailed microstructural analysis with conventional etching methods and optical metallography, rather than the more costly processes of electropolishing, or electron metallography (e.g., SEM, EBSD, and TEM).

2. Materials and Methods

For the weld trials, rolled plates of AA6082-T6 (Al–Si–Mg–Mn family) were used as the workpiece. The chemical composition of the AA6082-T6 aluminium alloy as the base metal is reported in Table 1.

Table 1. Element composition of the AA6082-T6 aluminium alloy (wt %). Data from [29].

AA6082-T6 Aluminium Alloy	
Chemical Element	% Present
Silicon (Si)	(0.70–1.30)
Magnesium (Mg)	(0.60–1.20)
Manganese (Mn)	(0.40–1.00)
Iron (Fe)	(0.0–0.50)
Chromium (Cr)	(0.0–0.25)
Zinc (Zn)	(0.0–0.20)
Titanium (Ti)	(0.0–0.10)
Copper (Cu)	(0.0–0.10)
Other (Each)	(0.0–0.05)
Other (total)	(0.0–0.15)
Aluminium (Al)	Balance

BFSW tests were conducted using a geometrically full-featured (including threads, flats and scrolls) bobbin tool manufactured from H13 tool steel with hardness of 560 HV. The butt-joint weldments comprised two pieces of similar plates with dimensions of 250 mm (length) × 75 mm (width) × 6 mm (thickness). The welding trials involved two sets of operation speeds; spindle rotational speed (ω) and weld travel rate (V).

The aim of the research was to identify the flow features of the BFSW weld in a defect-free structure compared with the presence of defects. A sample of each condition was procured. The defective weld was produced with welding parameters of ω = 400 rpm, V = 350 mm/min. The good weld was produced at ω = 650 rpm and V = 400 mm/min. The same tool was used in both cases. These process parameters were identified from previous work [30]. That reference also describes the tool geometry in detail, see also Table 2.

Table 2. Parameters of the BFSW process for the AA6082-T6 weld trials.

Work Temp (°C)	$D_{Shoulder}$ (mm)	D_{Pin} (mm)	$D_{Shoulder}/D_{Pin}$	Plate Thickness (mm)	Compression Ratio	Feed ω (rpm)	Speed V (mm/min)	Thread Pitch (mm)	Number of Threads
18 °C	21	7	3	6	3.75%	400 650	350 400	1.2	5

The BFSW experiments were performed on a 3-axis CNC machining centre (2000 Richmond VMC Model, 600 Group brand, Sydney, Australia) with a Fanuc control unit and 14-horsepower spindle motor capacity. There were no preheating or post-weld processes before or after the welding process. The direction of tool rotation was clockwise relative to the advancing direction of the welding (Figure 1). Table 2 gives more details of the bobbin-tool and the welding operation. After welding the quality of joints were first checked by visual examination, and then cross sectioned by an electro-discharge machine wire cut through the middle of the weld seam (perpendicular to the welding direction). The resulting surfaces were subjected to metallographic measurements. For metallographic analysis, firstly the specimens were prepared using standard mechanical polishing with different grades of SiC sand papers (600-grit, 800-grit and 1200-grit). To achieve a mirror surface, the micro-polishing step was conducted on a micro-cloth pad with a 3 μm diamond paste, and finally a 0.05 μm colloidal silica solution. The etching process was designed to remove the oxide film, and then delineate the flow lines. The polished specimens were first pre-etched for 3 min in a solution of (5 g NaOH + 1 g NaCl + 80 mL H_2O) at 70 °C. The developer mixtures are shown in Table 3, which describes the composition and other conditions of the chemical solutions (time and temperature). After completion of the etching process, the samples were rinsed in ethanol, and then dried with warm air. The flow patterns of the BFSW weld region were studied using stereoscopic and light optical microscopes (Olympus Metallurgical Microscope, Tokyo, Japan). In some cases, where the microscopic features needed to be clarified in more detail, the cross-section sample was re-polished and re-etched with other reagents. For some metallurgical validating, the etched samples were also subjected to elemental mapping using the scanning electron microscope (SEM) (JEOL 6100, JEOL Inc., Peabody, MA, USA) equipped with energy-dispersive X-ray spectrometer (EDS) detector (Oxford Instruments plc, Abingdon, UK).

Table 3. Different reagent compositions with separate sequences of processing.

Name of Etchant	Etchant Composition
A	2.5 mL HF + 2.5 mL HCl + 95 mL H_2O (30 s, 50 °C), then: 15 mL H_3PO_4 + 85 mL H_2O (30 s, 70 °C)
B	0.5 g $(NH_4)_2MoO_4$ + 3.0 g NH_4Cl + 1 mL HF + 18 mL HNO_3 + 80 mL H_2O (90 s, 70 °C)
C	10 g CrO_3 + 2 g Na_2SO_4 + 10 mL HNO_3 + 10 mL CH_3COOH + 1 mL HF + 80 mL H_2O (60 s, 70 °C)

3. Results

3.1. Macrostructure of the Cross-Section

Figure 2a,b reveal the macrostructural examinations of the AA6082-T6 BFSW joints (etched by Reagent A) for two different sets of welding speeds (longitudinal or rotating), refer to Table 1. Both joints exhibit a distinctive zone in the middle, processed by thermo-mechanical plastic deformation induced by the bobbin-shaped tool. This hourglass-shaped region—different from the basin-shaped SZ in CFSW [13]—is discernible in the middle of the weld from the base metal (BM) by the symmetrically curved borders at the both advancing side (AS) and retreating side (RS).

It is observed that the SZ, at the middle of the weld, is larger than the diameter of the pin. This is attributed to: (a) the pin recruiting a wider volume of matter through frictional contact with the substrate, and (b) recruitment of the substrate material via frictional heating from the shoulders. Both the pin and shoulders create frictional heat.

Basically, in BFSW the fully contained pin has eliminated the incomplete root penetration of CFSW, but in Figure 2a, a macro-size tunnel void has emerged at the bottom surface towards the AS, while Figure 2b shows a defect-free weld with an integrated structure. The origins of tunnel void emergence are not fully understood in the literature; however, it is clear that the formation of this macro-size defect in a solid state process does not seem to be more definite to have a metallurgical explanation. Existing theory attributes the tunnel voids to incomplete backfilling [8,31], that arises from incompatible welding speed parameters (ω, V), or dynamic interaction between the tool and workpiece [31,32]. It appears that there is insufficient material. As the metallography samples were cross sectioned from the middle of the weldment, the amount of the material loss can be considered similar to the size of the sprayed tail defect at the entry zone. The continuous tunnel void shows that the plates were not fully butted.

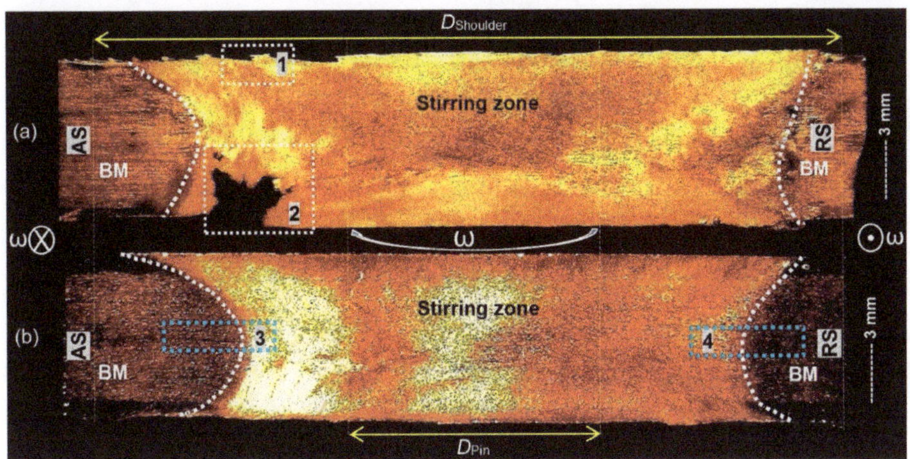

Figure 2. Macrostructure of the cross-section of the BFSW joint for two different welding speeds (etched by reagent A), (**a**) (w = 400 rpm, 350 mm/min), and (**b**) (w = 650 rpm, 400 mm/min). Arrows show proposed internal flows. Dashed white lines show the SZ borders. (1) weld crown, (2) tunnel void, (3) AS border, and (3) RS border.

From observation of the macrostructure, we infer the existence of the internal flow directions. This can affect the weld quality, when the failure of the flow regimes leads to the emerging of the defects (e.g., tunnel voids or hair-line micro-cracks). To provide a better explanation for the origins of the defects based on a flow-based observation, the metallographic measurements needed to be conducted in more depth with a focus on microstructural and flow feature integrity.

3.2. Microstructure Evolution

The transverse section of a BFSW weld-seam should reveal three typical zones; HAZ, TMAZ and SZ, distinguished from the BM towards the centre of the weld. However, macro-sections of the BFSW welds (Figure 2) only indicated a primary macrostructural region for the stir zone (SZ) at the centre of the weld. The other transitional zones could not be identified using macro etching. To make the TMAZ and HAZ regions visible we re-polished, used etchant B and higher magnification for the micrographs and then montaged them together.

Figure 3 shows a photomontage of the micrographs taken from the transverse cross-section of the BFSW weld (selected from the sample of Figure 2b) close to AS and the RS hourglass boundaries, etched by Reagent B. This reveals the deformation-induced grain refinement at the transition region (HAZ/TMAZ) on either side of the SZ.

The latter region exhibits a grain orientation transformation. The transformation is represented in grain size and morphology. The transition region has rotated-elongated grains morphology detectable with a distortion in grains compared with the parent material. Compared to the large columnar grain structure of the BM, the microstructure of the SZ reveals a homogenous distribution of the fine equiaxed grains. This abrupt change in structure separates the TMAZ/HAZ region from the plastically deformed finer-grained SZ, and the large directional grains observed in the BM structure. This is attributed to dynamic recrystallization (DRX), which in turn arises from the thermomechanical (heating & strain) mechanisms during stirring.

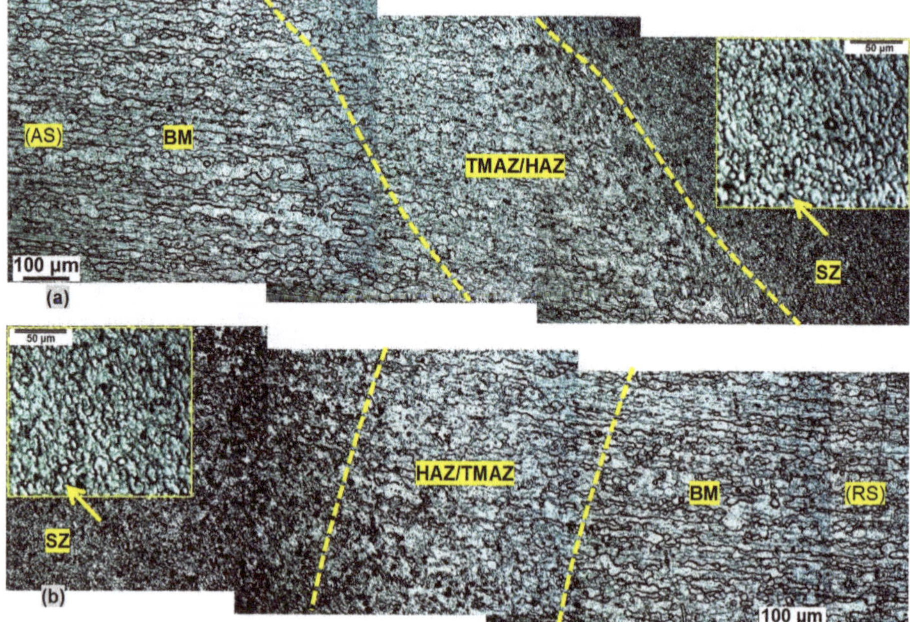

Figure 3. Photomontage of the microscopic features for the BFSW weld structure, near the AS and RS borders (etched by Reagent B). (**a**) transition region in AS border (regions 3 in Figure 2b), (**b**) transition region in RS border (regions 4 in Figure 2b).

3.3. Material Flow Features in the Weld Cross-Section

The magnified micrographs in Figure 3 from the interface of the SZ and the external region illustrate a metallurgical bonding between the processed weld region and the parent metal. To distinguish the flow patterns from the grains structure, etchant reagent C was used, giving the results in Figure 4.

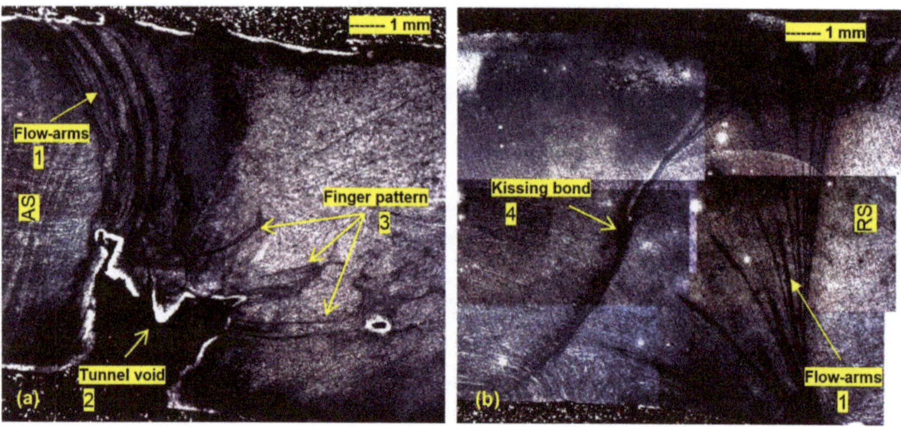

Figure 4. Photomontage microfeatures of the BFSW weld at the transverse cross-section (revealed by Reagent C). (**a**) AS border, (**b**) RS border. (1) flow-arms, (2) tunnel void, (3) finger patterns, and (4) kissing bond.

For this micro-flow evaluation, the sample of Figure 2a was chosen to reveal more details because of the existence of defects (e.g., tunnel void). As is shown in Figure 5a,b, the presence of the flow arms (banded patterns) is evident in the SZ at the hourglass borders for both sides of the weld. The flow arms of the AS are more compacted to each other.

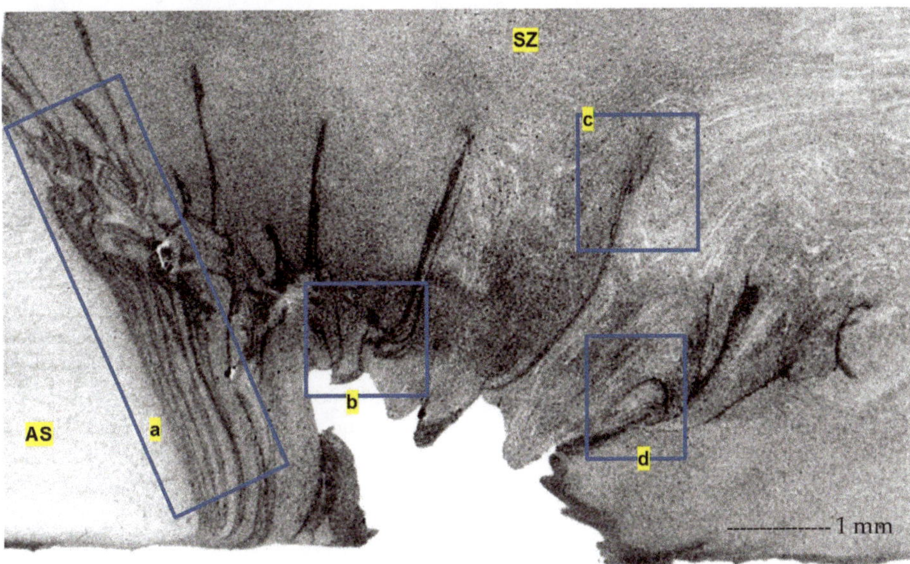

Figure 5. Micrograph of the tunnel void and the microflow features around the defect at the circumferential side of the tunnel void (revealed by Reagent C). (**a**) Cascade flow bands, (**b**) hook lines, (**c**) finger pattern, (**d**) swirling lines.

The flow arms become dispersed when reaching the tunnel void (Figure 4a). In this case, the flow arms near the bottom surface of the weld follow a finger pattern, stretched from the hourglass boundary (at the circumferential edge of the tunnel void) towards the centre of the weld. It is more evident that the compaction effect from the top shoulder was higher than the bottom shoulder, while their diameters and contact surface conditions are the same. The branching pattern can also be observed when the plastic flow behaviour—between the bottom and top surfaces—is affected by the geometrical features of the tool (threads, flats, scrolls) during the stirring action. We attribute this to an unsteady internal flow, in turn caused by the inability of the flow to completely fill the void left by the tool moving forward. Behind the tool, the material moves from the RS towards the AS, and is deposited at the AS. If there are any insufficiencies in the flow, due perhaps to previous loss of material at entry, or early cooling of the material, then the forward movement of the tool creates a gap that is difficult for the flow to fill. We attribute the striation layers themselves to batches of material cut by the flats out of the base material.

The macro-section of the bobbin weld in Figure 4b reveals another distinct flow-based problem. The appearance of this linear discontinuity is similar to the kissing bond defect in CFSW; solid-state bonding with poor or no metallurgical adhesion. One of the major causes of kissing bond defects in friction stir welds is insufficient engagement of the tool pin into the plasticized material. We interpret the current findings as due to a recoil/backlash between the pin and the material during the stirring, (such as might be caused by vibration – which was prevalent), causing incomplete joining. It should be noted the kissing bond defect is more prominent in samples produced at lower rotational speeds which can be attributed to the inconsistency in strain distribution due to occurred flow-based condition.

Furthermore, the position of the kissing bond defect is located exactly at the position of change in the flow direction from leading edge to the trailing edge of the pin.

It is very difficult to detect or accurately characterize the nature of this defect by the typical analysis methods, e.g., metallography or non-destructive testing (NDT). However, in Figure 4, we tried to delineate the exact location of the discontinuity line in the microstructure by Reagent C. As it is observable in Figure 4b, a continuous and uniform discontinuity flow was vertically delineated between the top surface and bottom surface through the SZ. Furthermore, this microscopic view confirms that the nature of the defect line is not a grain structure.

We propose that it is a kind of oxidation formed during the dynamic recrystallization of the deformed grains. The proposed explanation of is as follows: As the pin possesses a symmetrical thread-flats feature acting through the stirring zone, the rotating angle between the flat surface and the thread surface creates an empty space to allow air to enter into the weld region. During the stirring, which is at considerable temperature, this air causes a surface oxidation between the layers of stirred mass from the AS. Possibly the adiabatic compression of air pockets adds further heating.

The darker colour of the flow arms -visible in hourglass borders and around the tunnel void- is also attributed to this oxidation behaviour. In addition, it should be noted that this oxidized layer was formed in the sample containing the tunnel void defect, but not in defect-free samples (no tunnel void). Therefore, another flow-based relation can be assumed between the emergence of the tunnel void in the AS position, and formation of this oxidation bond at the RS proximity of the stirring zone.

While the pin-driven pressure force in the tunnel void region is insufficient to refill the position by a compensating flow, in the RS the pressure is high, which consequently squeezes the flow to the trailing edge of the tool. By revolution of the tool these oxidation bonding layers can be stretched/transferred also towards the RS. However, there is less compaction between the flow-arms at the RS and the refilling action and deposition of the stirred mass at the RS does not lead to a discontinuity at the RS.

We suggest the defects and oxidation layers have flow-based origins. Due to the clockwise rotation of the tool, the plastic flow behaviour differs in AS and RS of the SZ. As in the AS the rotation of the tool was in the direction of welding advance, the plastic deformation of the BM moves forward (and approaches the leading edge). Simultaneously in the opposite direction of the tool (RS), the extrusion nature of the process causes the plasticized flow to become squeezed through the region between the tool and base metal. This forms a channelized flow, which is transferred backwards on the trailing edge of the pin.

Similar to a kissing bond, this oxidation bonding likely causes a negative effect on mechanical properties of FSW joints. Typically, this brittle-nature defect can affect the fatigue/tensile strength of the weld as it is a stress raiser and eventually a place for initiating the fracture. It is suggested that in the presence of this oxidation layer, the crack initiation location and the crack growth would be mainly along the boundaries of this defect. This can cause a possibility for a failure mode, different to typical fracture in fusion welds.

One of the outcomes of the instability of the internal flow regimes is the emergence of tunnel void. Figure 5 shows micrographs of the AS region to evaluate the flow behaviour around the tunnel void defect.

The corresponding flow arms around the tunnel void and in proximity of AS border are shown in Figure 6 which reveals an intersecting layout from the top surface stretching downwards. The striation diagonal lines around the bay-shaped tunnel geometry (Figure 5) show the simultaneous effects of the pin (thread marks) and shoulders on the plastic deformation. The reason for the tunnel defect being at the bottom surface can be attributed to the action of the right-handed threads on the pin, which pump material upwards. There is also a mass deficit caused by loss of material at edge-entry. This mass deficit persists as a continuous void on the bottom surface.

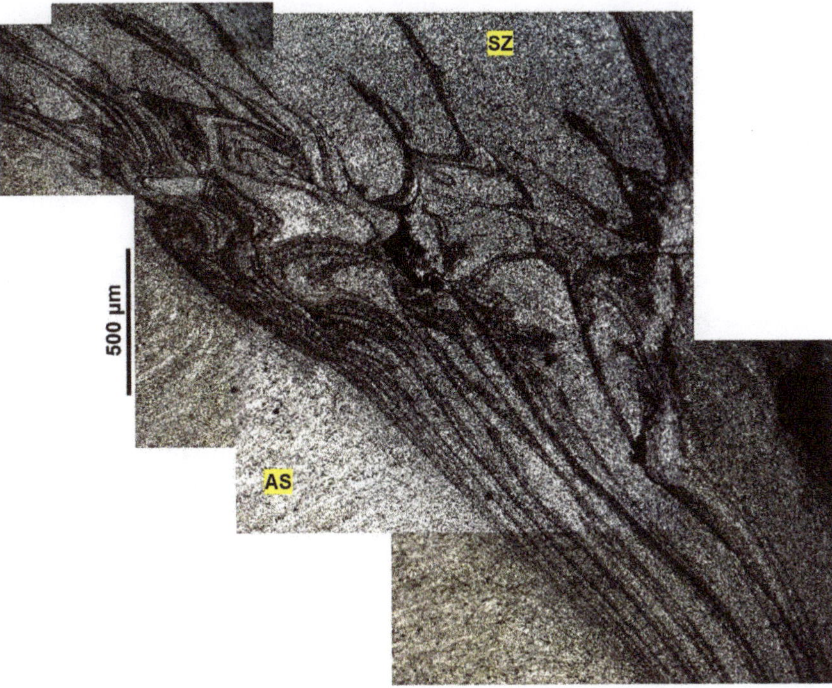

Figure 6. Interlaced flow patterns at the AS border of SZ as the following of the cascade microflow (Reagent C).

Direction of the scrolls compared with the tool rotation provides a circulation towards the centre – this is deliberate and is intended to provide a dynamic sealing of the weld. In an ideal condition, threads and flats on the pin and scrolls on shoulders improve the uniformity of stirring by driving the lateral motion and pumping of material inwards, this reduces spilling [30]. To provide a uniform stirring condition in the sub-shoulder area, the spiral scrolled features started from the edge of the shoulders and ended at the proximity of the pin location. Upward pumping of material (driven by the thread effect) also increases the curvature of hourglass border close to the top shoulder. This shows the compaction of the plasticized flow in this region is more than by the bottom shoulder region.

This upwards flow potentially creates an additional frictional contact between the upper shoulder and the workpiece. If so, we would expect that the temperature on the top shoulder to be higher than that of the bottom shoulder. Consideration of grain sizes suggests that this indeed the case [21]. This may further contribute to cooling and stiffening of the flow at the bottom surface.

The aim of Figure 7 is to clarify the role of the pin features (threads/flats) and the welding parameters (ω/V ratio; rev/mm) in occurrence of the tunnel void. As is shown in Figures 5 and 6, a collection of striation lines accumulated at the AS border. These can be attributed to the simultaneous interaction of the pin threads/flats and the speed ratio (ω/V) to form the flow lines pattern. As a rough measure, the speed ratio by rev/mm is a unit of magnitude of the distance between the flow lines which can form the size of each nib in a saw-tooth like pattern (Figure 5) with the same distance between the flow arms. However, the flow complexities cause an interlaced condition for the flow arms which makes it unreliable as an accurate measurement. The pin threads/flats can cause the localization of deformation through the narrow shear zone to form this saw-tooth like pattern at the internal/inward bay of tunnel defect (Figures 5 and 7d).

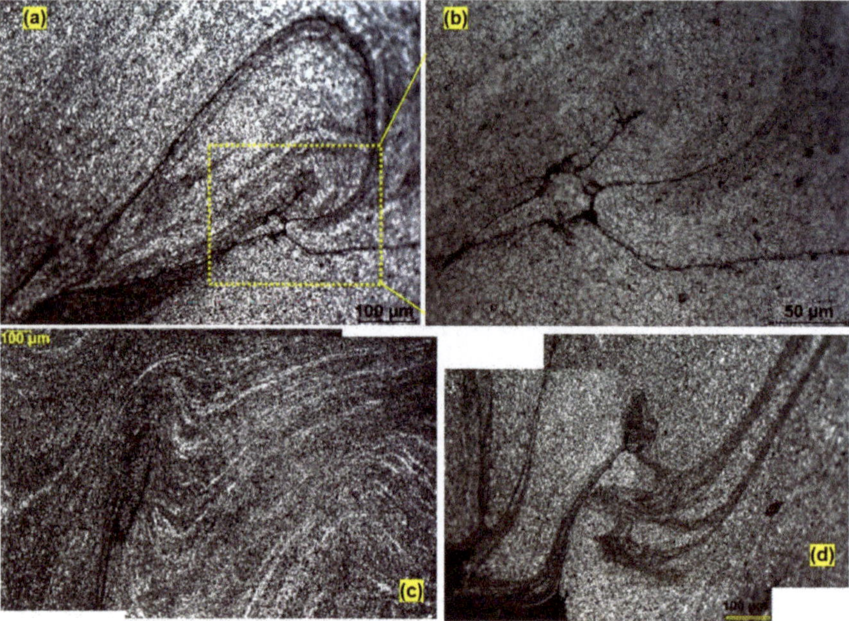

Figure 7. Microflows, tangled in different region of the cascade flow region of the SZ (Reagent C). (**a**) swirling line (region d in Figure 5), (**b**) higher magnification of (**a**); hair-line micro-crack, (**c**) magnified finger pattern and the wavy flow lines (region c in Figure 5), and (**d**) hook line flow at the jagged edge of the tunnel void (region b in Figure 5).

It is clear that the tunnel void geometry emerged due to a large lack of bonding, presumably due to the influence of insufficient pushing of material flow during the stirring. Hence, the tunnel void formation is attributed to insufficiency in compaction of the material flow at the position of the defect. This indicates that for a good weld the material must have good flow forced by the tool geometry and process variables.

The typical periodic striation lines with the finite width at the border of the AS show the typical forging zone around the pin. These flow lines form a projected area at the circumferential side of the tunnel void where the bonding patterns interlaced abruptly. Our interpretation is that, as the bonding lines are separating, the flow velocity gradually decreases which eventually comes to a sharp decrease at the interface of the tunnel defect. This can also change the orientation of the material flow lines, as the interval of the flow lines is maximally separated (compared to the compacted banding pattern in the proximity of the AS border in Figure 6).

By comparing the lack of bonding defect with the characteristics of a proper joint geometry, the main reason for formation of the tunnel void appears to be insufficient frictional heat and integrity of material flow, due to the internal force and rotating speed during the process. Poor material flow is attributed to an insufficient heat input, which leads to more bonding defects. Also, the compression ratio is not enough to create a consistent forging/pressure force to extrude the plastic mass and fully/decently compensate/refill the defect position.

The curved striation lines in Figure 7a (region d in Figure 6) elucidate these flow-based problems where the connection of the flow lines with the main regime has been disrupted. This leads to formation of a tangled flow around the saw-tooth like pattern at the circumferential side of the tunnel void. As shown in Figure 7b, this tangled flow is a suitable place for stress concentration within the layers of the mass flow, in which may cause microcracking. The propagation of microcracks could eventually lead to the failure of the weld by coalescence of the macrosize voids. The interlaced flow-lines in

Figure 7c show a stretched finger pattern (region c in Figure 6) which was evident before in Figure 5a. This flow defect, similar to the curved flow pattern shown in Figure 7a (semi-circular swirling band), can affect the flow integrity as a suitable position for imitation of the micro-crack.

Another flow feature, the enlarged micro-flow shown in Figure 7d, reveals the hook line flow patterns positioned at the internal edge of the tunnel void. We interpret this feature as a lack of material consolidation during stirring, similar to the root flaw in CFSW.

From a metallurgical viewpoint, there is some debate about the nature of the flow-arms as the elongated flow bands with a different colour compared to the matrix. They are not metallographic stains or over-etching. Nor are they some type of intrinsic defects of the base metal formed during the rolling procedure. Rather they are specifically associated with the tunnel defect, and visible across multiple different welds. The literature generally terms them flow lines, i.e. attributes them to internal flow of the material. To further understand the nature of the flow-arms we applied elemental mapping via EDS to determine the composition of the texture. To observe the exact position of the flow-pattern, the samples after etching were observed by Backscattering Electron (BSE) imaging via SEM microscopy, also supported by the elemental mapping via the energy-dispersive X-ray spectrometer (EDS) detector. The scanning analysis was done for the flow-arms region, also the selected location of the hair-line micro-crack, both shown in Figure 8.

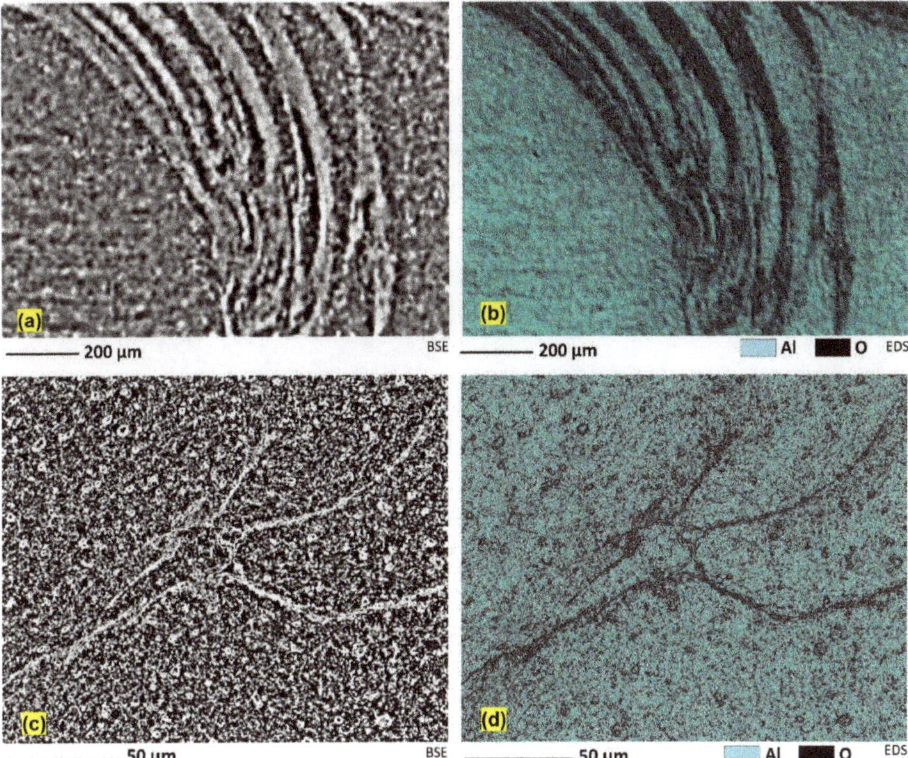

Figure 8. Analysis of the flow-arms region (region 1, Figure 4a) using (**a**) BSE, and (**b**) EDS elemental map, and similarly for (**c**,**d**) hair-line micro-crack position (demonstrated in Figure 7b). The oxygen-rich areas of the EDS elemental map have been delineated in dark (**b**,**d**).

In general, the BSE imaging can be used to show the different elements present in the sample. The microscopic analysis of the flow-arms region in Figure 8a,b confirms that the elongated flow-arms are rich in oxygen. This region is delineated as the darker area at the OM etched samples. Additionally, the EDS analysis of the location of the hair-line micro-crack shows that the edge of the crack is rich in oxygen (Figure 8c,d).

A further SEM study for the Flow-arms pattern in Figure 4b were demonstrated in Figure 9. The scanned area in Figure 9a belongs to the tip of the one of flow-arm branched at the RS of the weld region. The bunch of elongated grains in Figure 9a all belong to one flow line pattern in Figure 4b. The scanned area reveals that the elongated grains are distinctly separated from each other by main grain boundaries; however there are some sub-grain details inside of the grains. The higher magnification of the inside of the grain (Figure 9b) shows that there is a high density of the sub-grain boundaries within the elongated grain. This is a main thermomechanical characteristic of dynamic recrystallization, suggesting that the shearing within the texture can activate the sub-grain boundaries during the re-cooling process after stirring. It should be noticed that the nature of the sub-grain boundaries can be analysed further by EBSD and TEM techniques, which is beyond the scope of the current work. From this result it is concluded that the presence of the shearing during the stirring induced a stored strain within the compacted deposited layers of the plasticized mass at the back of the tool. This activated the formation of the microscopic features at the hourglass border, which in RS is revealed as the sub-grain boundaries, appearing as the elongated flow patterns similar to the oxidation layers at the AS. The high density of the sub-grain boundaries can cause a darker band during the etching, delineating a flow-arm shaped pattern similar to the oxidation patterns at the AS of the hourglass border.

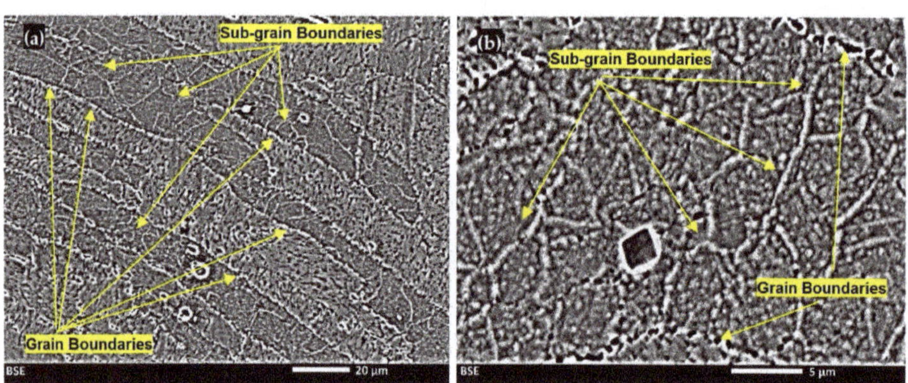

Figure 9. SEM analysis of the flow arm shaped branches at the RS. (**a**) A bunch of elongated grains at the position of the flow pattern, (**b**) higher magnification of the inside of the grain, representative of the high density of the sub-grain boundaries as a response to the stored strained during the DRX.

3.4. Sub-Surface Features of the Weld

Similar to the transition region HAZ/TMAZ, the surface region of the weld also experiences plastic deformation. The saw-tooth feature at the weld crown (region 1, Figure 2a) corresponds to the plastic deformation and the dynamic slip-stick engagement between the material and the shoulder. Figure 10 illustrates a gradient of the size of grains laterally (Regions A-B-C).

3.4.1. Cross-Section under the Weld Shoulder

The wavy pattern visible in Figure 9a represents the large deformation in the weld crown. The microstructure exhibits a morphological variation from a coarse grain structure at the top (Region A, Figure 10b), altered to an elongated pattern (Region B, Figure 10c) which eventually leads to a large number of globular equiaxed grains observable at the inside region (region C, Figure 10d). This microstructural subdivision with a gradual trend from the edge of the sub-shoulder region towards the inner parts of the SZ can be attributed to the grain shape adjustment during the continuous recrystallization induced by the large plastic deformation.

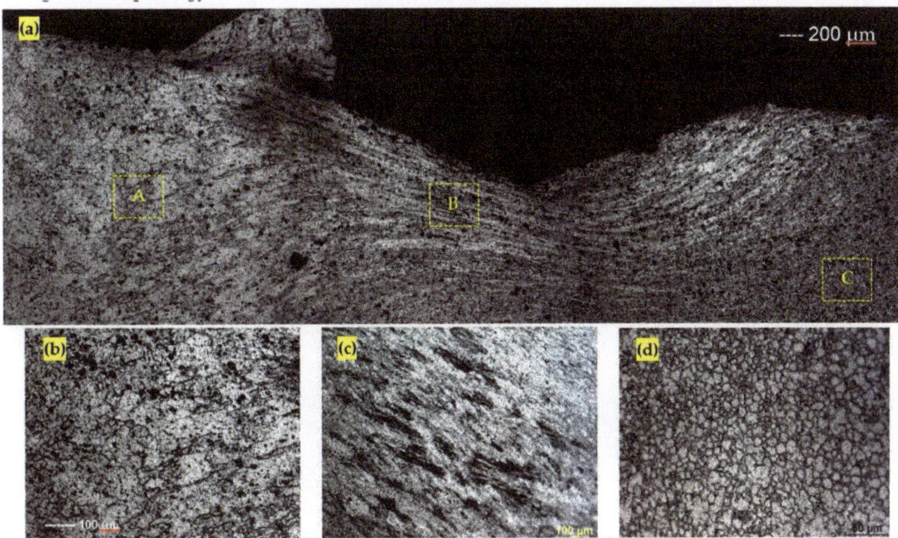

Figure 10. Grain and flow details of the sub-shoulder region of the weld crown (Reagent B). (**a**) macrograph of the sub-shoulder region; different microscopic features are enlarged as Regions A, B, C in Figure 10b–d, respectively; (**b**) Regions A; coarse grain structure, (**c**) Regions B; elongated grains, (**d**) Regions C; ultrafine grained microstructure.

The curving area near the top surface exhibits a random distribution of the coarse grains (Region A, Figure 10b). This can be caused by the thermal dissipation gradient near the surface, similar to the chill zone in casting.

At region B (Figure 10c) a lamellar morphology is initiated at the crown region where the slope increases at the edge and a large shear strain is imposed through the sub-shoulder area which simultaneously experiences the maximum compression rate. In this situation the dynamic recrystallization takes place very fast which pins the grain boundaries and rearrange the grains to an elongated morphology.

Stirring is inherently classified as a large strain warm deformation. During the stirring action at the inner layers of mass located far away from the top surface, the pin-driven plastic deformation prevails over the shoulders. This can also be considered a severe grain deformation due to the mechanical stirring action. Consequently, the elongated grains undergo a fragmentation and the subsequent formation of an ultrafine grained microstructure (Region C, Figure 10d). On the other hand, because this location is far from the surface, rapid cooling is unlikely. Therefore, the grains distribution is homogeneously equiaxed. This is also the region where other abnormal microscopic features have been observed, such as metallic-glass amorphous structure and localized shrinkage [33].

3.4.2. Plan View under the Shoulder (Longitudinal Axis)

To explore the relationship between the pin and sub-shoulder flow, we investigated the flow features of the weld in the plan view under the shoulder in the longitudinal axis of the weld-line. The results of the sub-shoulder flow observation are illustrated in Figures 11 and 12.

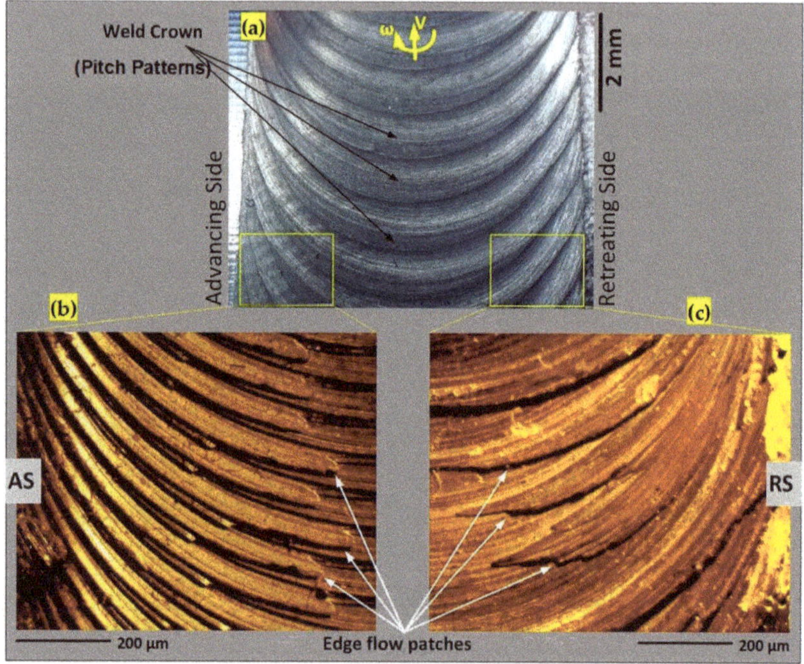

Figure 11. Surface flow patterns at the centre of the weld-line. (**a**) Plan view of the weld surface, (**b**) Macro-etched sub-shoulder (0.5 mm deep) flow patterns at the AS, and (**c**) Macro-etched sub-shoulder (0.5 mm deep) flow patterns at the RS.

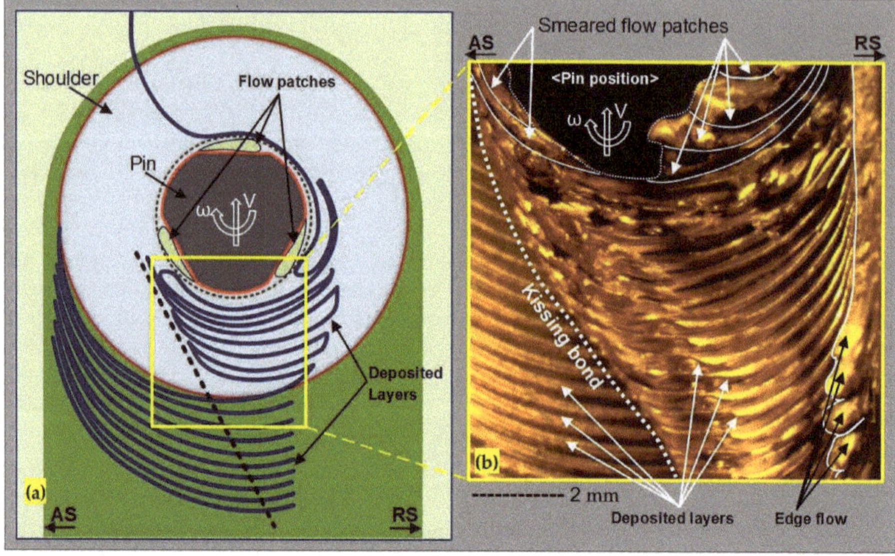

Figure 12. The top view of the sub-shoulder region with a proposed model for the flow-lines 0.5 mm underneath the shoulder during the stirring action. (**a**) the flow model of the sub-shoulder area, (**b**) the macro-etched sub-shoulder region including the actual flow patterns, at the position of the exit zone.

Figure 11 shows the plan view of the sub-shoulder flow for the weld-seam at the centre of the weld-line, where the tool was half way through the weld trial. Figure 11a shows a plan view of the actual weld-line, including periodic weld pitches at the weld crown, representative of the deposited stirred mass layers at the trailing edge of the tool. Figure 11a,b show the macro-etched flow patterns of the weld-seam for the same position of the weld in Figure 11a. The samples were polished to approximately 0.5 mm in depth, and were macro-etched to show the surface flow pattern both in the AS (Figure 11b) and RS (Figure 11c).

As shown in Figure 11b,c, the flow patches at the plan view of the weld-line exist at both sides; AS and RS. This is consistent with our interpretation that the sub-shoulder flow lines arise from the deposition of the stirred layers of mass at the trailing edge of the tool. The periodic deposition of the mass flow creates a pitch pattern for the surface flow which extends to some depth, driven by the action of the shoulder during the stirring. These features are not created by the pin, but rather the shoulder.

The features in Figure 11a,b are smeared material that extends to some depth below the surface, reveal that the flow lines at the edge of the weld-line, and different features are evident at the AS and RS.

3.4.3. Exit Zone

The region where the tool exits the weld provides a unique opportunity to examine the location of flow features without the complication of the packing and deposition that occurs elsewhere in the weld. The surface and subsurface (approximate depth of 0.5 mm) results for this region is shown in Figure 12.

The tentative explanation is that this region shows the combined effect of smeared flow patches from the shoulder (to the left of Figure 12b), and the flow patches from the pin (centre-right of Figure 12b). In between the two there appears to be a region of disturbed flow.

3.5. Building a Model of Internal Flow and Defect Formation

We propose a number of new mechanisms that affect flow, and the formation of defects.

3.5.1. Proposed Composition of Flow Layers

The metallurgical analysis in Figures 8 and 9 shows that the flow line patterns evident in optical microscopy are in fact oxidation effects in the AS, and the accumulation of the shearing and formation of high-density sub-grain boundaries during the DRX procedure in the RS. Thus, we propose that what are conventionally called flow arms are better understood as 'flow layers'.

We believe the more accurate interpretation, for the AS, is that they represent the oxidised fronts between packets of material that have been transported through the weld by tool motion and rotation. Where they occur, they are evident in all cross sections along the weld. An oxidization explanation is feasible considering the high temperature of the stirring process, though an explanation is needed for how the air enters the weld - more on this below. We interpret the layers as being a packing (or stacking) mechanism in the AS, i.e., the flow motion is orthogonal to the flow layer in this region. It is interesting to note that the metallographic macro/micrographs showed that the defect-free samples (see Figure 2) do not reveal flow-arms to the same extent.

For the RS the flow layers represent shearing between different layers, i.e., the flow motion may be in/out of the page. This is inferred from by the microstructure in Figure 9, which shows elongated grains with internal accumulation of sub-grain boundaries that have been subject to self-rearrangement during dynamic recrystallization. More specifically, the proposed DRX mechanism is that the stirred flow packets experience severe shearing, and hence stored strain is induced. By using the heat generated during re-cooling, the elongated grains cause a self-rearrangement to relieve the stored strain, leading to a high-density accumulation of sub-grain boundaries inside the grain. As the elongated grains have similar crystallographic orientation, so too the newly formed sub-grain boundaries have similar crystallographic direction. Consequently, both the grain boundaries and sub-grain boundaries etch similarly, and more so than the matrix. The dark colour of these layers under optical microscopy arises from the etching of these sub-grain boundaries with the characteristic darker contrast compared to the matrix.

3.5.2. Proposed Principles of Motion of Flow Layers

In this regard, and based on the metallurgical improvements, the suggested model in Figure 13 provides an explanation for the entering of the air into the weld region and beginning of the oxidation in flow layers. We introduce the concept of a flow patch. This is a chip of material that is excised from the base material, transported around the weld as a lump, and deposited in the wake of the weld as a flattened layer.

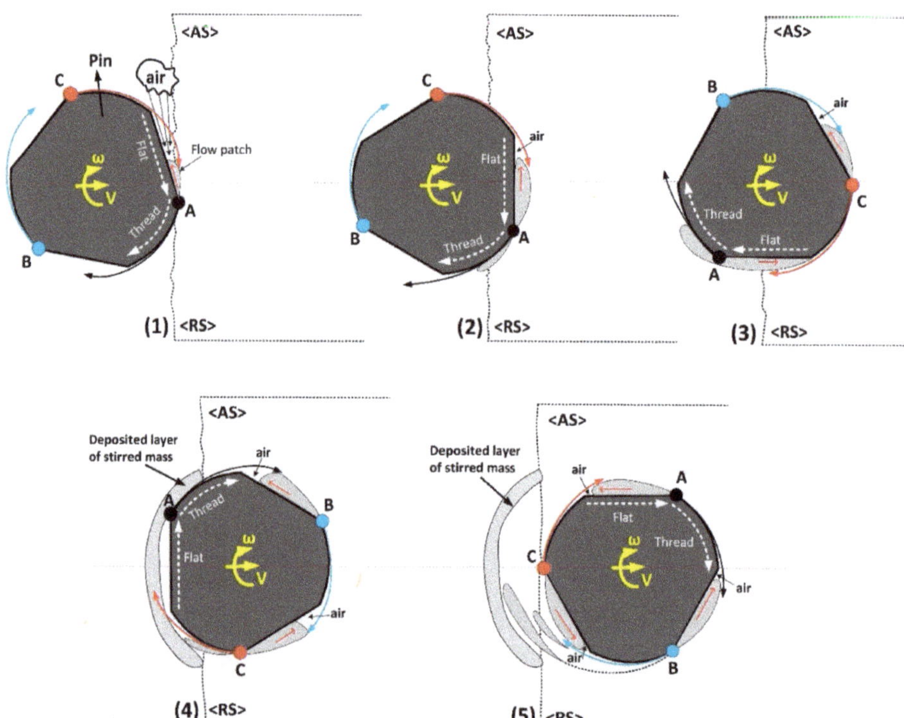

Figure 13. Schematic of the entering of the tool into the workpiece, with a proposed model for the entering of the air into the weld region and formation of the oxidation layer between the deposited mass layers at the trailing edge of the tool (Steps **1–5** explain the details of the discontinuous flow mechanism).

As shown in step 1 in Figure 13, the entering of the tool into workpiece initiates the excision of the flow patches on the tool, as chip formation on the flat area of the tool. These flow patches are softened by the heat generated from the friction, and in contact with the air form an oxide surface layer. By the further embedding of the tool into the workpiece, the flow patch is also pushed into the stirring zone between the AS and the RS (step 2). Simultaneously, a volume of the air also enters into the material which is mostly because of the angled geometry of the flat-thread configuration (steps 2 and 3). Air is readily available in the weld due to the shaking and vibration of the tool-substrate. This trapped air forms as an air-pocket at the corners of the flat-thread geometry and intensifies the oxidation layers of the plasticized mass during the stirring (steps 3 and 4). Eventually, the stirred mass is transported to the trailing edge of the tool, where they plough into the RS and are deposited as the flow layers of the weld (step 5). The oxidized surfaces of these flow patches are the flow layers that are observed in the cross section. They are deposited as periodic layers, corresponding to multiple flow patches.

- EXTRACTION AND FRAGMENTATION OF SUBSTRATE

The first step of the stirring action is the fragmentation of the solid grains, from the body of the workpiece. The frictional heat from this grain fragmentation can soften the mass and form the plasticized mass which enters into the stirring zone. Hot material is broken into large fragments by the tool features (threads and flats). These fragments may be partially attached to the weld mass, or they may be completely severed but retained within the confines of the weld. These fragments are evident in the lacerations on the internal weld surfaces at the entry of the weld, and as loose granules that fall out of the weld. The fragments are cut from the workpiece in an arc from the leading edge of the tool, from the advancing to retreating sides. We propose that the fragments are delineated by the flow layers, i.e. the region between any two flow arms represents a fragment of material that once had exposed surfaces in the region of the flow arms.

- TRANSPORT AND FORCED DEPOSITION

The weld mass, including attached and detached fragments, is moved by the rotation of the tool into the trailing edge space behind the tool. The motion is from the RS towards the AS and results in deposition of the material. This deposition process involved squeezing of the weld bulk and the fragments. This involves shear and forging-like processes, whereby the fragments are squashed back together in a plastic process. The shear adds a heating effect which softens the material and facilitates their solid state bonding. As the plasticised mass moves closer towards the AS, so the fragments are flattened and squashed up against the AS—this creates the hourglass border. We propose this is the reason for the increasingly fine spacing between the flow arms closer to the AS compared to the centre of the weld. The final result is thin streaks of flow arms, hence the observed interlaced flow.

3.5.3. Proposed Principles of Formation of Surface Features

This next explanation addresses the correspondence between internal and surface features. Figure 14 presents a flow model to explain this transport deposition. By revolution of the tool the transportation and deposition of the mass layers happens by the speed ratio (ω/V). The final flow patterns are observable as the flow-arms at the cross-section of the weld, and the related weld pitch patterns at the surface of the weld-line.

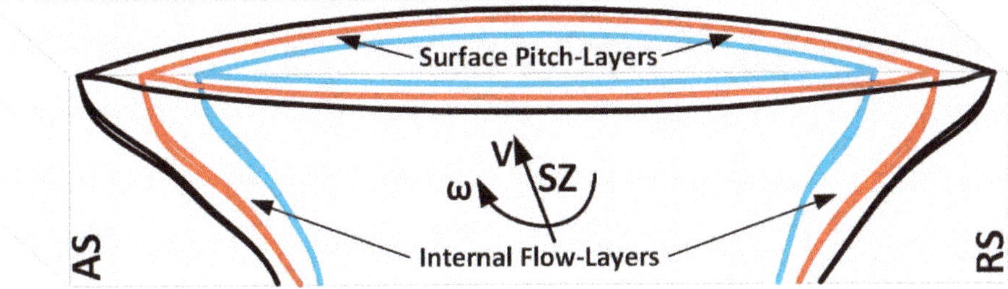

Figure 14. The proposed flow patterns for the flow-arms and the weld pitch patterns at the cross-section and surface of the weld region, respectively. The model aims to reveal the flow lines during the revolution of the tool and deposition of the layered mass at the trailing edge of the tool.

In general, the weld pitch patterns at the surface of the weld are directly dependent on the speed ratio between the rotation and advancing actions of the tool, simply as the number of revolution per forward movement, in which can directly influence the formation of the deposited layers at the stirring zone. It should be noted that the pressure imposed underneath the shoulder can affect the shape and size of these bonded layers and compact them together. Also, adding of the new deposited layers into the previous layers leads to increase in compaction condition at the back of the tool.

3.5.4. Integrated Weld Transport Model

In Figure 15, the simultaneous formation of the surface flow patches (pitch layers) and the internal flow layers has been demonstrated as a discontinuous flow model for the plasticized batches of the mass which are formed by the revolution of the tool. As the shoulder action is directly responsible for the formation of the surface flow patches at the sub-shoulder region, the rotation and the advancement of the pin at the mid-SZ area generates the plasticized flow batches at the proximity of the tool, at the breadth of AS-RS. The deposited flow layers at the trailing edge of the tool form the bended flow layers at the borders of the AS and RS, which are revealed as the hourglass-boundaries within the cross-section of the weld.

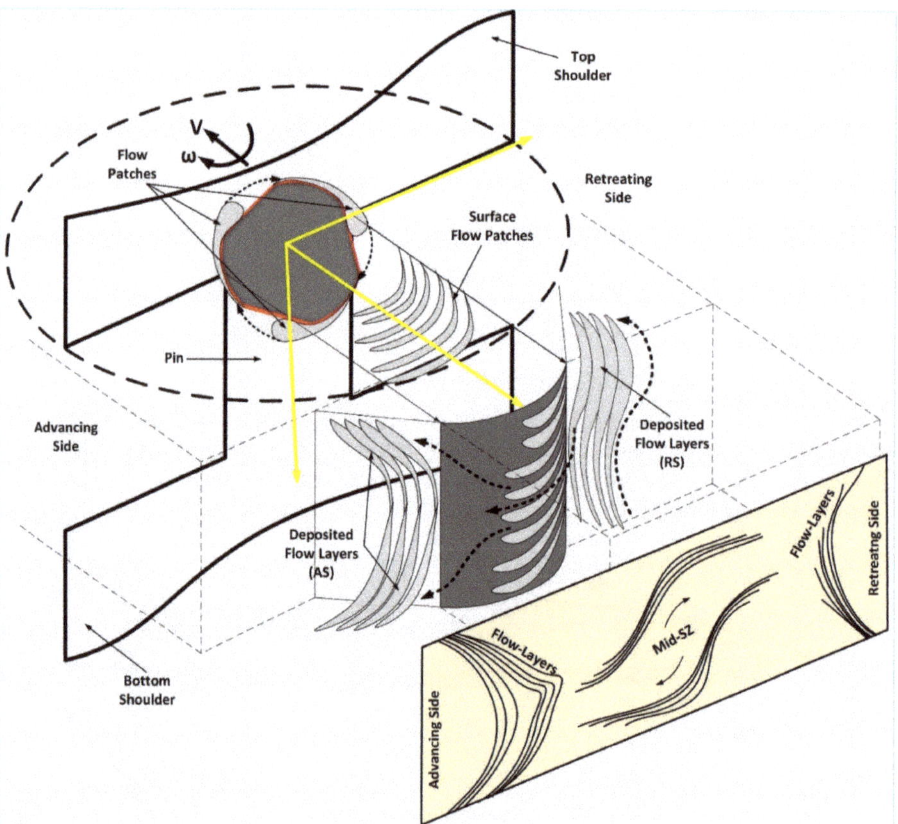

Figure 15. Schematic of the cross-section of the weld with a proposed model for the formation mechanism of the flow-arms at the breadth of the stirring zone, between the AS and RS.

4. Discussion

4.1. Originality

The work makes the following original contributions to the understanding of the welding of AA6082-T6.

4.1.1. Characterization of the Flow Layers at the Hourglass-Borders

The combination of the metallographic delineation in optical microscopy and the metallurgical analysis by the SEM and EDS, confirms that the dark curved patterns in the AS and RS hourglass borders of flow-arms, being oxidation layers (AS) and elongated grains with a high-density of accumulation of the sub-grain boundaries as the result of the shearing and DRX (RS).

The original aim of this work was to understand the relationship between flow regimes and physical defects. The results show that the flow layers are important in understanding the internal flow, and the resulting defects. Failed welds show more flow lines, and show in specific locations in the cross section. Also these flow lines are associated with oxidisation. OM methods are relatively quick and simple to apply, but identification of these characteristics of the flow lines required electron microscopy. Now that the metallurgical nature has been elucidated, it is potentially possible that the simpler OM method could be used in future as an industrial quality control tool.

4.1.2. Visualization of Flow Layers

In this work, based on the microscopic observations, we elucidated the internal flow features of the plastic deformation in different regions of the stirring zone during the bobbin friction stir processing. Key concepts that emerge are:

- Packets of material ('flow patches') are transported around the pin.
- Flow patches are transported round the RS to the back of the tool, where they experience high localized shearing at their mutual boundaries, as evidenced in high density of sub-grain boundaries.
- Flow patches are transported all the way round the tool to the AS, where they are stacked on each other and flattened in the process. Air enters at or before this stage and causes oxidization of the boundaries, as evidenced in EDS elemental mapping.

4.1.3. Proposing a Model of the Internal Flow Processes

Many other studies have been based on the assumption of continuous flow within the stirring zone. However, such approaches have not had very much success in predicting the actual flow. It has been particularly difficult to model the tunnel void as a flow-based defect using continuous flow assumptions. The present work puts this into context, by finding that the flow is highly discontinuous, based on observation of the microstructure. In turn this has been facilitated by discovery of a suitable reagent and metallographic measurement. Consequently, the present results imply that methods based on continuous flow computational fluid dynamics (CFD) are unlikely to be successful in explaining flow defects. In CFSW the flow lines ("onion rings") are attributed to the tool-material interaction [18]. However, these flow features are not defects per se. Likewise for BFSW the tool-material interaction has been identified as contributing to discontinuous flow as a bulk effect [34]. The present work is consistent with these findings, but extends them to the defect situation. In particular here it is proposed that the thread-flat features of the pin further to a finer scale discontinuity of flow, in terms of packets of material. These are deposited at the trailing edge of the tool, and squashed in the process.

4.2. Implications for Practitioners

Industry users of AA6082-T6 should note that tunnel defects are associated with internal oxidization of the weld cross section. This has the potential to cause reduced mechanical properties, corrosion resistance and the fatigue strength as the tunnel void as a macro-size discontinuity can deteriorate the integrity-related properties of the weld.

4.3. Limitations of this Work and Implications for Future Research

Our analyses of the flow layers were limited in the number of such layers investigated. We cannot exclude the possibility that flow layers may have features of both oxidization and DRX sub-grain boundary formation. A potential future research project could examine multiple flow layers, at different positions along their length, and evaluate both microstructure (e.g., using SEM or TEM) and elemental composition.

3D visualization of the flow layers was not possible during optical and electron microscopy. A possible research question could be to progressively re-polish the surface in a controlled manner to build up a 3D representation of the flow layers.

Another possible future line of research could be measurement of the mechanical properties of the weld in correlation with the flow features. Progressive loading of a sample of the cross section, with microscopy inspection in between, could identify the evolution of the crack propagation and failure mechanisms. Regarding the oxide layers, it may be useful to apply fractography or creep-fatigue tests to examine crack propagation.

It is to be expected that the formation of the macro-size tunnel void and the micro-cracks during stirring action would extensively affect the strength of the final weldment. Macroscopic defects are likely to be unacceptable to industry users, and hence preventing these is the first priority. Even if there is no tunnel defect, the internal micro-cracks are likely to cause adverse outcomes in the three-point bending and other tests used. Hence there could be value in further research into welding process settings, including tool features, that minimise these defects.

5. Conclusions

The internal flow features of the AA6082-T6 BFSW were evaluated using a set of developed reagents and optical microscopy. The key findings are that the dark curved patterns (conventionally called "flow-arms"), are actually oxidation layers at the advancing side, and at the retreating side are elongated grains with a high-density of accumulation of sub-grain boundaries due to DRX. A model of discontinuous flow within the weld is proposed, based on the microscopic observations. It is inferred that the internal flow is characterized by packets of material ("flow patches") being transported around the pin. At the RS they experience high localized shearing at their mutual boundaries, as evidenced in high density of sub-grain boundaries. Flow patches at the AS are stacked on each other and exposed to oxidization.

Author Contributions: Conceptualization, A.T., D.J.P.; methodology and formal analysis, A.T.; supervision, D.J.P., D.C., K.S.; validation, A.T., D.J.P.; writing–original draft, A.T.; writing–review & editing, A.T., D.J.P., D.C., K.S.

Funding: This research received no external funding.

Acknowledgments: Thanks are extended to Mike Flaws and Kevin Stobbs for assistance with electron and optical microscopy, respectively.

Conflicts of Interest: The authors declare no conflict of interest.

References

1. Thomas, W. Friction Stir Welding. International Patent Application No. PCT/GB92/02203, 6 December 1991.
2. Thomas, W.; Wiesner, C.; Marks, D.; Staines, D. Conventional and bobbin friction stir welding of 12% chromium alloy steel using composite refractory tool materials. *Sci. Technol. Weld. Join.* **2009**, *14*, 247–253. [CrossRef]
3. Threadgill, P.L.; Ahmed, M.; Martin, J.P.; Perrett, J.G.; Wynne, B.P. The use of bobbin tools for friction stir welding of aluminium alloys. *Mater. Sci. Forum* **2010**, *638*, 1179–1184. [CrossRef]
4. Thomas, W.; Wiesner, C. Recent developments of FSW technologies: Evaluation of root defects, composite refractory tools for steel joining and one-pass welding of thick sections using self-reacting bobbin tools. In *Trends in Welding Research, Proceedings of the 8th International Conference, Pine Mountain, GA, USA, 1–6 June 2008*; ASM International: Novelty, OH, USA, 2009; p. 25.
5. Thomas, W.; Nicholas, E. Friction stir welding for the transportation industries. *Mater. Des.* **1997**, *18*, 269–273. [CrossRef]
6. Svensson, L.E.; Karlsson, L.; Larsson, H.; Karlsson, B.; Fazzini, M.; Karlsson, J. Microstructure and mechanical properties of friction stir welded aluminium alloys with special reference to AA 5083 and AA 6082. *Sci. Technol. Weld. Join.* **2000**, *5*, 285–296. [CrossRef]

7. Russell, M.; Shercliff, H. Analytical modelling of microstructure development in friction stir welding. In Proceedings of the 1st International Symposium On Friction Stir Welding, Thousand Oaks, CA, USA, 14–16 June 1999.

8. Sued, M.; Tamadon, A.; Pons, D. Material flow visualization in bobbin friction stir welding by analogue model. *Proc. Mech. Eng. Res. Day* **2017**, *2017*, 368–369.

9. Waldron, D.J.; Roberts, R.W.; Dawes, C.J.; Tubby, P.J. Friction Stir Welding—A Revolutionary New Joining Method. *J. Aerosp.* **1998**, *107*, 1247–1252.

10. Xu, H.; Tang, H.; Liu, Z.; Xie, M.; Jiao, J. Microstructure and Mechanical Properties of 6082 Aluminum Alloy Joints Welded by MIG. *Hot Work. Technol.* **2010**, *1*, 42.

11. Zhang, H.; Wang, M.; Zhang, X.; Yang, G. Microstructural characteristics and mechanical properties of bobbin tool friction stir welded 2A14-T6 aluminum alloy. *Mater. Des.* **2015**, *65*, 559–566. [CrossRef]

12. Martin, J.; Wei, S. Friction Stir Welding Technology for Marine Applications. In *Friction Stir Welding and Processing VIII*; Springer: Berlin/Heidelberg, Germany, 2015; pp. 219–226.

13. Esmaily, M.; Mortazavi, N.; Osikowicz, W.; Hindsefelt, H.; Svensson, J.; Halvarsson, M.; Martin, J.; Johansson, L. Bobbin and conventional friction stir welding of thick extruded AA6005-T6 profiles. *Mater. Des.* **2016**, *108*, 114–125. [CrossRef]

14. Chen, C.; Kovacevic, R. Thermomechanical modelling and force analysis of friction stir welding by the finite element method. *Proc. Inst. Mech. Eng. Part C J. Mech. Eng. Sci.* **2004**, *218*, 509–519. [CrossRef]

15. Chen, Z.; Pasang, T.; Qi, Y. Shear flow and formation of Nugget zone during friction stir welding of aluminium alloy 5083-O. *Mater. Sci. Eng. A* **2008**, *474*, 312–316. [CrossRef]

16. Gopi, S.; Manonmani, K. Microstructure and mechanical properties of friction stir welded 6082-T6 aluminium alloy. *Aust. J. Mech. Eng.* **2013**, *11*, 131–138. [CrossRef]

17. Sued, M.; Pons, D.; Lavroff, J.; Wong, E.H. Design features for bobbin friction stir welding tools: Development of a conceptual model linking the underlying physics to the production process. *Mater. Des.* **2014**, *54*, 632–643. [CrossRef]

18. Krishnan, K. On the formation of onion rings in friction stir welds. *Mater. Sci. Eng. A* **2002**, *327*, 246–251. [CrossRef]

19. Liechty, B.; Webb, B. Modeling the frictional boundary condition in friction stir welding. *Int. J. Mach. Tools Manuf.* **2008**, *48*, 1474–1485. [CrossRef]

20. Tamadon, A.; Pons, D.; Sued, M.; Clucas, D.; Wong, E. Analogue Modelling of Bobbin Tool Friction Stir Welding. In Proceedings of the International Conference on Innovative Design and Manufacturing, Auckland, New Zealand, 24–26 January 2016.

21. Tamadon, A.; Pons, D.J.; Sued, K.; Clucas, D. Development of Metallographic Etchants for the Microstructure Evolution of A6082-T6 BFSW Welds. *Metals* **2017**, *7*, 423. [CrossRef]

22. Prangnell, P.; Heason, C. Grain structure formation during friction stir welding observed by the 'stop action technique'. *Acta Mater.* **2005**, *53*, 3179–3192. [CrossRef]

23. Fonda, R.; Knipling, K.; Bingert, J. Microstructural evolution ahead of the tool in aluminum friction stir welds. *Scr. Mater.* **2008**, *58*, 343–348. [CrossRef]

24. Fonda, R.; Reynolds, A.; Feng, C.; Knipling, K.; Rowenhorst, D. Material flow in friction stir welds. *Metall. Mater. Trans. A* **2013**, *44*, 337–344. [CrossRef]

25. Coelho, R.S.; Kostka, A.; Dos Santos, J.; Pyzalla, A.R. EBSD Technique Visualization of Material Flow in Aluminum to Steel Friction-stir Dissimilar Welding. *Adv. Eng. Mater.* **2008**, *10*, 1127–1133. [CrossRef]

26. Hilgert, J.; Schmidt, H.; Dos Santos, J.; Huber, N. Thermal models for bobbin tool friction stir welding. *J. Mater. Process. Technol.* **2011**, *211*, 197–204. [CrossRef]

27. Hilgert, J.; Hütsch, L.L.; dos Santos, J.; Huber, N. Material flow around a bobbin tool for friction stir welding. In Proceedings of the COMSOL Conference, Paris, France, 17–19 November 2010.

28. Hilgert, J.; Dos Santos, J.; Huber, N. Shear layer modelling for bobbin tool friction stir welding. *Sci. Technol. Weld. Join.* **2012**, *17*, 454–459. [CrossRef]

29. Davis, J.R. *Aluminum and Aluminum Alloys*; ASM International: Novelty, OH, USA, 1993.

30. Sued, M.K. Fixed Bobbin Friction Stir Welding of Marine Grade Aluminium. Ph.D. Thesis, University of Canterbury, Christchurch, New Zealand, 2015.

31. Tamadon, A.; Pons, D.J.; Sued, K.; Clucas, D. Formation Mechanisms for Entry and Exit Defects in Bobbin Friction Stir Welding. *Metals* **2018**, *8*, 33. [CrossRef]

32. DS, K.C.; Tamadon, A.; Pons, D.; Sued, M.; Clucas, D.; Wong, E. Preparation of Plasticine Material for Analogue Modelling. In Proceedings of the International Conference on Innovative Design and Manufacturing (ICIDM 2016), Auckland, New Zealand, 24–26 January 2016.
33. Tamadon, A.; Pons, D.; Sued, K.; Clucas, D. Thermomechanical grain refinement in AA6082-T6 thin plates under Bobbin friction stir welding. *Metals* **2018**, *8*, 375. [CrossRef]
34. Sued, M.K.; Pons, D.J. Dynamic Interaction between Machine, Tool, and Substrate in Bobbin Friction Stir Welding. *Int. J. Manuf. Eng.* **2016**, 1–14. [CrossRef]

Article

Effects of a Post-Weld Heat Treatment on the Mechanical Properties and Microstructure of a Friction-Stir-Welded Beryllium-Copper Alloy

Yeongseok Lim [1,2], Kwangjin Lee [2,*] and Sangdon Moon [1]

[1]   Department of Mechanical Design Engineering, Chonbuk National University, Jeonju 54896, Korea; dudtjr1215@gmail.com (Y.L.); msd@jbnu.ac.kr (S.M.)

[2]   Carbon & Light Materials Application R&D Group, Korea Institute of Industrial Technology, 222, Palbok-ro, Deokjin-gu, Jeonju-City 54853, Korea

*   Correspondence: kjlee@kitech.re.kr; Tel.: +82-63-210-3711

Received: 11 March 2019; Accepted: 17 April 2019; Published: 19 April 2019

Abstract: This paper investigated the microstructure and mechanical properties of a friction-stir-welded beryllium-copper alloy, which is difficult to weld with conventional fusion welding processes. Friction stir welding (FSW) was successfully conducted with a tungsten-carbide (WC) tool. Sound joints without defects were obtained with a tool rotational speed of 700 RPM and tool travel speed of 60 mm/min. A post-weld heat treatment (PWHT) of the FSW joints was performed to analyze the evolution of the microstructure at 315 °C for a half, one, two, three, four, five and eight hours, respectively. The microstructures of the joints were observed using an optical microscope (OM), a scanning electron microscope (SEM) and a transmission electron microscope (TEM). Observed softening of microstructure is suggested to be due to the dissolution of the strengthening precipitates during the FSW process, whereas the strength of the joints was recovered via the formation of the CuBe (γ') phase during the post-weld heat treatment. However, the strength was decreased upon an excessive post-weld heat treatment exceeding three hours. It is considered that the formation of the γ phase and the coarse γ' phase contributed to the reduction in the strength.

Keywords: friction stir welding; beryllium-copper alloy; mechanical properties; microstructure; post-weld heat treatment

---

1. Introduction

Cu alloys have been widely used in the aerospace, transportation and electric power industries due to their reasonable strength, excellent conductivity and good corrosion resistance [1]. Moreover, giga-grade high strength beryllium-copper alloy is used to manufacture several components, such as anti-galling cylinders for undersea cable communication system repeater housings, undersea pressure vessels, valves and gimbals, as well as connectors and drill components due to the high strength and hardness of this material and its excellent fatigue, corrosion and wear resistance capabilities [1,2]. Beryllium-copper alloys can be welded using conventional fusion welding methods; however, problems such as the formation of inclusions, blow holes, porosity and solidification cracking in the heat-affected zone (HAZ) can arise. However, this alloy has poor weldability and can induce softening when welded owing to the dissolution of strengthening precipitates [1,3]. Another major concern is the lack of suitable flux/core wire materials for beryllium-copper alloys [3].

A previous study has shown that the copper alloy joints fabricated by laser beam welding (LBM), which is most generally used welding method, show several challenges. For copper alloys with high thermal conductivity and high laser beam absorption, process stability and high laser power are required as a solution of spattering and high porosity problem [4,5]. Furthermore, when welding thicker plates of 3 mm or more, the higher laser power and the lower welding speed are required; however, defects, spatter and fluctuations due to unstable molten weld pool can be easily formed [6,7]. Despite many laser welding techniques suggested to improve the weldability of copper alloys, the complicated devices used, high cost, laborious set-up and the possible impurities inclusions limits their widespread industrial applications [8]. Especially in case of the welding of beryllium copper alloy, the laser beam welding butt joint of the 0.2 mm of thickness of beryllium copper plate shows sound joint strength as 90% of base metal; however, the liquation crack in HAZ involved in fracture was observed [9]. In solid state joining method, Lap joint of beryllium-copper alloy was fabricated using diffusion brazing with filler metal of Ag without defect at 750 °C for 1200 s. However, the tensile strength of the joint was comparatively low value as 173 MPa [10]. Thus, in order to fabricate sound thick beryllium copper alloy joint without defect caused by melting state, new welding methods are needed to overcome the above limits.

Friction stir welding is a solid-state joining process invented by The Welding Institute (TWI) in the UK [11]. It is an effective method due to the use of a low heat input welding process and eliminates melting and solidification associated problems [12], such as liquidation and solidification cracking [13]. It was well known that the FSW process produces high-quality welded region with a homogeneously refined microstructure and better mechanical properties than those yielded by conventional welding processes [14]. In particular, FSW is an effective technique for joining stain-less steel due to advantages such as low distortion and residual stress [15]. In recent days, studies about high strength steel [16], advanced high strength steel [17] and Ti-alloys [18] were performed. Murugan et al. showed that friction stir welding condition can get sound joints without defects in copper and bronze plates [19]. Sun et al. reported for the formation of denser twins in the stir zone during friction stir welding [20]. Guoliang et al. analyzed the precipitation behavior of Cu-2.0Be alloys in detail, but did not consider the FSW process [21].

In the present study, the authors tried to conduct friction stir welding of Giga-grade high strength beryllium-copper alloy plates and sound joints were successfully manufactured. Post-weld heat treatment (PWHT) was performed to improve the mechanical properties of the joints and the microstructural evolution was investigated during the PWHT. Based on the results, the authors discussed about the relationship between microstructure and mechanical properties during FSW and consequent PWHT. To conclude, the authors revealed the mechanism for dissolution and re-precipitation of strengthening γ' precipitates (CuBe) during FSW and consequent PWHT. In addition, this paper optimized the PWHT condition can obtain acceptable mechanical properties.

2. Materials and Methods

2.1. Materials and FSW Conditions

The material selected in this work is a 3-mm-thick commercial beryllium-copper alloy forged plate, referred to here as "C17200". The chemical composition of the plate is given in Table 1. These plates, with a length of 260 mm and a width of 120 mm, were bead-on-plate welded. Based on the preliminary welding test at various conditions as 600, 700, 800 RPM of a tool rotational speed and 50, 60, 70 mm/min of a tool traveling speed to figure out sound joint condition without impact on the friction stir welding tool and micro defects such as tunnel, crack and void of the joint, Friction stir welding was conducted at a tool rotational speed of 700 RPM and a tool traveling speed of 60 mm/min with an angular advance of 2 degrees as a standard condition. In this study, as shown in Figure 1, a 2.7-mm-long tungsten carbide (WC) tool with a hemispherical probe (6 mm diameter) and a concave shoulder (15 mm diameter) was selected.

2.2. Methods and Corresponding Conditions of Analysis

Figure 2 shows schematic representations of tensile and Charpy test specimens. This figure also shows the geometry of the tensile and Charpy testing following ASTM: E8 (gauge length: 40 mm) and

KS B 0809 standard, respectively. The specimens were cut perpendicular to the welding direction, with the stir zone (SZ) being centered within the gage length and notch. The FSW specimens were subjected to PWHT at 315 °C for durations ranging from a half hour to eight hours. A salt bath furnace was used for the PWHT process. The heat-treated specimens were subsequently cooled to room temperature. Figure 3 shows the solvus temperature and age-hardening temperature of the precipitates in a phase diagram of the beryllium-copper alloy. The post-weld heat-treatment temperature was determined to be 315 °C after referring to the phase diagram and to previous study results [1,2,15–17]. Figure 4 shows a process flow chart describing the FSW and PWHT condition. The microstructure of the transverse section of the FSW specimens was observed by OM using an Olympus (GX51, Tokyo, Japan) and a field emission scanning electron microscope (FE-SEM, JEOL, JSM-700F, Tokyo, Japan). The observed samples were prepared according to the standard procedure for specimen preparation, including grinding, polishing and etching. Surfaces were observed after immersion for 20 s using 40 vol% of HNO_3 and 60 vol% of a CH_3OH solution. Precipitation behavior in the SZ following PWHT was investigated by a high-resolution transmission electron microscope (HR-TEM, JEOL, JEM-2010, Tokyo, Japan) at 200 kV and a cs-corrected-field emission transmission electron microscope (CS-corrected-FE-TEM, JEOL, JEM-ARM-200F, Tokyo, Japan) at 200 kV. The thin foils used for the TEM observation were obtained from the SZ center and from base metal in the plane perpendicular to the welding direction. They were prepared by twin-jet electro polishing in 40 vol% of HNO_3 and 60 vol% of a CH_3OH solution at −20 °C. For the microhardness measurements, the samples were cut from the SZ center. Microhardness levels were measured on the welding direction plane of the SZ using a Vickers microhardness test machine (Daekyung Tech, DTR-300N, Incheon, Korea) with a load of 300 g and a dwell time of 10 s. The hardness value was determined by taking an average of five readings and excluding the maximum and minimum values. The tensile test (Instron, 8801, Norwood, MA, USA) was conducted at room temperature at a strain rate of 1.0 mm/min, and the tensile strength values under various PWHT conditions were determined by taking the mean of three test results. The yield strength was determined at an offset of 0.2 percentage.

Table 1. Chemical composition of the beryllium-copper alloy used in this work.

Element	Be	Si	Al	Cu
Mass%	1.9	0.2	0.2	Bal.

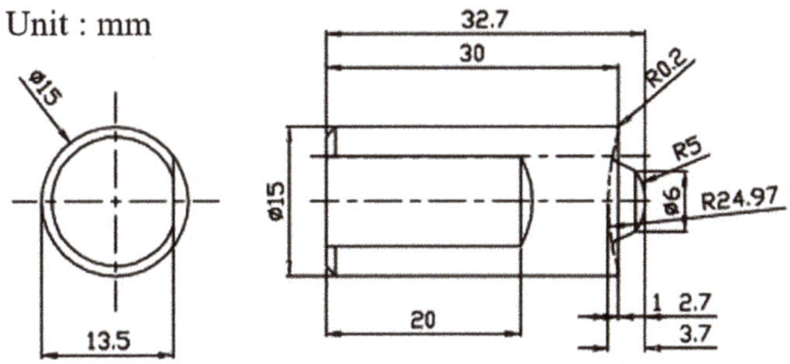

Figure 1. Illustraion of Tool geomatric.

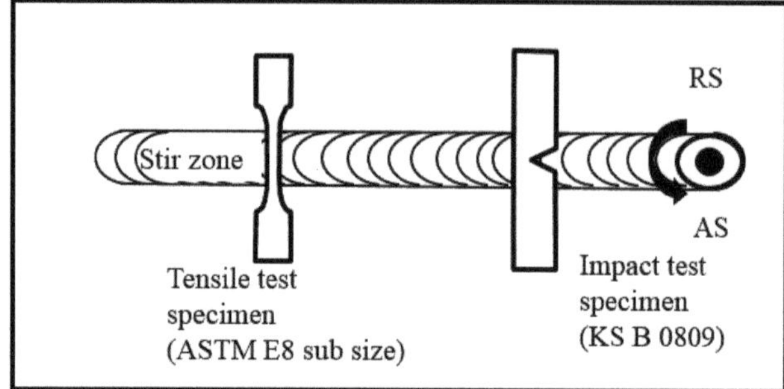

Figure 2. Schematic illustration of the friction-stir-welded plate and test specimen preparation.

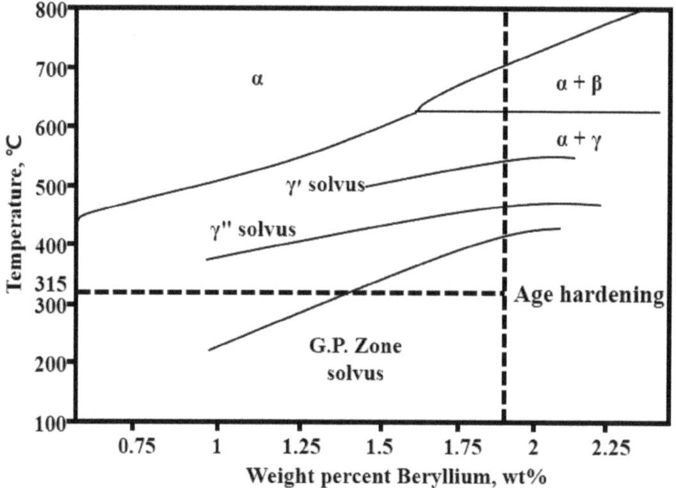

Figure 3. Phase diagram and metastable solvus of the beryllium-copper alloy.

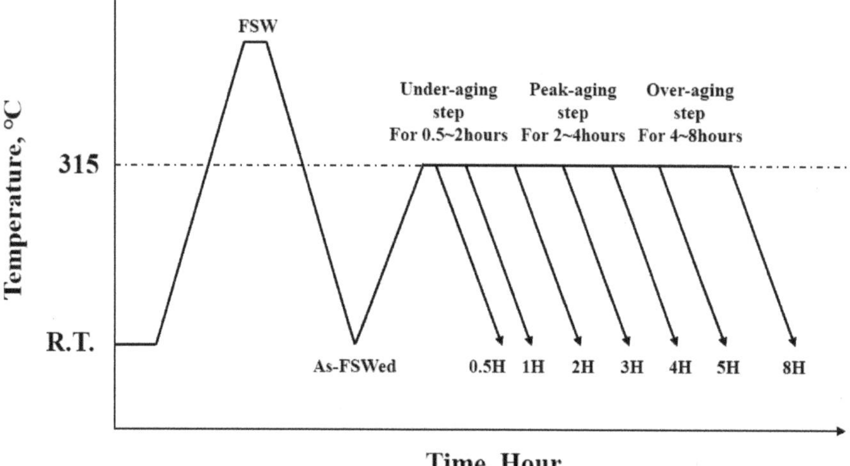

Figure 4. Process flow chart including the friction stir welding (FSW) and post-weld heat treatment conditions.

3. Results

3.1. Microstructure and Mechanical Properties of the FSW Joints and the Base Metal

The mechanical properties of the base metal and the joints are listed in Table 2. The tensile strength and hardness of the base metal were 1162 MPa and 380 HV, respectively, showing toughness values of 4.6 J and elongation of 9.4%. After FSW, both the tensile strength and hardness decreased down to 461 MPa and 162 HV, respectively. However, the toughness and elongation showed significant corresponding increases of 92.2 J and 53%. It was revealed that FSW has a considerable effect on the mechanical properties of beryllium-copper alloys implying that microstructural changes occur in the welding region during FSW. An optical micrograph and a TEM bright-field image of the base metal are shown in Figure 5a,b, respectively. From Figure 5a, it can be seen that the equiaxed α matrix ranged from 40 to 150 μm. In Figure 5b, a dense γ' (CuBe) needle-like precipitates can be observed. Figure 6 shows the overall cross-section of the FSW joints. It can easily be identified from Figure 6a that the welding zone of the FSW joint is divided into three regions: the stir zone (SZ), the thermo-mechanical-affected zone (TMAZ) and the heat-affected zone (HAZ). The coarse grain structure in the base metal (BM) is replaced by introducing the SZ containing refined grains (as shown in Figure 6b). This is attributed to the dynamic recrystallization caused by severe thermal plastic deformation in the SZ. A mixture of coarsened, refined and elongated grains in the TMAZ (a transitional region between the SZ and the HAZ) was typical due to the incomplete dynamic recovery and recrystallization in this zone (Figure 6c). The grain size in the HAZ (Figure 6d) was similar than that of the BM (Figure 5a). Figure 7 shows the microstructures of the BM and SZ. Compared with the BM (Figure 7a) and the SZ (Figure 7c) under low magnification, it is apparent that the grains are significantly refined due to the dynamic recrystallization stemming from the frictional heat and plastic flow of the softened material during FSW. It could also be observed that the microstructure consisted of discontinuous precipitation (DP) cells (γ) located at the grain boundaries of α matrix, as presented in Figure 7b. As shown in Figure 7d, the grain boundaries in the SZ do not have DP cells due to their dissolution into the BM during FSW. Figure 8 shows a TEM bright-field image of the SZ. The microstructure consists of only α phase, without γ' precipitates. This suggests that the strengthening γ' precipitates dissolved into the BM during the FSW heat cycle. Such dissolution of the γ' precipitates can be related to decreases in the hardness and tensile strength of the SZ.

Table 2. Mechanical properties of the base metal and the FSW joints.

Specimen	Tensile Strength (MPa)	Yield Strength (MPa)	Failure Strain (%)	Hardness (HV)	Toughness (J)
Base metal	1162	673	9.4	380	4.6
FSW joints	461	258	53	162	92.2

Figure 5. Microstructural analysis of the base metal: (**a**) optical microscopy, (**b**) transmission electron microscopy.

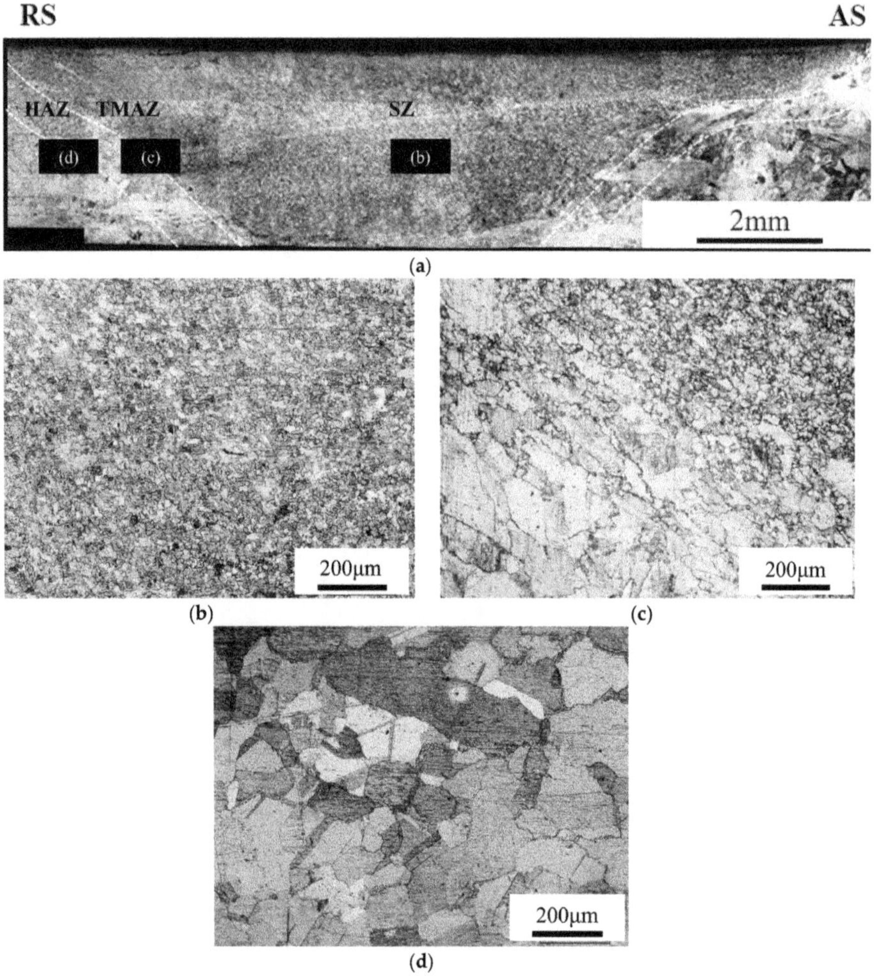

Figure 6. Cross-sectional macrograph of (**a**) the FSW region and (**b**) micrographs of the stir zone (SZ), (**c**) thermo-mechanical-affected zone (TMAZ) and (**d**) heat-affected zone (HAZ).

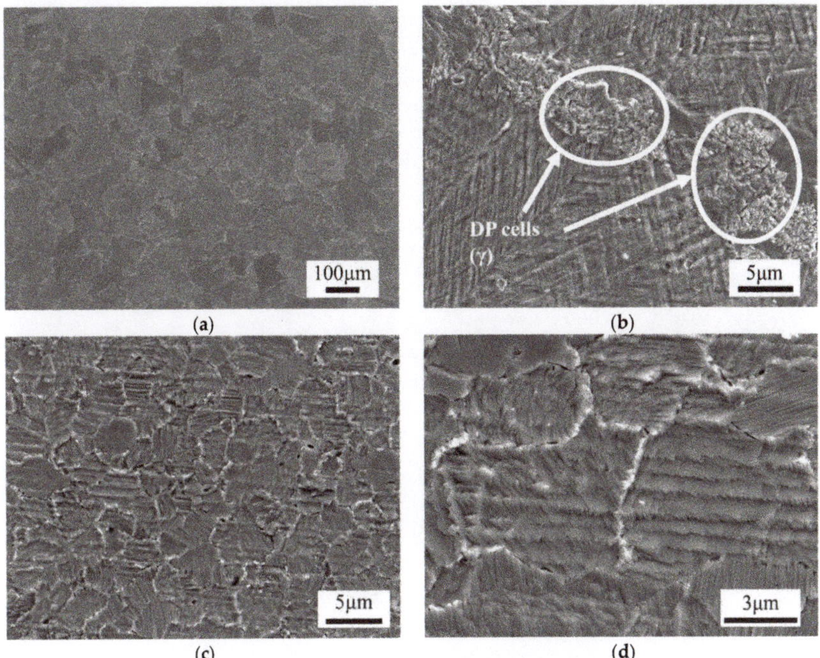

(a)

(b)

(c)

(d)

Figure 7. Scanning electron microscope (SEM) Images of the base metal: (**a**) 100×, (**b**) 3000× and stir zone (**c**) 3000×, and (**d**) 7000×.

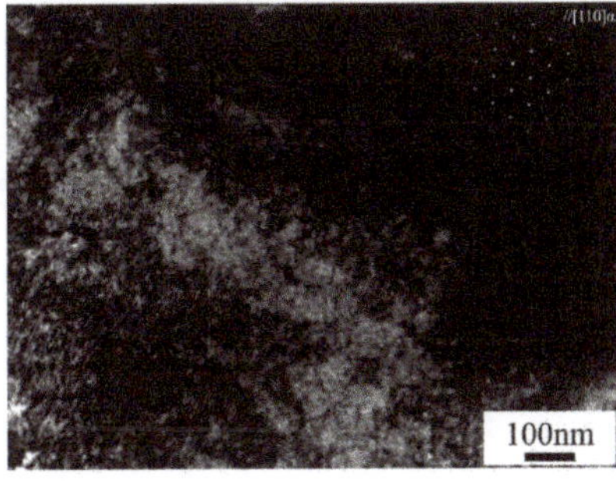

Figure 8. Transmission electron microscope (TEM) bright-field images in the stir zone.

3.2. Behavior of Mechanical Properties during PWHT

Figure 9 presents the outcomes of tests of the Vickers hardness and Charpy impact absorption energy in the stir zones of FSW joints subjected to a post-weld heat treatment. When the FSW was performed, the hardness decreased significantly to 162 HV from 380 HV. This provides evidence that the hardness was decreased in the SZ due to the dissolution of the strengthening precipitates, as shown in Figure 8. When the joint was subjected to PWHT at 315 °C, the microhardness increased steeply during the under-aging step (US) of the PWHT process. Furthermore, as PWHT lasted for approximately three hours, the results for the joint showed that the hardness values recovered similarly to the BM outcome. In other words, this was the peak-aging step (PS). When PWHT lasted for more than four hours, the hardness decreased slightly. Thus, this was the over-aging step (OS). The Charpy impact absorption energy increased up to 92.2 J from 4.6 J, as listed in Table 2, as the FSW was conducted. However, it decreased severely during the under-aging step. Subsequently, the decrease lasted for four hours. However, the Charpy impact absorption energy value increased slightly during the over-aging step. Figure 10 shows the behavior of the tensile properties of the FSW joints after PWHT. All of the tensile test specimens were fractured at the BM. Both the tensile and yield strength increased early during the under-aging step. Subsequently, these increases lasted for four hours of PWHT. After the peak-aging step, these values decreased gradually. On the other hand, the strain decreased abruptly early in the under-aging step. Additionally, it showed little change during the peak-aging and over-aging steps. Hence, PWHT has a decisive effect on the recovery of mechanical properties such as the hardness, yield strength and tensile strength of FSW joints. These results also indicate that microstructural evolution such as precipitation can occur during PWHT.

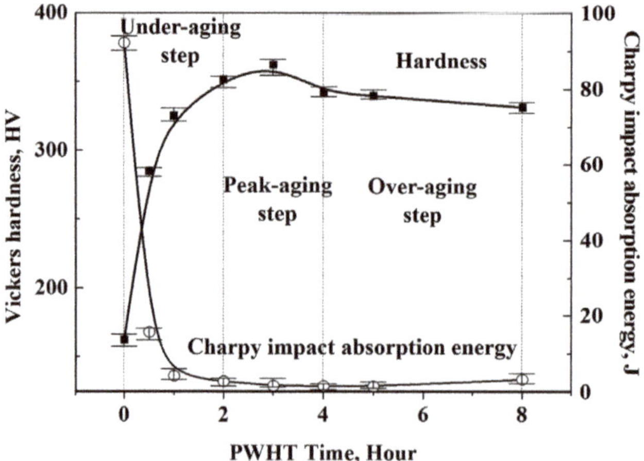

Figure 9. Behavior of the Vickers hardness and Charpy impact absorption energy in the stir zone of FSW joints subjected to a post-weld heat treatment. PWHT: post-weld heat treatment.

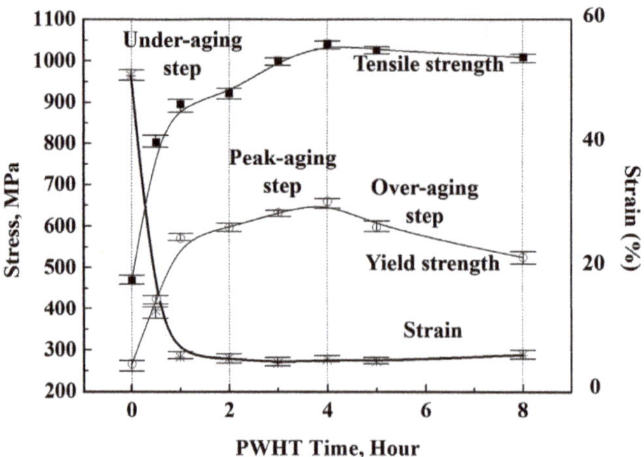

Figure 10. Behavior of the tensile properties of FSW joints after a post-weld heat treatment.

3.3. Microstructural Evolution during PWHT

Figure 11 shows SEM images of DP cells formed at the grain boundaries of the stir zone after the samples were subjected to PWHT for 30 min, three hours and eight hours. In the early stage of the under-aging step, very few and relatively small (Figure 11a) DP cells were observed. When the

PWHT process was extended to three hours, the number of DP cells increased (Figure 11b). After eight hours, many colonies of DP cells formed (Figure 11c). The DP cells are involved in the formation of a solute-depleted matrix phase (α') and a precipitate phase (γ) as a duplex transformation product, typically with nucleation at the grain boundaries with growth into one side of the supersaturated matrix (α phase) [22–25]. The DP cells consumed numerous solute atoms and diminished the precipitation hardenability; in this case, the γ phase readily formed instead of the finer metastable γ' and γ'' phases. TEM observations were used to characterize the precipitates which formed during PWHT, and these results are shown in Figure 12. As shown in Figure 8, no precipitates were found at the foil prepared from the FSW joints. However, numerous nano-scale sphere-type precipitates were observed in the foil prepared from the sample which underwent PWHT for 30 min (Figure 12a). Many γ' phase precipitates arose depending on the duration of PWHT (Figure 12b). When the PWHT process lasted for eight hours, coarsened γ' phases were observed (Figure 12c).

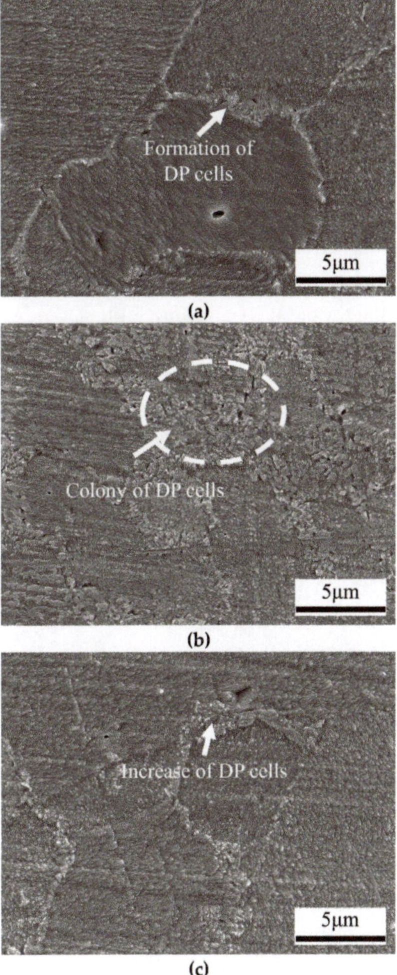

Figure 11. SEM images of discontinuous precipitation (DP) cells formed at the grain boundaries of the stir zone subject to PWHT: (**a**) 0.5 hour, (**b**) three hours and (**c**) eight hours.

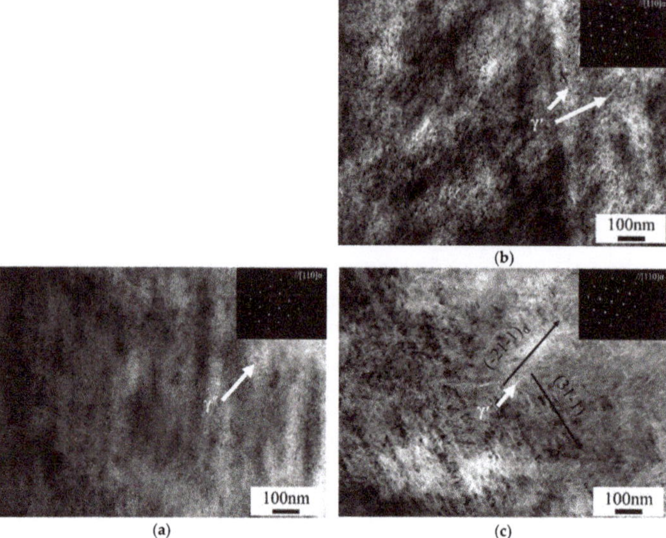

Figure 12. High-resolution transmission electron microscope (HR-TEM) images and selected area diffraction pattern (SADP) outcomes of the post-weld heat treated foils: (**a**) 0.5 hour, (**b**) three hours and (**c**) eight hours.

Figure 13 shows high-magnification CS-TEM bright-field images and corresponding SAD patterns of samples after PWHT for 30 min (a) and three hours (b) with the axis parallel to $[001]_\alpha$. In Figure 13a, the γ'' precipitates were nucleated and grew to a length of 5~15 nm along the $(-100)_\alpha$ direction with three to six layers of Be. With a PWHT duration up to three hours, the γ'' precipitates grew and coarsened to γ' in the same direction and more than eight Be layers formed. The images indicate the distance between the atoms with a straight line in two directions, showing results of a = 0.236 nm, b = 0.232 nm (Figure 13a), and a = 0.270 nm, b = 0.279 nm (Figure 13b). These outcomes were fairly close to the phase parameter of γ' as reported by Guoliang et al. [21].

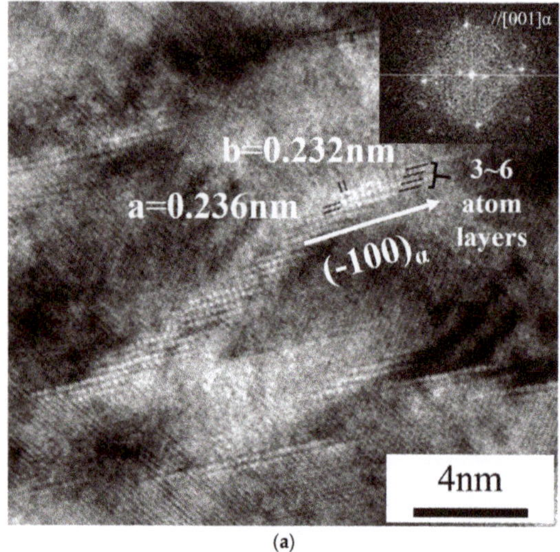

(**a**)

Figure 13. *Cont.*

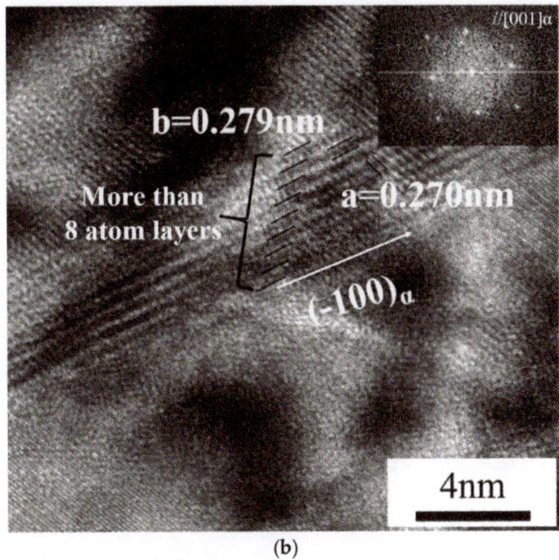

(b)

Figure 13. CS-TEM images and SADP outcomes of precipitates consisting of (**a**) γ'' and (**b**) γ' phases in PWHTed foils.

4. Discussion

Relationship between the Mechanical Properties and Microstructure during FSW and Consequent PWHT

In case of laser beam welding butt joint of beryllium-copper alloy, the joint shows liquation crack in HAZ involved in fracture. Also it is possible to get sound joint properties of 0.2 mm of thickness plate after welding; however, there is currently limited research on more than 3 mm of thickness [9].

In the case of solid-state joining process, which use not only thermal factor but secondary factors such as mechanical or chemical factor, the problems caused by solidification from melting state is expected to be suppressed.

Esmati et al. attempted diffusion brazing lap joint of beryllium copper alloy with filler metal of Ag content. The brazing joint was successfully fabricated without defect at 750 °C for 1200 s. However, maximum tensile strength was comparatively low value as 173 MPa [10].

Especially, this study was conducted to get a sound thick plate joint that has good joint properties without solidification defects by proposing the properly controlled friction stir welding condition.

Based on preliminary welding test in several friction stir welding conditions, the authors were able to figure out optimal condition to control the heat input properly. When the welding condition accompanied higher heat input, since the microstructure of the joint is expected to exhibit greater dissolution of the γ' and γ precipitates, the faster and the stronger hardening effect is expected in the same post-weld heat treatment condition due to the higher driving force. Meanwhile, when the heat input during the welding process was lower, since the less γ' and γ precipitates are dissolved, it is assumed to show the slower and the weaker hardening effect due to the less driving force to form precipitates.

The effect of a post-weld heat treatment on the mechanical properties and microstructures of friction-stir-welded beryllium-copper alloy were investigated, as presented in the results section. In this section, the relationship between the mechanical properties and precipitation behavior during FSW and consequent PWHT is discussed. The process flow from BM to PWHT has five steps based on the alternation of the mechanical properties and microstructures. This process is shown in Figure 14. Initially, the base metal has acceptable hardness and tensile strength, as a considerable amount of γ' precipitates, which are the primary strengthening mechanism of the present alloy, were present in the BM, as shown in Figure 5b. Interestingly, in the second step just after FSW, the tensile strength and hardness decreased abruptly, while the toughness and ductility increased sharply. The non-existing presence of strengthening γ' precipitates is considered to be the primary cause of the fall in both the hardness and tensile strength. The authors believe that the γ' precipitates dissolved into the base metal owing to the frictional heat generated by the rotation of the tool. A significantly refined grain microstructure was formed in the stir zone. Both the refined grains and softening through the dissolution of the γ' precipitates had an effect on the upsurge of the toughness. The PWHT period includes steps 3, 4 and 5. During the third step, called the under-aging stage of PWHT within a half hour (PWHT, US), a recovery of the hardness and tensile strength was obtained. In contrast, the toughness and ductility expressed a large decline. This outcome appears to be related mainly to the formation of globular γ'' precipitates, as shown in Figures 12a and 13a. During the fourth stage of the peak-aging step of PWHT from two hours to four hours (PWHT, PS), the gradual strengthening tendency continues as the aging time is extended. The peak tensile strength and hardness could be determined from microstructures mainly containing γ' precipitates which grew from γ'' precipitates. Finally, during the fifth stage of the over-aging step of PWHT from four hours to eight hours, a slight decrease in hardness and tensile strength occurred. The growth and coarsening of γ' precipitates are considered to be the dominant causes of this finding. If PWHT involves longer aging times up to a few days, the low mechanical properties through the softening of the microstructure are suggested to appear. It was clearly revealed that when the Giga-grade high-strength beryllium-copper alloy, which shows mechanical properties mainly due to precipitation of the γ' phase is subjected to FSW, the hardness and tensile strength decrease sharply because the γ' precipitates dissolve into the base metal due to the frictional heat generated during the FSW process. It was also found that PWHT is indispensable to recover the hardness and tensile strength of FSW joints. However, excess aging times exceeding three to four hours at 315 °C bring about a decline in the hardness and tensile strength. In conclusion, it is crucial to control metastable precipitates such as γ'' and γ' to secure reliable FSW joints with beryllium-copper alloys.

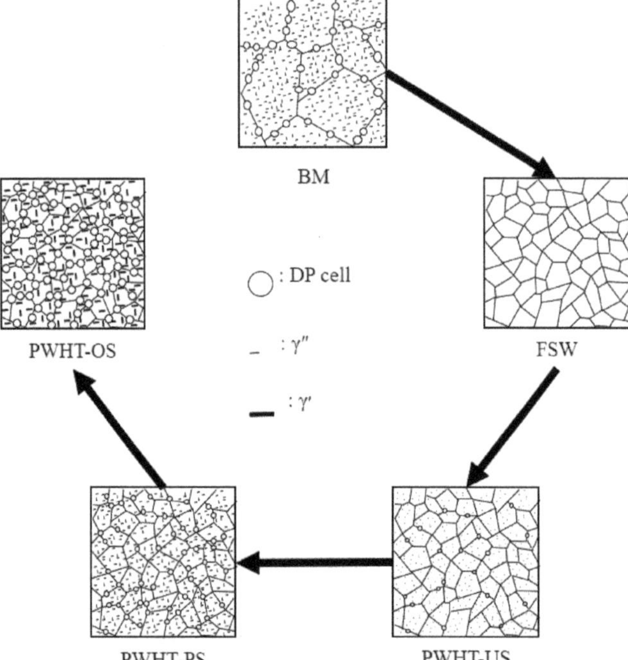

Figure 14. Schematic diagram of the evolution of the microstructure during the FSW and PWHT processes.

5. Conclusions

To understand the microstructural evolution and behavior of the mechanical properties during FSW with a consequent PWHT process, FSW was conducted using a beryllium-copper alloy and the resulting joints were subjected to PWHT at 315 °C for up to eight hours. The authors determined the microstructural and mechanical behaviors and the relationship between both outcomes. The results from this study are summarized below.

1. Friction stir welding of a Giga-grade high-strength beryllium-copper alloy was successfully conducted with a tool rotational speed and travel speed of 700 RPM and 60 mm/min, respectively. Sound joints without defects could be obtained.
2. A refined grain microstructure was formed in the stir zone. The strengthening γ' (CuBe) precipitates dissolved into the base metal due to the frictional heat generated during the FSW process.
3. After FSW, the hardness and tensile strength decreased significantly, whereas the toughness and ductility increased sharply. It was found that the dissolution of γ' precipitates is the dominant cause of the mechanical property changes during FSW.
4. When PWHT is conducted with the FSW joints, γ'' precipitates forms at an early stage within a half hour during the under-aging step. As the PWHT process is maintained, γ' precipitates, which are the primary strengthening phase of this alloy, forms in the stir zone. This increased with an increase in the PWHT time. They became coarsened when the PWHT process exceeded four hours.
5. When the FSW joints are subjected to PWHT, the hardness and tensile strength increase remarkably at an early stage of the under-aging step. The gradual increase is maintained for up to four hours of PWHT. Past this time, however, the hardness and tensile strength gradually decrease. The toughness and ductility expressed crosscurrents during the same PWHT process.
6. To conclude, to obtain sound joints with acceptable mechanical properties, PWHT of 3~4 h at 315 °C is essential and the precipitation of the γ' phases must be controlled.

Author Contributions: Conceptualization, K.L.; Investigation, Y.L.; Project administration, K.L.; Validation, S.M.; Writing–original draft, Y.L.; Writing–review & editing, K.L.

Funding: This work was financially supported by R&D program of the Korea Institute of Industrial Technology (JA180012). The authors would like to express their gratitude to KITECH.

Acknowledgments: The FSW tools used in this study were supported by R&D program of Ministry of Trade, Industry and Energy (NK180051). The authors would like to express their gratitude to MOTIE.

Conflicts of Interest: The authors declare no conflict of interest.

References

1. Davis, J.R. *ASM Specialty Handbook: Copper and Copper Alloys*; #06605G; ASM international: Metals Park, OH, USA, 2001.
2. Crone, W.C. Compositional variation and precipitate structures of copper-beryllium single crystals grown by the Bridgman technique. *J. Cryst. Growth* **2000**, *218*, 381–389. [CrossRef]
3. Tech Breifs Materion. 'Welding Copper Beryllium'. Available online: https://materion.com/resource-center/technical-papers/copper-beryllium-wrought-alloys (accessed on 2 November 2018).
4. Zhang, L.J.; Ning, J.; Zhang, X.J.; Zhang, G.F.; Zhang, J.X. Single pass hybrid laser-MIG welding of 4-mm thick copper without preheating. *Mater. Des.* **2015**, *74*, 1–18. [CrossRef]
5. Engler, S.; Ramsayer, R.; Poprawe, R. Process studies on laser welding of copper with brilliant green and infrared lasers. Lasers in manufacturing 2011: Proceedings of the Sixth International Wlt Conference on Lasers in Manufacturing. *Phys. Proced.* **2011**, *12*, 339–346. [CrossRef]
6. Miyagi, M.; Zhang, X.D. Investigation of laser welding phenomena of pure copper by X-ray observation system. *J. Laser Appl.* **2015**, *27*, 042005. [CrossRef]

7.   Reisgen, U.; Olschok, S.; Turner, C. Welding of thick plate copper with laser beam welding under vacuum. *J. Laser. Appl.* **2017**, *29*, 022402. [CrossRef]

8.   Auwal, S.T.; Ramesh, S.; Yusof, F.; Manladan, S.M. A review on laser beam welding of copper alloys. *Int. J. Adv. Manuf. Technol.* **2018**, *96*, 475–490. [CrossRef]

9.   Mousavi, S.A.A.A.; Niknejad, S.T. Study on the microstructure and mechanical properties of Nd: YAG pulsed laser beam weld of UNS-C17200 copper beryllium alloy. *J. Mater. Process. Technol.* **2010**, *210*, 1472–1481. [CrossRef]

10. Esmati, K.; Omidvar, H.; Jelokhani, J.; Naderi, M. Study on the microstructure and mechanical properties of diffusion brazing joint of C17200 Copper Beryllium alloy. *Mater. Des.* **2014**, *53*, 766–773. [CrossRef]

11. Mishra, R.S.; Ma, Z.Y. Friction stir welding and processing. *Mater. Sci. Eng. R* **2005**, *50*, 1–78. [CrossRef]

12. Park, S.H.C.; Sato, Y.S.; Kokawa, H.; Okamoto, K.; Hirano, S.; Inagaki, M. Corrosion resistance of friction stir welded 304 stainless steel. *Scr. Mater.* **2004**, *51*, 101–105. [CrossRef]

13. Santos, T.F.A.; Hermenegildo, T.F.C.; Afonso, C.R.M.; Marinho, R.R.; Paes, M.T.P.; Ramires, A.J. Fracture toughness of ISO 3183 X80M (API 5L X80) steel friction stir welds. *Eng. Fract. Mech.* **2010**, *77*, 2937–2945. [CrossRef]

14. Sato, Y.S.; Kurihara, Y.; Park, S.H.C.; Kokawa, H.; Tsuji, N. Friction stir welding of ultrafine grained Al alloy 1100 produced by accumulative roll-bonding. *Scr. Mater.* **2004**, *50*, 57–60. [CrossRef]

15. Lim, Y.S.; Kim, S.H.; Lee, K.J. Effect of residual stress on the mechanical properties of FSW joints with SUS409L. *Adv. Mater. Sci. Eng.* **2018**, *2018*, 9890234. [CrossRef]

16. Das, H.; Lee, K.J.; Hong, S.T. Study on microtexture and martensite formation of friction stir lap-welded DP 590 steel within A1 to A3 temperature range. *J. Mater. Eng. Perform.* **2017**, *26*, 3607–3613. [CrossRef]

17. Das, H.; Mondal, M.; Hong, S.T.; Lim, Y.S.; Lee, K.J. Comparison of microstructural and mechanical properties of friction stir spot welded ultra-high strength dual phase and complex phase steels. *Mater. Charact.* **2018**, *139*, 428–436. [CrossRef]

18. Ko, Y.J.; Lee, K.J.; Baik, K.H. Effect of tool rotational speed on mechanical properties and microstructure of friction stir welding joints within Ti-6Al-4V alloy sheets. *Adv. Mech. Eng.* **2017**, *9*, 1–7. [CrossRef]

19. Murugan, R.; Thirumalaisamy, N. Experimental and numerical analysis of friction stir welded dissimilar copper and bronze plates. *Mater. Today Proc.* **2018**, *5*, 803–809. [CrossRef]

20. Sun, Y.F.; Xu, N.; Fujii, H. The microstructure and mechanical properties of friction stir welded Cu–30Zn brass alloys. *Mater. Sci. Eng. A* **2014**, *589*, 228–234. [CrossRef]

21. Guoliang, X.; Qiangsong, W.; Xujun, M.; Baiqing, X.; Lijun, P. The precipitation behavior and strengthening of Cu-2.0wt% Be alloy. *Mater. Sci. Eng. A* **2012**, *558*, 326–330. [CrossRef]

22. Williams, D.B.; Butler, E.P. Grain boundary discontinuous precipitation reactions. *Int. Met. Rev.* **1981**, *26*, 153–183. [CrossRef]

23. Manna, I.; Pabi, S.K.; Gust, W. Discontinuous reactions in solids. *Int. Mater. Rev.* **2013**, *46*, 53–91. [CrossRef]

24. Bonfield, W.; Edwards, B.C. Precipitation hardening in Cu 1.81 wt% Be 0.28 wt% Co Part 2 Discontinuous precipitation. *J. Mater. Sci.* **1974**, *9*, 409–414. [CrossRef]

25. Bonfield, W.; Edwards, B.C. Precipitation hardening in Cu 1.81 wt% Be 0.28 wt% Co Part 1 Continuous precipitation. *J. Mater. Sci.* **1974**, *9*, 398–408. [CrossRef]

 *metals*

Article

The Influence of Pre- and Post-Heat Treatment on Mechanical Properties and Microstructures in Friction Stir Welding of Dissimilar Age-Hardenable Aluminum Alloys

Yang Jia [1], Sicong Lin [1], Jizi Liu [1,2,*], Yonggui Qin [1] and Kehong Wang [1,*]

[1]   School of Materials Science and Engineering/Herbert Gleiter Institute of Nanoscience, Nanjing University of Science and Technology, Nanjing 210094, China; jiayang@mail.njust.edu.cn (Y.J.); 117116022355@njust.edu.cn (S.L.); 116116001365@njust.edu.cn (Y.Q.)

[2]   State Key Laboratory of Advanced Design and Manufacturing for Vehicle Body, Hunan University, Changsha 410082, China

*   Correspondence: jzliu@njust.edu.cn (J.L.); wkh1602@njust.edu.cn (K.W.)

Received: 27 September 2019; Accepted: 25 October 2019; Published: 28 October 2019

Abstract: An Al-Mg-Si alloy 6061 and an Al-Zn-Mg alloy 7A52 were joined by friction stir welding successfully. Pre- and post- heat treatment were employed to improve the strength of the weld. The results show a best weld joint with the lowest hardness of 100 HV in 6061 matrix, being achieved by post-solid-solution and subsequent two-stage artificial aging for the whole weld joint of the 7A52 and 6061 solid solution. Under this condition, the weld nugget zone (WNZ) is stronger than 6061 matrix but it has lower hardness than 7A52 matrix. The hardness of WNZ is contributed by the combination of η' and L precipitates, dynamically changes along with the ratios between the number of η' and L precipitates. The higher the number density of η' precipitates, the hardness of WNZ is closer to that of the 7A52 matrix. Otherwise, the higher number density of L precipitates, the hardness of WNZ is closer to that of 6061 matrix. The coexistence of η' and L precipitates is a direct result from the mixture of 7A52 and 6061 alloys achieved by stirring. Precipitates identification and composition analysis reveal a dynamic WNZ with constituent transition in hardness and composition.

Keywords: friction stir welding; aluminum alloy; heterogeneity; mechanical; microstructure

---

1. Introduction

As high strength aluminum alloys, Al-Zn-Mg alloys of the 7000 series family and Al-Mg-Si alloys of the 6000 series family are widely used in the defense, aerospace, automotive, and structural applications. In these heat-treatable alloys, Mg combines with Zn or Si (Cu) to form a large number of strengthening precipitates during aging, such as the metastable precursors of η-MgZn$_2$ [1,2] and β''-Mg$_5$Si$_6$, L, and/or Q-phase [3,4], contributing to the high strength of these alloys. Presently, there is great interest in these alloys to advance strengthening to further reduce the weight of structures in high-speed train, aircraft, and other applications. However, to join components that were made by these alloys is a large challenge in assembly, because of the sensitivity of the strengthening metastable precipitates in these alloys to thermal cycles introduced by traditional fusion welding. The strength of the weld would be significantly reduced due to precipitates coarsening in the heat affected zone (HAZ) [5]. As with the fusion welding joint of Al-Mg-Si alloys, the strength of the HAZ is even lower than that of the weld zone, due to over-aged HAZ resulting from heat input. Moreover, the conventional fusion welding processes are not suitable for dissimilar welding applications, due to the different thermodynamic behavior of dissimilar materials.

Friction stir welding (FSW) is an effective method for joining many combinations of dissimilar materials at the solid state, being pioneered and practiced by Wayne Thomas in 1991 [6]. As a new family of solid-state joining techniques, benefitting from the decreasing of heat input and the introduction of deformation during FSW, the joint might be free from the dendritic structure typical of a fusion-weld joint, and it has finer microstructures and better mechanical properties than the respective fusion welding joint. These welding processes have been successfully applied to join dissimilar aluminum alloys [7–13]. In the literatures, investigators have paid more attention to study the effect of FSW parameters on microstructures, mechanical properties, and defects formation [7,9,11–24]. Researchers found that FSW parameters, such as tool design, tool rotation and traverse speed, depth of tool plunge, angle of tool tilt, tool pin offset, and welding gap, etc., may lead to defect formation, if they are not selected properly [8,25,26].

Beyond a weld joint without defects, engineers strive to pursue higher mechanical properties. Thus, researchers need to understand the microstructural evolution and the underlying physics of FSW. For precipitation-hardened Al alloys, such as 6000 and 7000 series, the thermos-mechanical process of FSW results in the change of precipitate structure, size, distribution, and density in different zones [27]. Generally, this change is negative for the strength of the FSW joint. In the nugget zone of the weld (WNZ), the pre-existed hardening precipitates will dissolve due to the sufficient high temperature that is generated during FSW. The region outside of the WNZ is termed the thermo-mechanical affected zone (TMAZ), where precipitates will be sheared or surrounded by dislocations introduced by severe plastic deformation. Outside of the TMAZ is the heat affected zone (HAZ), in which precipitates coarsen and transform into the stable partner, which makes HAZ over-aged and soften. Obviously, an important reason for the drop of strength in FSW joint is the loss of strengthening precipitates. Researchers naturally adopted post-heat-treatment to reform the strengthening precipitates. The effect of post-weld heat treatment on FSW joints of dissimilar Al alloys were investigated [28–31], they found that, after post-weld heat treatment, the strength of the weld can be improved if there are new precipitates forming during post heat treatment, the location and mode of fracture under tension are completely different from the as-weld joints [32–35]. However, not all post-weld heat treatment is effective. Mahoney et al. found that the post-weld T6 aging treatment would not help in the improvement of the mechanical properties of FSW joints of AA7075-T651 [36]. Sullivan and Robson even found that post weld heat treatment (T7 temper) coarsened nano-sized precipitates in matrix, WNZ, and TMAZ of FSW AA7449 joints, which results in the decrease of hardness in these area [37]. The above conflicting results revealed that the aging conditions of the as-received materials and the precipitates in the as-weld sample should be considered before the application of post-weld heat treatment.

In this work, the effect of pre- and post- heat treatment on the hardness of the weld joints between two technologically important age hardenable Al alloys (7A52 and 6061) was investigated from the point view of microstructural evolution at different scales. Moreover, the underlying physics of precipitation kinetics was revealed by high-resolution transmission electron microscopy (HRTEM) (Thermo Fisher Scientific, Waltham mass, Massachusetts, USA). A new strategy of two-stage aging was proposed to restore precipitation strengthening in the weld based on the above.

2. Material and Experimental Details

Two technologically important age hardenable Al alloys, 7A52 and 6061, were employed in this study. These alloys were received in the form of 5 mm thick sheets. The chemical compositions of these alloys were determined, as listed in Table 1.

Table 1. Chemical composition of 7A52 and 6061 aluminum alloy.

Alloys	Al	Zn	Mg	Mn	Cr	Zr	Ti	V	Cu	Fe	Si
7A52	Bal.	4.40	2.33	0.27	0.18	0.083	0.09	0.008	0.10	0.21	0.07
6061	Bal.	0.18	1.05	0.10	0.22	-	0.01	-	0.25	0.18	0.51

The FSW were carried out while using a commercial gantry type FSW machine (FSW-LM-025-2030,Jiangsu Ruicheng Machinery Co., Ltd, Yixing, China) with position control mode. The friction stir butt welds between the rolled plates of 7A52 and 6061 aluminum alloys were obtained at a welding speed of 90 mm/min. and rotational speed of 500–700 rpm, while employing a single pass welding procedure. The plates were held firmly, covered by thermal insulation board, and then preheated at 150 °C for 2 h before welding. Figure 1 shows the welding direction. The FSW tool was made of die steel and had flat shoulder with truncated conical pin having anticlockwise thread of 1 mm pitch. Figure 1 demonstrates the profile of the FSW tool. The depth of shoulder plunge was kept 0.2–0.3 mm from workpiece surface.

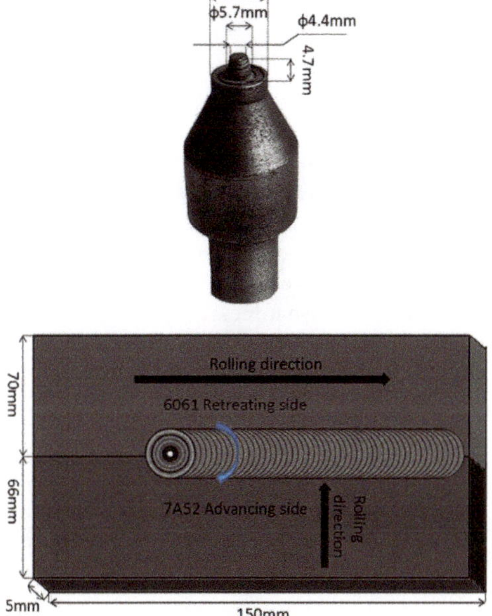

Figure 1. Schematic illustration of the friction stir butt welding of 7A52/6061 dissimilar Al alloys and an image of actual friction stir welding (FSW) tool used in all welds.

Before welding, the as-received alloy plates were treated as solid solution (SS): 475 °C for 1.5 h for 7A52 alloy and 535 °C for 1 h for 6061 alloy, or further treated as T6-state after SS: aging at 120 °C for 24 h for 7A52 alloy and 180 °C for 30 min. for 6061 alloy. The SS treatment was carried out in a box-type furnace and the age treatment was carried out in oil bath. The SS treated samples and T6 treated samples were joined by FSW, respectively. After welding, different heat treatments were employed to strengthen the weld joint. They were solid solution treated again or aged directly. The parameters in the post-heat-treatment are the same as the pre-heat treatment. If the two side of the weld joint were aged respectively, one side being immersed into hot oil and the other side is uncovered.

The visually sound joints were selected by visual inspection or while using optical microscopy for hardness test and microstructure analysis. Microhardness variations across the joint were measured on a HMV-G 21DT Micro Vickers Hardness Tester with an applied load of 100 g and a holding time of 10 s, each value was obtained by averaging at least five successive measurements. With the help of indentation of the hardness test, the specimens for microstructure characterization were accurately extracted from specific zones. A focal-ion beam (FIB)/SEM system (Carl Zeiss Auriga 45-66, Jena, Germany) was employed to examine grains of FSW joints in electron back-scattered diffraction (EBSD) (Oxford Instruments, Oxford, UK) mode. An FEI Titan G2 60-300 transmission electron microscope (TEM) (Thermo Fisher Scientific, Waltham mass, Massachusetts, USA) was used to examine the precipitates in different zones. Specimens for EBSD were prepared by mechanical polishing and then electrochemical polishing in a solution mixed of 10% perchloric acid and 90% ethyl alcohol at the voltage of 15 mV and at the temperature of −30 °C. The specimens for TEM were prepared as the following: the specimens were thinned down to 100 μm and then punched to form a disc with the diameter of 3 mm. These discs were mechanically ground and subsequently electro-polished while using a twinjet unit (TJ100-SE produced by LEBOscience) with an electrolyte of 30% nitric acid in methanol at −25 °C and 20 V. The specimens for microstructure examination were sectioned from the transverse section perpendicular to the welding direction, because the transverse section comprises all of the zones with different microstructures and associated mechanical properties.

3. Results

3.1. Hardness Evolution with Different Heat Treatment

The 7A52 and 6061 Al alloys were T6-heat-treated, respectively, with the average hardness of 165 HV and 115 HV, and then joined by FSW. Figure 2 presents the hardness variation across the weld joint along the central line of the transverse section with different rotational speeds, where the different zones are indicated. The hardness curve of FSW joint is asymmetrical with respect to the weld center. The lowest hardness (62 HV) appears near the center of the WNZ on the 6061 side, the hardness gradually rises from WNZ to the TMAZ, to the HAZ, till the 6061 matrix, or to the center line on the opposite side. A sudden increase in hardness occurs at the center line from 98 HV on the 6061 side to 138 HV on the 7A52 side. Subsequently, the hardness drops slowly to approximately 120 HV in the TMAZ on the 7A52 side, and the hardness afterwards increases slowly from HAZ to the 7A52 matrix. As shown in Figure 2, the width of HAZ on the 7A52 side is affected by the rotational speed, the higher rotational speed selected, the wider HAZ formed.

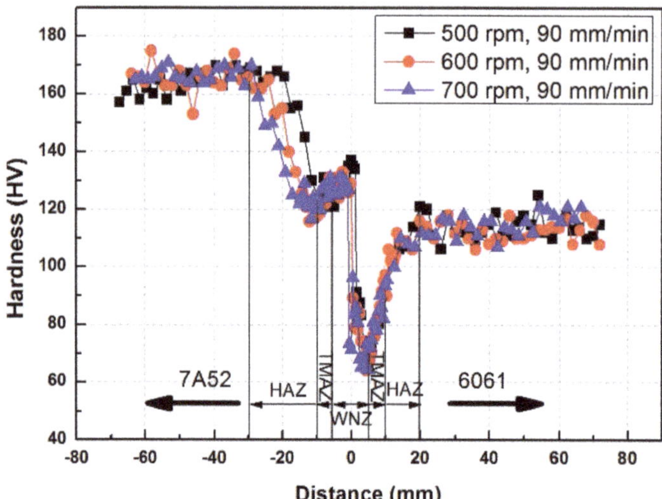

Figure 2. Microhardness profile along the middle section of the weld joint of T6-treated 7A52 and 6061 Al alloys at different rotational speed.

After solid solution, the 7A52 and 6061 Al alloys were joined by FSW at the rotation speed of 600 rpm, and post-aging was then employed; Figure 3a presents the hardness variation across the weld joint along the central line of the transverse section. It is very interesting that the WNZ and HAZ is stronger than the matrix on the 7A52 side, although a narrow area with low hardness near 78 HV was found to be in TMAZ, as shown in Figure 3a. On the 6061 side, the hardness keeps constant from the WNZ to the matrix. After welding, aging treatment was employed with three different heating methods, as shown in Figure 3b–d; the hardness change occurs on the 7A52 side, not on the 6061 side. After post-aging, on the 7A52 side, the hardness in TMAZ significantly increased, the highest hardness was tested in WNZ, and then successively drops from WNZ to the matrix, via TMAZ and HAZ. Aging 7A52 and 6061 alloys separately obtained the highest hardness of 150 HV in WNZ on the 7A52 side (Figure 3d). Unfortunately, the hardness of the matrix is too low after welding, even cannot be increased after post-aging, because the matrix of 7A52 and 6061 were over aged during pre-heating at 150 °C for 2 h before welding.

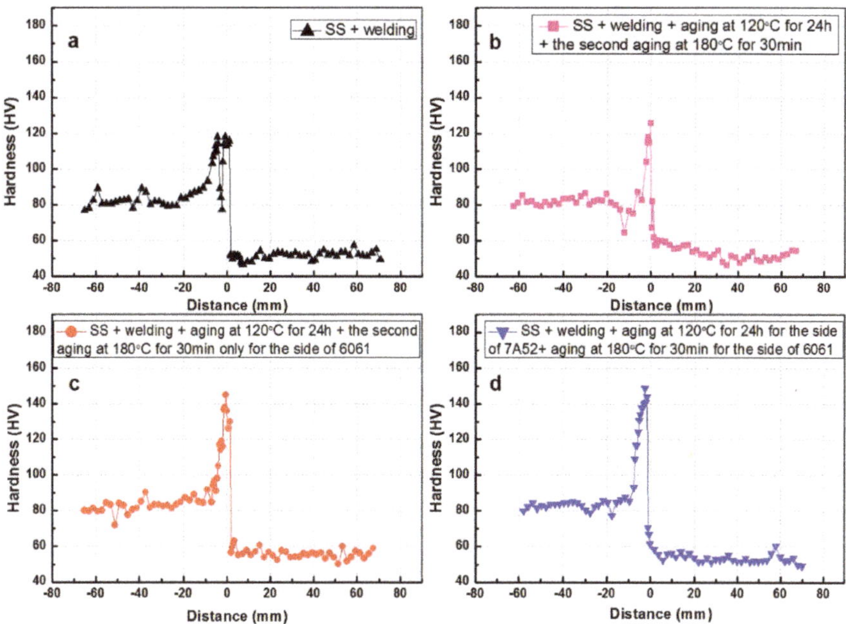

Figure 3. Microhardness profile along the middle section of the weld joint of 7A52 and 6061 Al alloys with pre-solid-solution and different post-aging.

When considering the negative effect of pre-heating on the hardness of the matrix, solid solution treatment was employed after welding and before post-aging. The solid solution temperature of 6061 alloy is higher than that of the 7A52 alloy, the 7A52 side was solid-solution treated first at 475 °C for 1.5 h, and then the 6061 side was treated at 535 °C for 1 h. Affected by the thermal transmission, precipitation occurs in the 7A52 side during the subsequent solid solution of the 6061 side. Hence, after post-solid-solution-treatment, the hardness of 7A52 matrix (near 120 HV in Figure 4a) is higher than that of the solid solution treated 7A52 alloy (about 80 HV). As shown in Figure 4a, after the post-solid-solution treatment, the hardness on both side of the central line keeps constant from the WNZ to the matrix. However, the hardness on the 7A52 side is much higher than that on the 6061 side and an abrupt increase appears near the central line from 6061 side to 7A52 side. After the subsequent post-aging, as shown in Figure 4b–d, the hardness of the weld joint was significantly improved, and still kept steady on both sides without signs of softening. With different heating procedures for post-aging, different hardness is achieved on both sides. As is well known, the strength of a weld joint is dependent upon the weakest region. Hence, the study focused on how to improve the lowest hardness. From Figure 4, it can be seen that the weld joint undergoing two-stage artificial aging (SS + welding + SS + aging at 120 °C for 24 h + the second aging at 180 °C for 30 min.) has the lowest hardness of about 100 HV on the 6061 side (Figure 4b), higher than that obtained by other post-aging treatments (Figure 4c,d), meanwhile the hardness of the 7A52 side is approximately 140 HV (Figure 4b). More importantly, the hardness evolution from the 7A52 side to 6061 side is gradual in Figure 4b–d, as opposed to the sudden shift seen in Figures 3 and 4a. This smooth change in hardness benefits the weld joint in stress concentration during service.

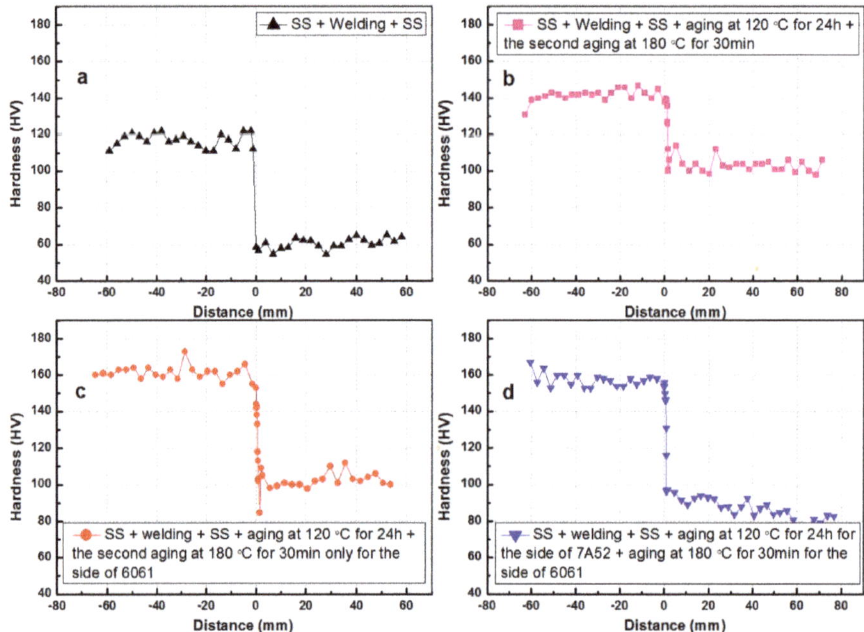

Figure 4. Microhardness profile along the middle section of the weld joint of 7A52 and 6061 Al alloys with pre-solid-solution and different post-heat-treatment (solid solution treatment and aging).

3.2. Microstructures

3.2.1. Alloying Elements Redistribution in WNZ

Figure 5 shows the second electron image and EDS mappings of the main alloying elements in two adjacent grains near the central line in the post-solid-solution treated sample corresponding to Figure 4a. As seen from Figure 5a, there are obviously a large number of the second phases at the grain boundary between these two grains. Figure 5b–d show the grains are composed of Zn, Mg, and Si, and Figure 6 shows that the right grain has a higher level of Mg and Zn, but a lower level of Si. In other words, the composition of grains in WNZ has been changed and the elements diffusion occurred during stirring or post-solid-solution. The line scanning in Figure 6 reveals the content of alloy elements is higher gradually from 6061 side to 7A52 side, which is the basic of the gradual increase of precipitation hardening in WNZ after post-aging from 6061 side to 7A52 side, as shown in Figure 4b–d.

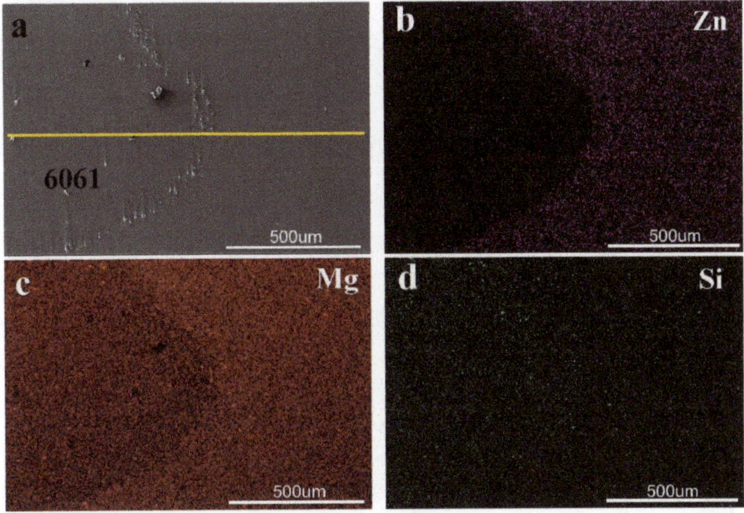

Figure 5. The second electron image of the weld joint and the corresponding EDS mappings. (**a**) The second electron (SE) image, (**b–d**) the mapping of Zn, Mg and Si.

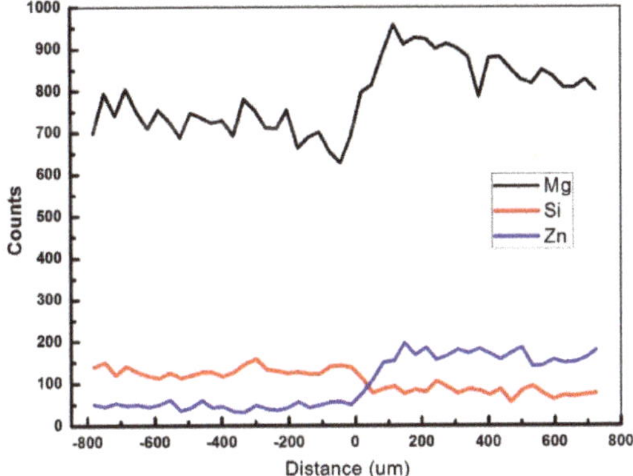

Figure 6. The line profiles of the main alloying elements (Zn, Mg, and Si) obtained along the yellow line in Figure 5a.

3.2.2. Grains

Figure 7 shows the orientation maps of grains in and adjacent to different zones locating by hardness measurement. The left is 6061 alloy and the right is 7A52 alloy. The as-rolled 6061 alloy plate comprises of recrystallized approximately equiaxed grains, and the as-rolled 7A52 alloy plate is composed of deformed elongate grains. After welding, grains in WNZ are equiaxed with a mean size of 5 µm, resulting from dynamic recrystallization, independent of the pre-heat-treatment. In TMAZ on the advancing side (6061), the as received equiaxed grains were severely elongated, and bended too far along the direction of metal flow. On the other hand, in TMAZ on the retreating side (7A52), there are more equiaxed recrystallized grain when compared with the 7A52 matrix. On the retreating side, it was found that the transition of grains is pretty smooth from TMAZ to WNZ, in comparison to the advancing side. The width of the TMAZ on both sides is approximately 1 mm.

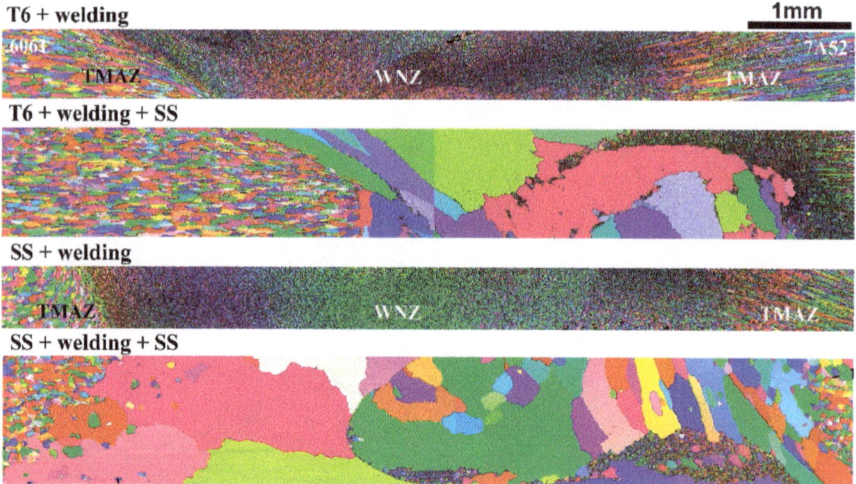

Figure 7. Electron back-scattered diffraction (EBSD) orientation map of grains in FSW joint of dissimilar 7A52 and 6061 alloys.

After post-solid-solution treatment, abnormal grain growth was found in both the WNZ for (T6 + welding + SS) sample and (SS + welding + SS) sample; this phenomenon has also been reported in literatures [32,33]. Obviously, the pre-treatment has significant influence on the region of abnormal grain growth. In (T6 + welding + SS) sample, abnormal grain growth only occurred in WNZ, but in (SS + welding + SS) sample, abnormal coarse grains were found in WNZ and TMAZ. The size of these coarse grains is of the order of millimeter or sub-millimeter.

3.2.3. Microstructures in the Interior of Grains

Figure 8 shows the TEM bright-field images and high-resolution TEM (HRTEM) images of intragranular precipitates in the T6-treated matrix. In the 6061 alloy, all of the images were acquired in $<001>_{Al}$ directions, there are three types of precipitates observed: L phase, β'' phase, and Q' phase. The overview of known precipitates in the Al-Mg-Si alloy is listed in Table 2 in Ref. [3]. These precipitates were distinguished from morphology (Figure 8a) or with the help of HRTEM images and the corresponding Fast Fourier Filtering transform (FFT) patterns (Figure 8b–d). The needle-like precipitates were identified as the β'' phase (Figure 8d), the lath-like precipitates with the cross-section elongated along $<510>$Al were the Q' phase (Figure 8b), or with the cross-section elongated along $<001>_{Al}$ were the L phase (Figure 8c). The L phase is a precursor of Q' phase. The addition of Cu has suppressed the formation of β'' by the formation of Cu-containing precipitates (L and Q'), which result in the coexistence of lath-like and needle-like precipitates. In 7A52 alloy, all of the images were taken along $<112>_{Al}$ orientations, where a large number of GPII zones can be observed in Figure 8d. Two crystallographic orientations, $<110>_{Al}$ and $<112>_{Al}$, are most suitable to identify disc-like precipitates in Al-Zn-Mg alloys, but along $<112>_{Al}$ more structural details can be revealed for the η'-phase [38,39]. As shown in Figure 8d, the disc-like GPII zones are fully coherent with the Al matrix, composed of Zn-rich layers on the Al-{111} atomic planes. The profiles of GPII zones are presented by stress field contrast, resulting from atomic size difference between Zn, Mg, and Al.

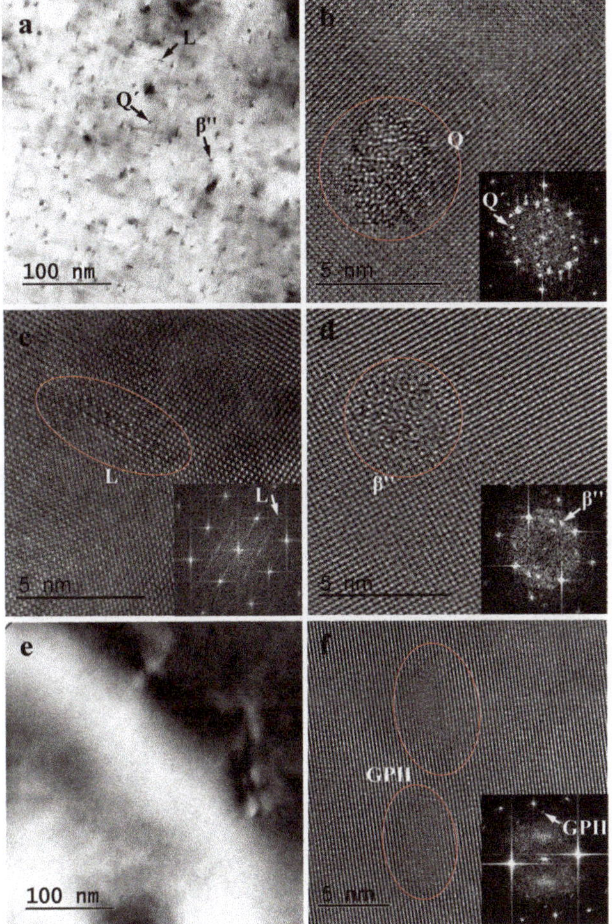

Figure 8. Transmission electron microscope (TEM) images of precipitates: (**a**) bright-field TEM images of the T6-treated 6061 matrix along <001>$_{Al}$ zone axis; (**b–d**) high-resolution transmission electron microscopy (HRTEM) images of different types of precipitates in (a), they are Q′ phase (**b**), L phase (**c**), β″ phase (**d**), respectively; (**e**) bright-field TEM images of the T6-treated 6061 matrix along <112>$_{Al}$ zone axis; and, (**f**) HRTEM image of GPII zones in (c). The inserts are the corresponding Fast Fourier Filtering transform (FFT) patterns.

Figure 9 shows the intragranular microstructure of WNZ in the weld joint of dissimilar T6-treated 6061 and 7A52. The corresponding hardness distribution across the weld joint is shown as the blue one in Figure 2. Figure 9a,b were taken from the area at the central line, where no precipitate is observed. Obviously, the pre-existing precipitates were resolved into the matrix during stirring, resulting in the lowest hardness being equal to the hardness of the solid solution state. The slight increase in hardness, from the 6061 side to the 7A52 side, might be due to solution strengthening increasing that results from alloying elements redistribution occurring during stirring. Over the lowest hardness on the 6061 side, the hardness increase is due to L precipitates forming, as shown in Figure 9c,d.

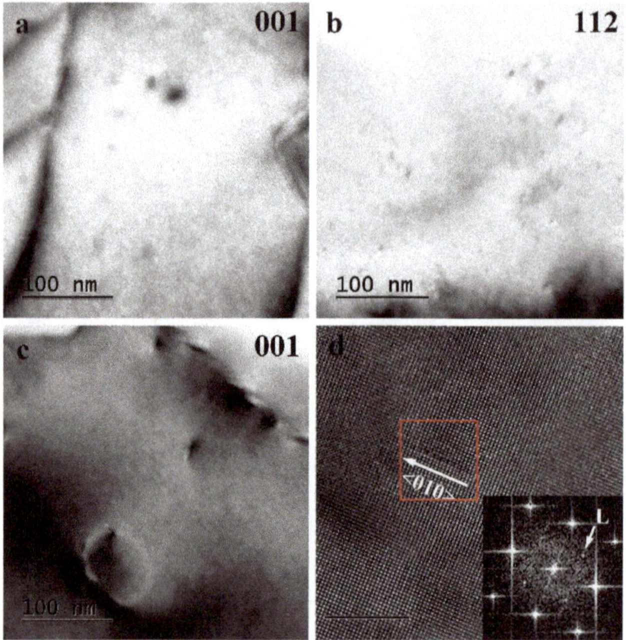

Figure 9. TEM bright-field images of the weld nugget zone (WNZ) along different zone axis: (**a**) on the 7A52 side along <001>$_{Al}$ zone axis; (**b**) on the 7A52 side along <112>$_{Al}$ zone axis; (**c**) on the 6061 side with the beam parallel to the <001>$_{Al}$ zone axis; (**d**) High-resolution TEM images and corresponding FFT patterns of L precipitate in (**c**) with the beam parallel to the <001>$_{Al}$ zone axis.

Figure 10 shows the microstructures after post-aging in the area at the central line in WNZ, where the hardness is in between that of the 7A52 and 6061 matrix, see Figure 4b. A large number of precipitates were observed along <001>$_{Al}$ and <112>$_{Al}$ orientations, as shown in Figure 10a,c, respectively. In Figure 10a, there are two types of precipitates with different size, the smaller with the size of about 1 nm are L phase, being identified from the HTTEM image in Figure 10b, the larger with the size of about 10 nm is difficult to identify from the <001>$_{Al}$ orientation. Projected along <112>$_{Al}$ (Figure 10c), the larger precipitates are also observed with distinct structural features, as shown in Figure 10d, they are identified as the η′ phase. Additionally, a few of stable η phase are also observed here. It must be noted that the coexistence of L phase and η′ phase in the same grain, which reveals the grain with the composition of Al-Zn-Mg-Si-Cu, it is the result of the fully mix of 7A52 and 6061 alloys by means of stirring. The above in agreement with the result revealed by the line scan in Figure 6.

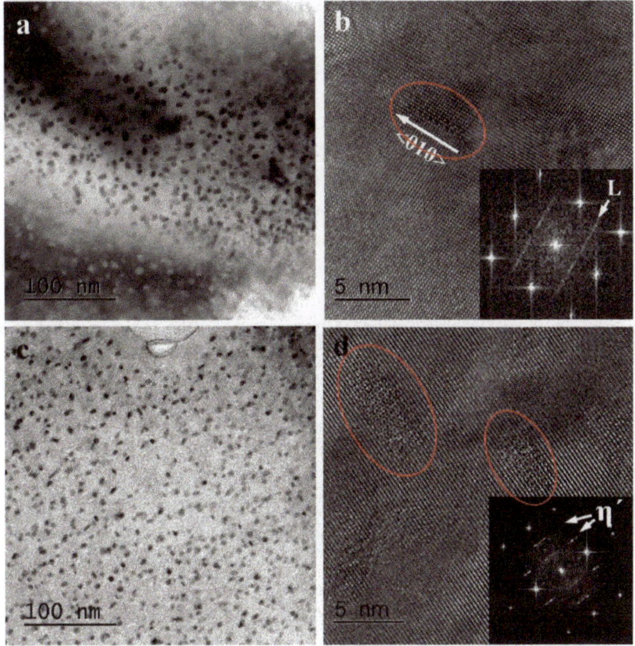

Figure 10. TEM bright-field images of the same area in WNZ after post-aging as the same as Figure 4b along different zone axis: (**a**) along <001>$_{Al}$ zone axis; (**c**) along <112>$_{Al}$ zone axis. High-resolution TEM images and corresponding FFT patterns of different precipitates: (**b**) L precipitate in (**a**) along <001>$_{Al}$ zone axis; (**d**) η′ precipitate in (**c**) along <112>$_{Al}$ zone axis.

4. Discussion

The T6-treated 7A52 and 6061 alloys were joined by FSW, a region with the lowest hardness (Figure 2), the same as a solid solution, greatly weakened the weld joint. The lowest hardness results from precipitate decomposition (Figure 9a,b) during FSW under the role of heat inputs and dislocation moving. The dissolution of precipitates in WNZ is due to the temperature in FSW process is about ~425–480 °C, which is high enough to cause the dissolution of strengthening precipitates in WNZ [40]. An effective approach for improving the hardness of this region is to make precipitates form again. The local heat treatment would be difficult to employ, because this region is very narrow and the heat inputs will soften the adjacent area via precipitates coarsening. Thus, this study employed the integral heat treatment. For cost saving and improving elements diffusion, after solid solution treatment, the alloys were directly welded by FSW, and the weld joints were then aged. The displayed results revealed that the hardness of the WNZ is much higher than the matrix, as shown in Figure 3, which is satisfactory for the weld. However, the matrixes were significantly softened, because over-aging occurred during pre-heating at about 150 °C before the weld. Hence, the weld joints should be solid solution treated again.

When considering the different age temperature for 7A52 and 6061 alloys, the former is approximately 120 °C, and the latter is about 180 °C. Thus, there are three approaches for aging after post-solid-solution: (1) the welding structure was aged at 120 °C for 24 h and then 180 °C for 30 min.; (2) both sides of the weld joint were respectively aged: the 7A52 side was aged at 120 °C for 24 h, and then the other side 6061 was aged at 180 °C for 30 min.; and, (3) both sides of the weld joint were aged at 120 °C for 24 h, and then the 6061 side was further aged at 180 °C for 30 min. The ideal weld joints were obtained via the approach (1) the hardness change across the weld is described as in Figure 4b, where the 6061 matrix was strengthened to be similar to the T6-state, the 7A52 matrix was over-aged, and the hardness slowly changed in WNZ from 7A52 side to 6061 side. Composition analysis (Figures 5 and 6) and precipitates identification reveal the WNZ with the composition of Al-Zn-Mg-Si-Cu. During the first-stage aging at 120 °C, η-$MgZn_2$-series precipitates form, and most of Zn and Mg moved from solid solution to form precipitates. The decrease of Mg in the matrix of WNZ inhibited the nucleation of β'' phase, thus, after the second-stage aging, none of the β'' precipitates were detected (Figure 10), unlike T6-treated 6061 alloy (Figure 8). In addition, Cu suppresses the formation of β'' by the formation of Cu-containing L precipitates [3]. The hardness of WNZ is lower than that of 7A52 matrix and higher than that of 6061 matrix; it is dependent upon the mixed composition by 7A52 and 6061 alloys. In other words, the hardness is contributed by η' and L precipitates. The number density of η' precipitates is higher, the hardness of WNZ is closer to that of 7A52 matrix. Otherwise, the number density of L precipitates is higher, the hardness of WNZ is closer to that of 6061 matrix. Obviously, precipitates influence the hardness of the WNZ more.

The tensile properties of the weld that correspond to Figure 4 are shown in Table 2. After the post-weld heat treatment, the ultimate tensile strength of the weld significantly increased, in comparison to that of the weld without post-weld heat treatment; however, the elongations of all samples are very poor. The fracture topographies in Figure 11 shows a large number of coarse second phases (arrowed in Figure 11b,d) within dimples. In Figure 5, an amount of second phase is also observed on the grain boundary in WNZ. Obviously, these second phases formed during FSW, not being caused by the post-weld heat treatment. The micro-hardness is more dependent upon precipitates, while the tensile properties are usually affected by more complicated factors, such as defects, interfaces, the second phases, and so on.

Table 2. Tensile properties of the weld.

Samples	Ultimate Tensile Strength (MPa)	Elongation (%)
SS + welding	174	5.66
SS + welding + aging at 120 °C for 24 h + the second aging at 180 °C for 30 min	260	4.89
SS + welding + aging at 120 °C for 24 h + the second aging at 180 °C for 30 min. only for the side of 6061	250	4.73
SS + welding + aging at 120 °C for 24 h for the side of 7A52 + the second aging at 180 °C for 30 min. for the side of 6061	210	4.99

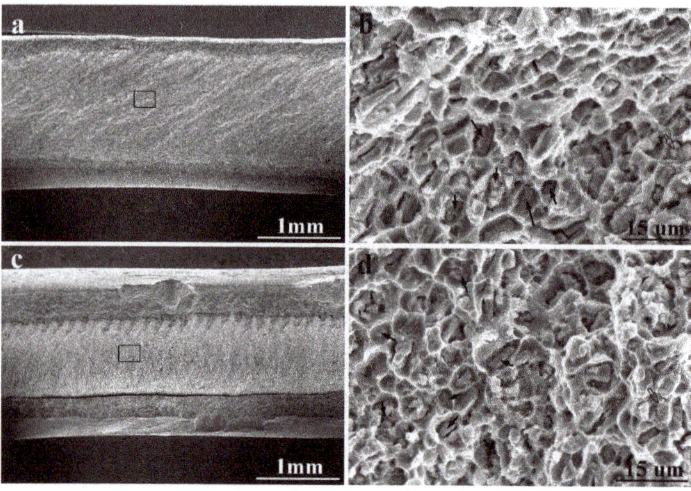

Figure 11. Fractured surface of the joint failed in the WNZ (SEM): (**a**) and (**b**) the sample of (SS + Welding); (**c**) and (**d**) the sample of (SS + welding + aging at 120 °C for 24 h + the second aging at 180 °C for 30 min.). (**b**) and (**d**) are the enlarged image of the boxed area in (**a**) and (**c**), respectively.

5. Conclusions

FWS joints of dissimilar 7A52 and 6061 Al alloys were achieved by post- integral-heat-treatment, the effect of heat treatment on the hardness and the microstructures of the weld were systematically investigated. The results are summarized, as follows.

(1) The T6-treated 7A52 and 6061 alloys were joined by FWS, a weakest region was revealed in WNZ, with the lowest hardness of 62 HV being similar to the solid solution of 6061 alloy.

(2) The solutionized 7A52 and 6061 alloys were joined by FWS, the WNZ can be strengthened by the post-aging, but the hardness of the matrixes cannot be improved by the post-aging.

(3) The weld joints of solutionized 7A52 and 6061 alloys were achieved by post-solid-solution and subsequent artificial aging. The best post-aging is designed as two-stage for the whole weld joint: 120 °C for 24 h and then 180 °C for 30 min. The hardness of the 7A52 matrix reaches approximately 140 HV, and the hardness of the 6061 matrix is higher than 100 HV. The hardness of WNZ slowly drops from 7A52 side to 6061 side, due to the precipitation of η′ phase and L phase.

Author Contributions: Conceptualization, J.L.; Data curation, Y.J., S.L. and Y.Q.; Formal analysis, J.L.; Funding acquisition, K.W.; Investigation, Y.J., S.L., J.L. and Y.Q.; Methodology, S.L. and J.L.; Supervision, J.L. and K.W.; Writing—original draft, Y.J. and J.L.; Writing—review & editing, J.L. and K.W.

Funding: This research was funded by the National Key R&D Program of China, grant number 2017YFA0204403; the National Natural Science Foundation of China, grant number 51301064; the open fund supported by State Key Laboratory of Advanced Design and Manufacturing for Vehicle Body at Hunan University, grant number 31715009; the Fundamental Research Funds for the Central Universities, grant number 30918011341.

Acknowledgments: All TEM experiments were performed at the Materials Characterization and Research Center of Nanjing University of Science and Technology.

Conflicts of Interest: The authors declare no conflict of interest.

References

1. Hornbogen, E. Hundred years of precipitation hardening. *J. Light Met.* **2001**, *1*, 127–132. [CrossRef]
2. Lendvai, J. Precipitation and Strengthening in Aluminium Alloys. *Mater. Sci. Forum* **1996**, *217*, 43–56. [CrossRef]
3. Ding, L.; Jia, Z.; Nie, J.-F.; Weng, Y.; Cao, L.; Chen, H.; Wu, X.; Liu, Q. The structural and compositional evolution of precipitates in Al-Mg-Si-Cu alloy. *Acta Mater.* **2018**, *145*, 437–450. [CrossRef]
4. Murayama, M.; Hono, K. Pre-precipitate clusters and precipitation processes in Al–Mg–Si alloys. *Acta Mater.* **1999**, *47*, 1537–1548. [CrossRef]
5. Murr, L.E. *Friction-Stir Welding and Processing, Handbook of Materials Structures, Properties, Processing and Performance*; Springer International Publishing: Cham, Switzerland, 2017; pp. 1–18.
6. Thomas, W.M.; Nicholas, E.D.; Needham, J.C.; Nurch, M.G.; Temple-Smith, P.; Dawes, C.J. Friction Stir Butt Welding. GB Patent Application No. 9125978.8, 6 December 1991.
7. Jafari, H.; Mansouri, H.; Honarpisheh, M. Investigation of residual stress distribution of dissimilar Al-7075-T6 and Al-6061-T6 in the friction stir welding process strengthened with SiO_2 nanoparticles. *J. Manuf. Process.* **2019**, *43*, 145–153. [CrossRef]
8. Khan, N.Z.; Siddiquee, A.N.; Khan, Z.A.; Shihab, S.K. Investigations on tunneling and kissing bond defects in FSW joints for dissimilar aluminum alloys. *J. Alloy. Compd.* **2015**, *648*, 360–367. [CrossRef]
9. Jenarthanan, M.P.; Varma, C.V.; Manohar, V.K. Impact of friction stir welding (FSW) process parameters on tensile strength during dissimilar welds of AA2014 and AA6061. *Mater. Today Proc.* **2018**, *5*, 14384–14391. [CrossRef]
10. Moradi, M.M.; Aval, H.J.; Jamaati, R. Effect of pre and post welding heat treatment in SiC-fortified dissimilar AA6061-AA2024 FSW butt joint. *J. Manuf. Process.* **2017**, *30*, 97–105. [CrossRef]
11. Avinash, P.; Manikandan, M.; Arivazhagan, N.; Ramkumar, K.D.; Narayanan, S. Friction Stir Welded Butt Joints of AA2024 T3 and AA7075 T6 Aluminum Alloys. *Procedia Eng.* **2014**, *75*, 98–102. [CrossRef]
12. Barbini, A.; Carstensen, J.; Santos, J.F.D. Influence of a non-rotating shoulder on heat generation, microstructure and mechanical properties of dissimilar AA2024/AA7050 FSW joints. *J. Mater. Sci. Technol.* **2018**, *34*, 119–127. [CrossRef]

13. Sidhar, H.; Mishra, R.S. Aging kinetics of friction stir welded Al-Cu-Li-Mg-Ag and Al-Cu-Li-Mg alloys. *Mater. Des.* **2016**, *110*, 60–71. [CrossRef]

14. Eskandari, M.; Aval, H.J.; Jamaati, R. The study of thermomechanical and microstructural issues in dissimilar FSW of AA6061 wrought and A390 cast alloys. *J. Manuf. Process.* **2019**, *41*, 168–176. [CrossRef]

15. Choudhary, S.; Choudhary, S.; Vaish, S.; Upadhyay, A.K.; Singla, A.; Singh, Y. Effect of welding parameters on microstructure and mechanical properties of friction stir welded Al 6061 aluminum alloy joints. *Mater. Today Proc.* **2013**, *50*, 872–878. [CrossRef]

16. Elnabi, M.M.A.; Elshalakany, A.B.; Abdel-Mottaleb, M.M.; Osman, T.A.; el Mokadem, A. Influence of friction stir welding parameters on metallurgical and mechanical properties of dissimilar AA5454–AA7075 aluminum alloys. *J. Mater. Res. Technol.* **2019**, *8*, 1684–1693. [CrossRef]

17. Milagre, M.X.; Mogili, N.V.; Donatus, U.; Giorjão, R.A.R.; Terada, M.; Araujo, J.V.S.; Machado, C.S.C.; Costa, I. On the microstructure characterization of the AA2098-T351 alloy welded by FSW. *Mater. Charact.* **2018**, *140*, 233–246. [CrossRef]

18. Devaiah, D.; Kishore, K.; Laxminarayana, P. Optimal FSW process parameters for dissimilar aluminium alloys (AA5083 and AA6061) using Taguchi Technique. *Mater. Today Proc.* **2018**, *5*, 4607–4614. [CrossRef]

19. Mishra, R.S.; Sidhar, H. Chapter 4—FSW of Al–Cu and Al–Cu–Mg Alloys. In *Friction Stir Welding of 2XXX Aluminum Alloys Including Al-Li Alloys*; Mishra, R.S., Sidhar, H., Eds.; Butterworth-Heinemann: Oxford, UK, 2017; pp. 47–77.

20. Dehghani, M.; Amadeh, A.; Mousavi, S.A.A.A. Investigations on the effects of friction stir welding parameters on intermetallic and defect formation in joining aluminum alloy to mild steel. *Mater. Des.* **2013**, *49*, 433–441. [CrossRef]

21. Astarita, A.; Squillace, A.; Scala, A.; Prisco, A. On the Critical Technological Issues of Friction Stir Welding T-Joints of Dissimilar Aluminum Alloys. *J. Mater. Eng. Perform.* **2012**, *21*, 1763–1771. [CrossRef]

22. Khodir, S.A.; Shibayanagi, T. Friction stir welding of dissimilar AA2024 and AA7075 aluminum alloys. *Mater. Sci. Eng. B* **2008**, *148*, 82–87. [CrossRef]

23. Astarita, A.; Squillace, A.; Carrino, L. Experimental Study of the Forces Acting on the Tool in the Friction-Stir Welding of AA 2024 T3 Sheets. *J. Mater. Eng. Perform.* **2014**, *23*, 3754–3761. [CrossRef]

24. Astarita, A.; Squillace, A.; Armentani, E.; Ciliberto, S. Friction stir welding of AA 2198 T3 rolled sheets in butt configuration. *Metall. Ital.* **2012**, *7*, 31–40.

25. Kumar, S.R.; Rao, V.S.; Pranesh, R.V. Effect of Welding Parameters on Macro and Microstructure of Friction Stir Welded Dissimilar Butt Joints between AA7075-T651 and AA6061-T651 Alloys. *Procedia Mater. Sci.* **2014**, *5*, 1726–1735. [CrossRef]

26. Kalemba-Rec, I.; Kopyściański, M.; Miara, D.; Krasnowski, K. Effect of process parameters on mechanical properties of friction stir welded dissimilar 7075-T651 and 5083-H111 aluminum alloys. *Int. J. Adv. Manuf. Technol.* **2018**, *97*, 2767–2779. [CrossRef]

27. Jia, Y.; Qin, Y.; Ou, Y.; Wang, K.; Liu, J. The Influence of Microstructural Heterogeneity on Mechanical Properties of Friction Stir Welded Joints of T6-Treated Al-Zn-Mg Alloy 7A52. *Metals* **2018**, *8*, 527. [CrossRef]

28. Velotti, C.; Astarita, A.; Buonadonna, P.; Dionoro, G.; Langella, A.; Paradiso, V.; Prisco, U.; Scherillo, F.; Squillace, A.; Tronci, A. FSW of AA 2139 Plates: Influence of the Temper State on the Mechanical Properties. *Key Eng. Mater.* **2013**, *554*, 1065–1074. [CrossRef]

29. Prisco, U.; Squillace, A.; Astarita, A.; Velotti, C. Influence of Welding Parameters and Post-weld Aging on Tensile Properties and Fracture Location of AA2139-T351 Friction-stir-welded Joints. *Mater. Res.* **2013**, *16*, 1106–1112. [CrossRef]

30. Astarita, A.; Prisco, U.; Squillace, A.; Velotti, C.; Tronci, A. Mechanical characterization by DOE analysis of AA6156-T4 friction stir welded joints in as-welded and post-weld aged condition. *Mater. Test.* **2015**, *57*, 192–199. [CrossRef]

31. Singh, R.K.R.; Sharma, C.; Dwivedi, D.K.; Mehta, N.K.; Kumar, P. The microstructure and mechanical properties of friction stir welded Al–Zn–Mg alloy in as welded and heat treated conditions. *Mater. Des.* **2011**, *32*, 682–687. [CrossRef]

32. Safarbali, B.; Shamanian, M.; Eslami, A. Effect of post-weld heat treatment on joint properties of dissimilar friction stir welded 2024-T4 and 7075-T6 aluminum alloys. *Trans. Nonferrous Met. Soc. China* **2018**, *28*, 1287–1297. [CrossRef]

33. Sharma, C.; Dwivedi, D.K.; Kumar, P. Effect of post weld heat treatments on microstructure and mechanical properties of friction stir welded joints of Al–Zn–Mg alloy AA7039. *Mater. Des.* **2013**, *43*, 134–143. [CrossRef]

34. Sato, Y.S.; Kokawa, H. Distribution of tensile property and microstructure in friction stir weld of 6063 aluminum. *Metall. Mater. Trans. A* **2001**, *32*, 3023–3031. [CrossRef]

35. Bayazid, S.M.; Farhangi, H.; Asgharzadeh, H.; Radan, L.; Ghahramani, A.; Mirhaji, A. Effect of cyclic solution treatment on microstructure and mechanical properties of friction stir welded 7075 Al alloy. *Mater. Sci. Eng. A* **2016**, *649*, 293–300. [CrossRef]

36. Mahoney, M.W.; Rhodes, C.G.; Flintoff, J.G.; Bingel, W.H.; Spurling, R.A. Properties of friction-stir-welded 7075 T651 aluminum. *Metall. Mater. Trans. A* **1998**, *29*, 1955–1964. [CrossRef]

37. Sullivan, A.; Robson, J.D. Microstructural properties of friction stir welded and post-weld heat-treated 7449 aluminium alloy thick plate. *Mater. Sci. Eng. A* **2008**, *478*, 351–360. [CrossRef]

38. Liu, J.Z.; Chen, J.H.; Yang, X.B.; Ren, S.; Wu, C.L.; Xu, H.Y.; Zou, J. Revisiting the precipitation sequence in Al–Zn–Mg-based alloys by high-resolution transmission electron microscopy. *Scr. Mater.* **2010**, *63*, 1061–1064. [CrossRef]

39. Li, X.Z.; Hansen, V.; GjØnnes, J.; Wallenberg, L.R. HREM study and structure modeling of the η′ phase, the hardening precipitates in commercial Al–Zn–Mg alloys. *Acta Mater.* **1999**, *47*, 2651–2659. [CrossRef]

40. Rhodes, C.G.; Mahoney, M.W.; Bingel, W.H.; Spurling, R.A.; Bampton, C.C. Effects of friction stir welding on microstructure of 7075 aluminum. *Scr. Mater.* **1997**, *36*, 69–75. [CrossRef]

Article

Residual Stress, Microstructure and Mechanical Properties in Thick 6005A-T6 Aluminium Alloy Friction Stir Welds

Xiaolong Liu [1,†], Pu Xie [2,†], Robert Wimpory [3], Wenya Li [4], Ruilin Lai [5], Meijuan Li [1], Dongfeng Chen [1], Yuntao Liu [1] and Haiyan Zhao [6,*]

[1]   Department of Nuclear Physics, China Institute of Atomic Energy, Beijing 102413, China
[2]   CRRC Changchun Railway Vehicles Co. Ltd., Changchun 130000, China
[3]   Helmholtz-Zentrum Berlin für Energie und Materialien, Hahn Meitner Platz 1, 14109 Berlin, Germany
[4]   Shanxi Key Laboratory of Friction Welding Technologies, School of Materials Science and Engineering, Northwestern Polytechnical University, Xi'an 710072, China
[5]   Department of Mechanical and Electrical Engineering, Central South University, Changsha 410083, China
[6]   State Key Laboratory of Tribology, Department of Mechanical Engineering, Tsinghua University, Beijing 100084, China
*   Correspondence: hyzhao@tsinghua.edu.cn; Tel.: +86-10-6278-4578
†   These authors contributed equally to this work and should be considered as co-first authors.

Received: 3 July 2019; Accepted: 18 July 2019; Published: 21 July 2019

Abstract: Plates (37 mm thick) of 6005A-T6 aluminum alloy were butt joined by a single-sided and double-sided friction stir welding (FSW). The 3D residual stresses in the joints were determined using neutron diffraction. The microstructures were characterized by a transmission electron microscope (TEM) and electron backscatter diffraction (EBSD). In the single-sided FSW specimen, there were acceptable mechanical properties with a tensile strength of 74.4% of base metal (BM) and low residual stresses with peak magnitudes of approximately 37.5% yield strength of BM were achieved. The hardness is related to the grain size of the nugget zone (NZ), and in this study, precipitations were dissolved due to the high heat input. In the double-sided FSW specimen, there were good mechanical properties with a tensile strength of 80.8% of BM, but high residual stresses with peak magnitudes of approximately 70% yield strength of BM were obtained. The heat input by the second pass provided an aging environment for the first-pass weld zone where the dissolved phases were precipitated and residual stresses were relaxed.

Keywords: residual stresses; friction stir welding; neutron diffraction; aluminium alloys; hardness; precipitation

1. Introduction

Aluminium alloy 6005A, a β-precipitation (Mg_2Si) strengthened heat treatable alloy, is widely used in rail transportation industries due to its excellent corrosion resistance and extrusion characteristics [1,2]. The welding joints to connect the extruded sheets and plates are usually the weak regions since the welding heat input introduces the microstructural changes deteriorating the mechanical properties and residual stresses causing fatigue cracks and stress-induced corrosion [3,4]. Therefore, sound welding is required to join extruded aluminium alloy sheets and plates.

Friction stir welding is a solid-state joining technique, which involves both plastic and thermal deformations [5,6]. In this process, process parameters, such as tool rotation and traverse speeds, need to be optimized to get good quality joints. The effect of tool rotation and traverse speeds, on microstructural changes and residual stresses of friction stir welding (FSW) aluminium alloys, have been studied widely [1–4,7–34]. Simar et al. reported that the β″ originally present in the base metal

(BM) fully dissolved in the nugget zone (NZ) and coarsened in the heat affected zone (HAZ) of 6005A-T6 aluminium FSW joint [1]. They also explained the softened region around the weld center. Dong et al. studied the effect of welding speed on microstructures and the mechanical properties of 6005A-T6FSW joints, reporting an increased tendency of tensile properties with increasing the welding speed [4]. Wang et al. concluded that the 6061-T6 aluminum FSW joints made at low welding speed exhibited lower residual stress, due to a change in microstructure and stress relaxation that occurred as a result of the longer heating time associated with the low welding speed [8]. Therefore, microstructure, mechanical properties and residual stresses should be considered together to obtain the optimized FSW process. However, most studies have focused on the effects of welding speed and rotation speed on either microstructures and mechanical properties, or residual stresses in aluminium sheets and thin plates.

Nonetheless, FSW has been applied to the production of large prefabricated aluminium panels in high speed railcars, which helps to reduce the weight and improve the integrity of aluminium sheets. To join thick 6005A aluminium plates using FSW, a better understanding of residual stresses and microstructure is required. However, the residual stresses and microstructures have not been studied in thick 6005A-T6 aluminium alloy plates joined by single-sided and double-sided FSW.

Among the methods of residual stress measurement, neutron diffraction can nondestructively characterize the 3D residual stress distribution of engineering materials [35,36]. The deep penetration capability of neutrons into most metallic materials makes neutron diffraction a powerful tool for determining residual stresses through welds. The E3 residual stress diffractometer at HZB is one such high-performance neutron residual stress instrument, capable of experimental measurement of 3D residual stress distributions [37]. This instrument was chosen to characterize the residual stresses in thick 6005A-T6 aluminium alloy FSWs.

In this work, thick 6005A-T6 aluminium alloy plates (37 mm thick) were studied to determine the single-sided and double-sided FSW residual stress distributions using neutron diffraction. In addition, microstructures and mechanical properties were studied to characterize welding behavior. This experimental study was carried out to understand the 3D residual stress distributions and microstructural changes. This study can be helpful in the design of welding strategies.

2. Experimental Details

2.1. Materials

The base metal used in this investigation was 37 mm thick 6005A-T6 aluminium alloy with the chemical composition and mechanical properties shown in Table 1. The dimensions of the 6005A-T6 test plates are shown in Figure 1.

Table 1. The chemical composition and mechanical properties of 6005A aluminium alloy.

Chemical Composition (wt.%)									Mechanical Properties	
Mg	Si	Fe	Cu	Mn	Cr	Ti	Zn	Al	Yielding Strength (MPa)	Tensile Strength (MPa)
0.54	0.62	0.19	0.07	0.14	0.01	0.01	0.02	Bal.	250	200

Further, FSW was carried out on a LM-FSW-5025 machine (China FSW Center, Beijing, China). For the single-sided FSW, the tool consisted of a 35 mm diameter shoulder with a pin length of 36.5 mm and a root diameter of 18 mm. The single-sided FSW process was performed with a rotation speed of 350 rpm and a traverse speed of 40 mm/min. The down force was approximately 8–9 kN. For the double-sided FSW, the tool consisted of a 32 mm diameter shoulder with a pin length of 18 mm and a root diameter of 14.4 mm. To apply the second pass of the FSW, the plates were rotated around their welding axes after completing the first weld. All welding passes were conducted with a rotation speed of 650 rpm and a traverse speed of 200 mm/min. The down force was approximately 13–14 kN.

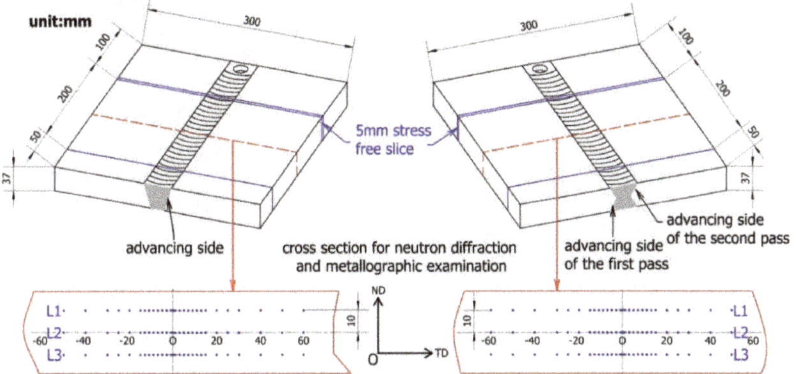

Figure 1. The layout of single-sided and double-sided friction stir welding (FSWs).

The temperature of the NZ was estimated using the following relation [38]:

$$\frac{T}{T_m} = \mathrm{K}\left(\frac{w^2}{v \times 10^4}\right)^{\alpha} \tag{1}$$

where T_m is the melting temperature of 6005A aluminium alloy 654 °C, exponent α is a dimensionless constant selected as 0.05, K is a dimensionless constant selected as 0.7, v is the traverse speed and ω is the rotation speed [39]. According to Equation (1), the temperature of the nugget zone in the single-sided FSW is 431.4 °C, while in the double-sided, FSW is 423.5 °C.

2.2. Neutron Diffraction

Due to the high penetration ability of thermal neutrons, neutron diffraction is an excellent engineering tool for providing 3D residual stresses nondestructively in bulk components. Therefore, neutron diffraction was applied to characterize the residual stresses in thick 6005A-T6 aluminium alloy FSWs, and was performed on the dedicated residual stress neutron diffractometer E3 at HZB, Germany.

As shown in Figure 1, the middle section with 200 mm in length was cut using electrical discharge machining (EDM, LA350, SSG, Suzhou, China) from the specimen for a neutron diffraction measurement. The longitudinal direction (LD), transverse direction (TD) and normal direction (ND) were assumed to be the principal directions in the bulk of the components and were measured as three orthogonal directions. The points along three lines, L1, L2 and L3, in the cross section were chosen to characterize the residual stresses, as shown in Figure 1. The diffraction peaks of the marked points were measured with the scattering vector parallel to three orthogonal directions. The peak positions, 2θ, were analyzed using the least square Gaussian fitting method [40].

The measurements were made using the Al(311) Bragg reflection, which is the strongest diffraction reflection and also is weakly affected by intergranular strains [36]. The E3 is equipped with a perfectly bent Si(400) crystal monochromator providing a wavelength of 1.47Å. Therefore, the Al(311) reflection was at a scattering angle of $2\theta{\sim}74°$. Figure 2 shows the experimental setup on E3 for the measurement of the transverse component. The gauge volume was defined by an incident primary slit with 3 mm and a secondary radial collimator with 2 mm. The height of the primary slit was set to 10 mm for transverse and normal measurements, and 3 mm for longitudinal measurements.

Figure 2. The experimental setup.

The elastic lattice strains ε_i in the *i*-direction (*i* = LD, TD, ND) were calculated using the following equation [36]:

$$\varepsilon_i = \frac{d_i - d_0}{d_0} \tag{2}$$

The elastic strains were converted to residual stresses (σ_{LD}, σ_{TD}, σ_{ND}) using the generalized Hooke's law [36]:

$$\sigma_i = \frac{E_{hkl}}{1 + v_{hkl}}\left[\varepsilon_i + \frac{v_{hkl}}{1 - 2v_{hkl}}(\varepsilon_{LD} + \varepsilon_{TD} + \varepsilon_{ND})\right] \tag{3}$$

where *i* is the LD, TD or ND component corresponding to the three orthogonal directions. The diffraction elastic constants (E_{311}) of 69 GPa and Poisson's ratio (v_{311}) of 0.35 were computed using the Kröner model via the software, IsoDEC [41].

To obtain a precise stress-free reference lattice parameter, d_0, is an important part of the diffraction-based, residual strain/stress experiment. To address a possible issue of d_0 variation due to microstructural changes, a full stress analysis was performed on 5-mm slices made by EDM from the specimen. The measurements were repeated in the same positions as for the specimen. Then, in an approximation of a biaxial stress state and the condition of the through thickness component (longitudinal in this case) to be zero ($\sigma_{LD} = 0$), the calculation of the stress-free parameters was made according to the following equation:

$$d_0 = \frac{1 - v_{hkl}}{1 + v_{hkl}}d_{LD} + \frac{v_{hkl}}{1 + v_{hkl}}(d_{TD} + d_{ND}) \tag{4}$$

As well as the TD and ND stress components in the slice as a by-product of the analysis,

$$\sigma_{TD} = \frac{E_{hkl}}{1 + v_{hkl}}[\varepsilon_{TD} - \varepsilon_{LD}], \ \sigma_{ND} = \frac{E_{hkl}}{1 + v_{hkl}}[\varepsilon_{ND} - \varepsilon_{LD}]. \tag{5}$$

2.3. Microstructure Characterization

The metallographic samples were cut perpendicular to the welding direction using EDM to avoid thermal degradation. The optical microstructure examination was performed on an Olympus microscope.

In order to compare the difference in precipitation, the microstructure examination in the NZ was conducted by a transmission electron microscope (TEM, Tecnai G2 F20, FEI, Hillsboro, OR, USA). The metallographic samples were polished down to a thickness of 80~100 μm. The final thickness reduction was obtained by electro-polishing with a HNO_3 solution (HNO_3 30% in volume in methanol at ~30 °C under 9V).

The electron backscatter diffraction (EBSD) samples were examined in a high resolution Philips XL30 field emission gun (FEG) SEM (Philips, Amsterdam, The Netherlands) interfaced to an HKL Channel EBSD orientation mapping system. The resulting EBSD orientation maps, with a step size of 0.1–0.25 μm and an area of 145 × 127 μm, were used to characterize the grain structures present in the NZ at three positions on cross sections along the weld center line, namely, the top (10 mm above the mid plane), the center (at the mid plane), and the bottom of the nugget (10 mm below the mid plane). The maps were processed using in house software (VMap) to determine the grain size.

2.4. Mechanical Properties

The hardness tests were performed using the Vicker hardness method, using a load of 500 g applied for 10 s. Three lines were chosen on the transverse section of the welds with a distance of 1 mm between neighboring measured points, as shown in Figure 1. The specimens were cut to 3-mm slices to release the residual stress, which minimized the influence of residual stress on the hardness test.

The tensile tests on welded specimens were performed on the transverse, i.e., perpendicular to the welding direction. Eight samples were cut using EDM from the single-sided FSW and double-sided FSW. The transverse cross-section of the tensile samples is shown in Figure 3. These tensile samples were 25 mm thick, 6 mm wide and 200 mm long, with an initial gauge length of 100 mm. The tensile test was carried out on a universal electronic tensile testing machine (MTS Landmark) with a crosshead moving speed of 2 mm/min. The range of load cell is ±100 KN.

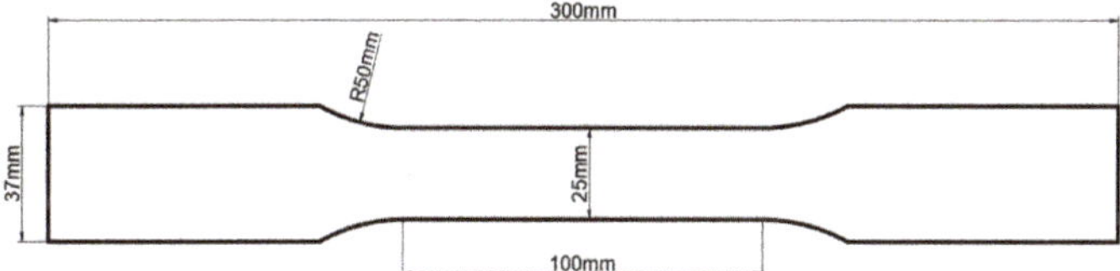

Figure 3. The layout of a tensile specimen.

3. Results and Discussion

Heat-treatable aluminium alloys derive much of their strength from the presence of fine precipitates. As residual stresses and microstructures are of great significance in determining weld performance, they were studied together for thick 6005A-T6 aluminium alloy FSWs.

3.1. Microstructural Evolution

Figure 4 shows the macrostructure of the 6005A-T6 single-sided and double-sided FSW joints. The NZ, thermo-mechanically affected zone (TMAZ), HAZ and BM are divided by the dotted lines. The TMAZ is more optically distinct on the advancing side and more diffuse on the retreating side in both specimens. The microstructures, including grain size and precipitation at the marked positions in Figure 4, are discussed further.

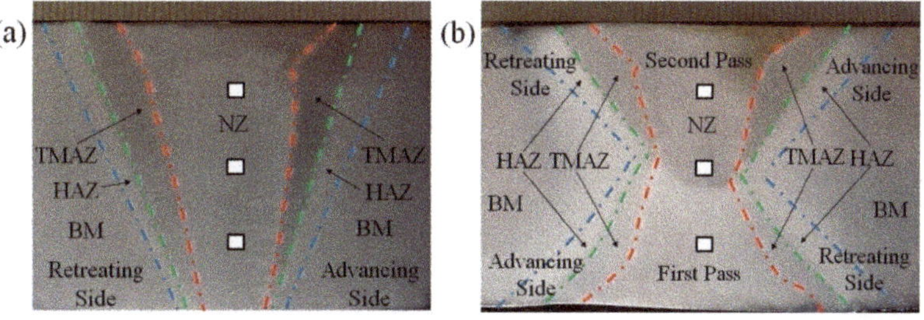

Figure 4. Macrographs (optical) of a single-sided FSW (**a**) and a double-sided FSW (**b**).

3.1.1. Features in the Single-Sided FSW

Figure 5 shows the grain map and the corresponding grain size distribution, TEM and its diffraction pattern in the single-sided FSW. The average grain size in the NZ was 9.16, 8.49 and 7.38 µm on L1, L2 and L3 line respectively. It is often suggested that the microstructure in the NZ is the result of continuous dynamic recrystallization supported by dynamic recovery [6]. The recrystallized grain size is mainly affected by the heat input in the welding process [6]. More heat input generated in the upper NZ and less heat input in the bottom NZ, which resulted in the larger grain size (9.16 µm) in the upper NZ and smaller grain size (7.38 µm) in the bottom NZ.

In order to study the microstructural evolution, the BM was observed by TEM, as shown in Figure 6. The BM contains a high density of fine hardening precipitates, as reported in the literature [6]. In the 6005A-T6 single-sided FSW, no precipitation was observed in the NZ. It is inferred that precipitations in the NZ were dissolved due to the friction heat and plastic deformation.

3.1.2. Features in the Double-Sided FSW

Figure 7 shows the grain map and the corresponding grain size distribution, TEM and its diffraction pattern in the double-sided FSW. The average grain size in the NZ was 7.48, 6.25 and 9.69 µm on L1, L2 and L3 lines respectively. The grain size in the first pass is much larger than that in the second pass. No precipitation was observed in the NZ of the second-pass region and the middle region. As compared to the BM, fewer precipitations were observed in the NZ of the first-pass region.

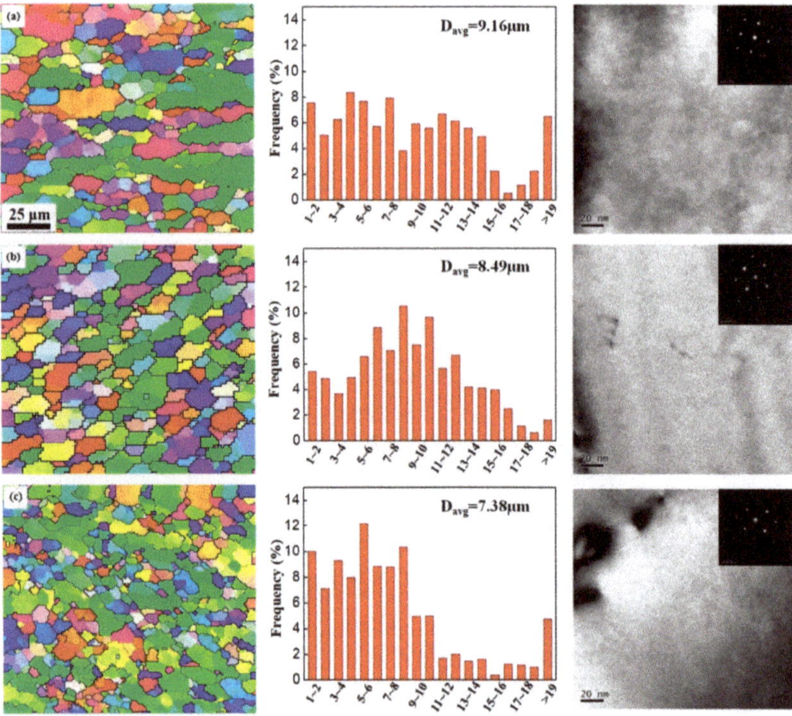

Figure 5. The grain map (on the left), grain size distribution (in the middle), TEM and diffraction pattern (on the right) of the NZ on L1 (**a**), NZ on L2 (**b**) and NZ on L3 (**c**) in the single-sided FSW.

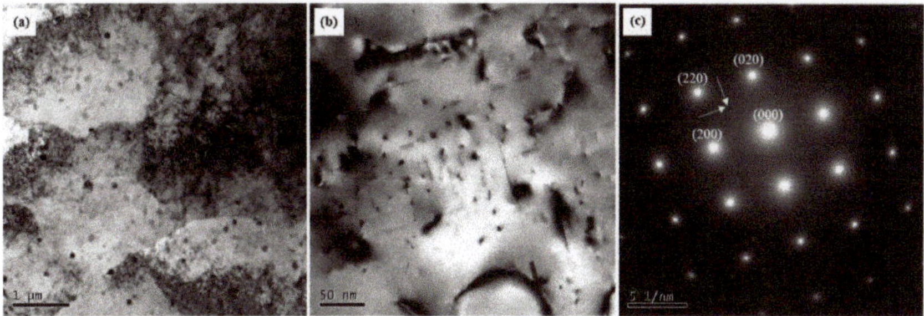

Figure 6. The microstructure of the base metal (BM) observed by SEM (**a**), TEM (**b**) and the diffraction pattern (**c**).

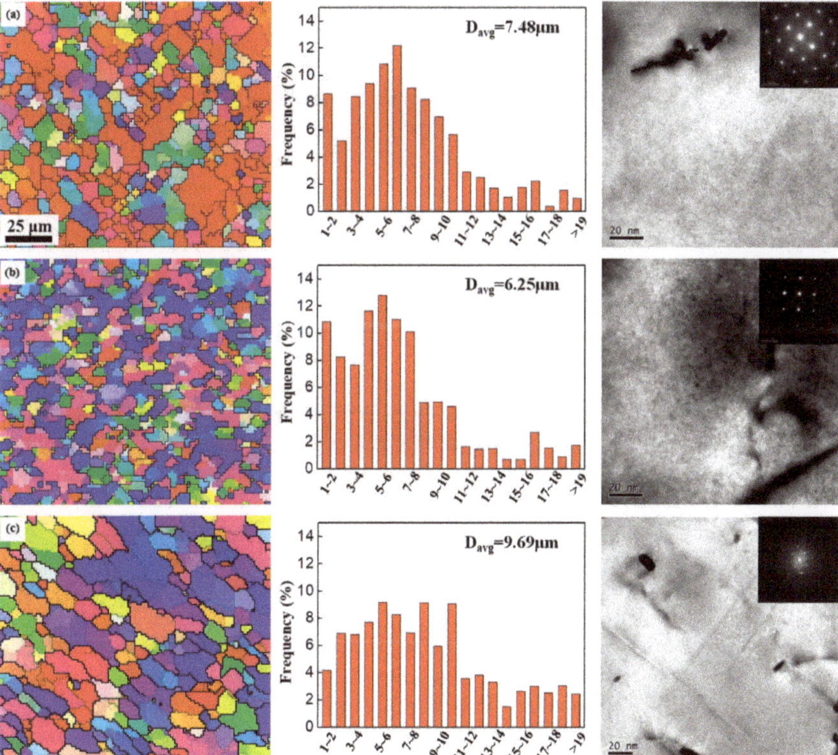

Figure 7. The grain map (on the left), grain size distribution (in the middle), TEM and diffraction pattern (on the right) of the NZ on L1 (**a**), NZ on L2 (**b**) and NZ on L3 (**c**) in the double-sided FSW.

The precipitations were dissolved in the second-pass NZ, while the precipitations were observed in the first-pass NZ. As the parameters were the same for the first and second pass, the authors inferred that the precipitations in the NZ were dissolved during the process of dynamic recrystallization in the first weld pass. The heat input by the second weld pass provided the aging temperature that the strengthening phases were precipitated in the first-pass NZ with grain growth occurring at the same time.

3.1.3. Comparison of Grain Size between Single-Sided and Double-Sided FSWs

The grain size of the NZ in the single-sided FSW is larger than the second pass of the double-sided FSW. The temperature of the NZ in the single-sided FSW is 7.9 °C higher than the double-sided FSW. The higher temperature is useful to activate the dynamic recrystallization and grain growth. This comparison verified the effect of the heat input on the recrystallized grain size.

3.2. Mechanical Properties

3.2.1. Hardness Distribution

Figure 8 shows the hardness profiles in the weld zone of single-sided and double-sided FSWs. The obtained profiles follow the general features with a central plateau and two valleys, which is a typical hardness behavior for all FSW heat treatable aluminium alloys [6].

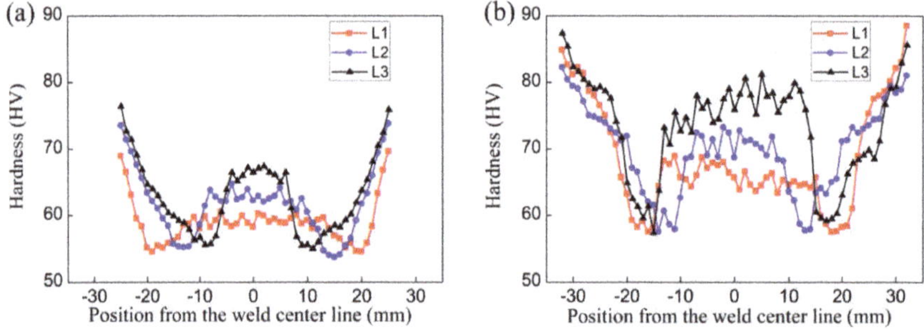

Figure 8. The hardness distribution in the single-sided FSW (**a**) and the double-sided FSW (**b**).

In the single-sided FSW, the minimum hardness on three lines is nearly the same with a value of approximately 54 HV and all appear in the HAZ. The valleys on L1 line, L2 line and L3 line are located approximately 20, 15 and 10 mm away from the weld center, respectively. The distance between two valleys decreases towards the bottom of the weld, which corresponds to the shape of the rotating pin. The average hardness of the NZ (central plateau) is 59.2, 62.8 and 66.4 HV on L1, L2, L3 lines, respectively. The hardness in the NZ increases in the order of the upper, middle and bottom regions. With the absence of precipitation in the NZ, fine grains usually benefit the strength [6]. Therefore, the hardness of NZ in the bottom region (grain size 7.38 μm) is 12.2% larger than that in the upper region (grain size 9.16 μm).

In the double-sided FSW, the hardness is generally larger than the single-sided FSW. The minimum hardness on three lines is nearly the same with approximately 58 HV and all appear in the HAZ. The valleys on L1 line, L2 line and L3 line are located approximately 18, 13 and 17 mm away from the weld center, respectively. The average hardness of the NZ (central plateau) is 66.0, 71.1 and 76.9 HV on L1, L2 and L3 lines respectively. The hardness of the middle NZ is 7.7% larger than that of the second-pass NZ. This is caused by the grain size in the middle NZ with 6.25 μm and in the second-pass NZ with 7.48 μm. The hardness of the first-pass NZ is approximately 16.5% larger than the second-pass NZ. The second-pass NZ has no precipitation and smaller grains, while the first-pass NZ has a small quantity of precipitations and larger grains. It is concluded that the strengthening precipitations overwhelm larger grain size in controlling the hardness.

3.2.2. Tensile Strength

The tensile strength of a single-sided FSW and a double-sided FSW was 186 ± 1 MPa and 202 ± 2 MPa, equivalent to 74.4% and 80.8% of BM (250 MPa), respectively. The larger tensile strength coincides with the larger hardness of the double-sided FSW. Figure 9 shows the fracture location of the single-sided FSW and the double-sided FSW. In the single-sided FSW, the fracture occurred in the HAZ of the advancing side. In the double-sided FSW, the crack initiated in the HAZ of the advancing side of the second pass and propagated to the HAZ of the retreating side of the first pass.

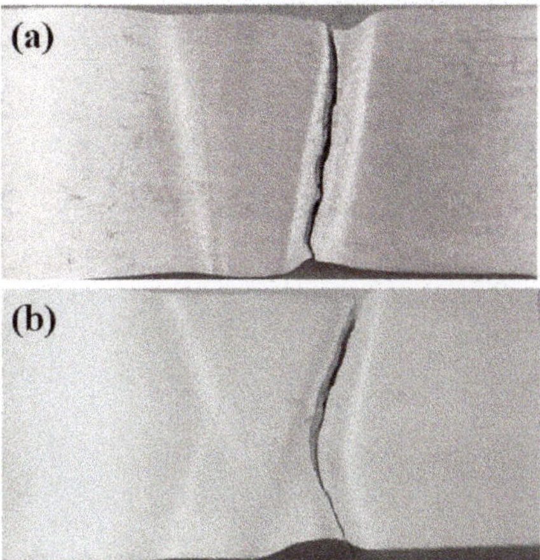

Figure 9. The fracture location of the single-sided FSW (**a**) and the double-sided FSW (**b**).

3.3. Residual Stresses

Figure 10 shows the residual stresses determined by neutron diffraction in the 6005A-T6 aluminium alloy single-sided and double-sided FSWs. In both specimens, the longitudinal and transverse residual stress distributions agreed with the typical 'M' shape pattern of FSW aluminum alloys [6]. The normal residual stresses were observed with little variations, especially on L3 line of the single-sided FSW which remained near to a value of zero. Notably, the average uncertainty of the obtained stresses is approximately ±13 MPa in these measurements.

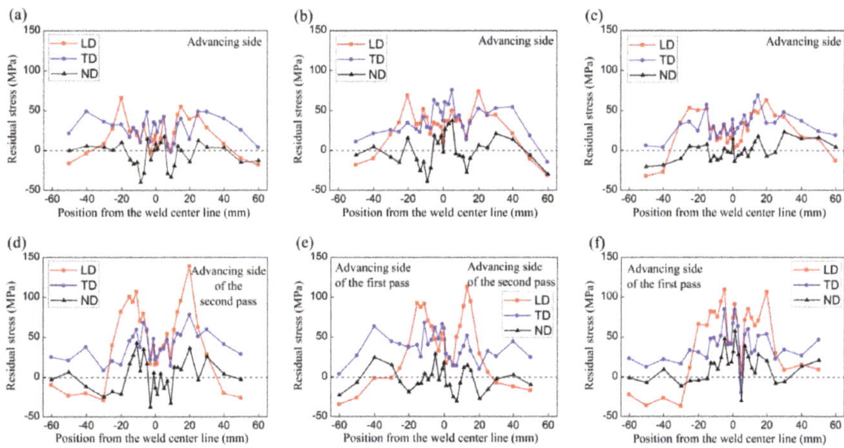

Figure 10. Residual stresses on L1 (**a**), L2 (**b**), L3 (**c**) in the single-sided FSW and L1 (**d**), L2 (**e**), L3 (**f**) in the double-sided FSW.

3.3.1. Features in the Single-Sided FSW

In the single-sided FSW, the residual stresses along the three lines have a similar distribution in both trend as well as magnitude. σ_{LD} is generally equal to σ_{TD} and larger than σ_{ND}. The maximum tensile σ_{LD} and σ_{TD} are nearly the same, approximately 75 MPa. This stress value amounts to 37.5% of the yield strength of Al-6005A-T6 alloy, which is 200 MPa at room temperature.

The introduction of residual stresses is inevitable due to severe thermomechanical deformation by FSW. Therefore, it is important to control the amount of residual stress. Hassan et al. reported an optimum combination of rotational and travel speeds that gave the best mechanical performance [11]. This optimum condition shifted to a higher rotational speed when the travel speed was increased. High heat input associated with low traverse and high rotation speeds leads to more extensive softening in the weld region, resulting in an overall reduction in the magnitude of the longitudinal residual stress. It seems difficult to get both optimized mechanical properties and residual stresses. However, the present 6005A-T6 single-sided FSW achieved an acceptable condition for comprising the mechanical properties with a tensile strength of 74.4% of BM and residual stresses with peak magnitudes of approximately 37.5% yield strength of BM.

3.3.2. Features in the Double-Sided FSW

In the double-sided FSW, σ_{LD} is generally larger than σ_{ND} and σ_{TD}. There is large difference between σ_{LD} magnitudes on the three lines, as shown in Figure 11. The maximum tensile σ_{LD} is approximately 140 MPa on L1 line ($xt = -10$), 115 MPa on L2 line ($xt = 0$) and 110 MPa on L3 line ($xt = 10$) respectively. These stress values amount to 70%, 57.5% and 55% of the yield strength of BM. The residual stress profiles of the L1 line and L2 line show obvious asymmetric distributions. The highest stresses of those lines occur on the advancing side of the NZ compared to the retreating side. The maximum residual stress in the upper weld zone is 140 MPa, 27% larger than the bottom weld zone, where the stress is 110 MPa. As the parameters were the same for the first and second pass, an inference was made that the peak magnitude of residual stress produced by the first weld pass was approximately 140 MPa. The heat input by the second weld pass provided the post-treatment environment of stress relaxation for the first-pass weld zone.

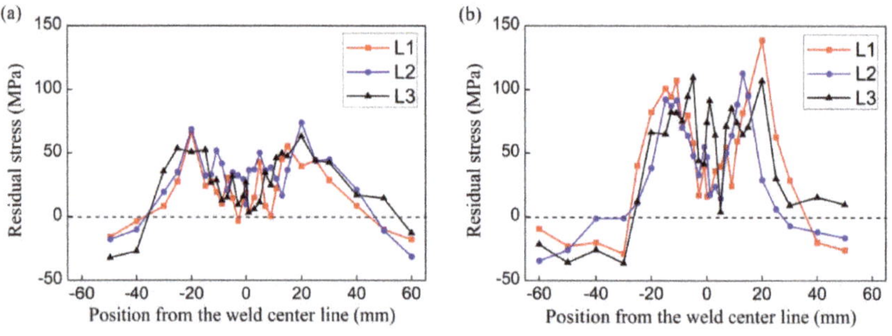

Figure 11. Replot of longitudinal residual stresses in a single-sided FSW (**a**) and a double-sided FSW (**b**).

Although the tensile strength of the double-sided FSW was a little higher than the single-side FSW, the longitudinal residual stresses in the double-sided FSW are much larger than the single-sided FSW, as shown in Figure 11.

4. Summary and Conclusions

In this study, residual stresses, microstructures and mechanical properties of the 6005A-T6 single-sided and double-sided FSWs were studied. The 3D residual stresses were characterized using neutron diffraction. The microstructures were observed by TEM and EBSD.

In the 6005A-T6 single-sided FSW, there were acceptable mechanical properties with tensile strength 74.4% of BM, and low residual stresses with peak magnitudes of approximately 37.5% yield strength of BM were achieved. This indicated a good quality of weld with such low residual stresses. However, good mechanical properties with a tensile strength of 80.8% of BM, but high residual stresses with peak magnitudes of approximately 70% yield strength of BM were obtained in the 6005A-T6 double-sided FSW.

In the 6005A-T6 single-sided FSW, there was no precipitation in the NZ. The hardness of the NZ is related to the grain size. The less hardness at the upper NZ and high hardness at the bottom NZ coincided with the larger grain size at the upper NZ and smaller grain size at the bottom NZ, respectively.

In the 6005A-T6 double-sided FSW, there were lower residual stresses and larger hardness in the first-pass weld zone, but higher residual stresses and smaller hardness in the second-pass weld zone. This was due to the heat input by the second weld pass which provided an aging environment for the first-pass weld zone, where the dissolved phases were precipitated and residual stresses were relaxed.

Author Contributions: Conceptualization, X.L. and H.Z.; investigation, P.X., R.W., R.L. and M.L.; Writing—Original Draft preparation, X.L.; Writing—Review and Editing, W.L., Y.L. and D.C.; funding acquisition, Y.L. and D.C.

Funding: This research activity was financially supported by Ministry of Science and Technology of China (grant 2017YFA0403704), National Natural Science Foundation of China (grants 11605293 and 51327902).

Conflicts of Interest: The authors declare no conflict of interest.

References

1. Simar, A.; Bréchet, Y.; De Meester, B.; Denquin, A.; Pardoen, T. Microstructure, local and global mechanical properties of friction stir welds in aluminium alloy 6005A-T6. *Mater. Sci. Eng. A* **2008**, *486*, 85–95. [CrossRef]
2. Ji, S.; Meng, X.; Liu, J.; Zhang, L.; Gao, S. Formation and mechanical properties of stationary shoulder friction stir welded 6005A-T6 aluminum alloy. *Mater. Des.* **2014**, *62*, 113–117. [CrossRef]
3. Lee, W.B.; Yeon, Y.M.; Jung, S.B. Evaluation of the microstructure and mechanical properties of friction stir welded 6005 aluminum alloy. *Mater. Sci. Technol.* **2003**, *19*, 1513–1518. [CrossRef]
4. Peng, D.; Hongmei, L.; Daqian, S.; Wenbiao, G.; Jie, L. Effects of welding speed on the microstructure and hardness in friction stir welding joints of 6005A-T6 aluminium alloy. *Mater. Des.* **2013**, *45*, 524–531.
5. Thomas, W.M.; Nicholas, E.D.; Needham, J.C.; Murch, M.G.; Temple-Smith, P.; Dawes, C.J. The Welding Institute, TWI. International Patent Application No. PCT/GB92/02203; GB Patent Application No. 9125978.8, 6 December 1991.
6. Threadgill, P.L.; Leonard, A.J.; Shercliff, H.R.; Withers, P.J. Friction stir welding of aluminium alloys. *Int. Mater. Rev.* **2009**, *54*, 49–93. [CrossRef]
7. Xu, W.; Liu, J.; Zhu, H. Analysis of residual stresses in thick aluminum friction welded butt joints. *Mater. Des.* **2011**, *32*, 2000–2005. [CrossRef]
8. Wang, X.L.; Feng, Z.; David, S.A.; Spooner, S.; Hubbard, C.R. Neutron diffraction study of residual stresses in friction stir welds. In Proceedings of the 6th International Conference on Residual stresses, Oxford, UK, 10–12 July 2000; pp. 1408–1414.
9. Haghshenas, M.; Gharghouri, M.A.; Bhakhri, V.; Klassen, R.J.; Gerlich, A.P. Assessing residual stresses in friction stir welding: Neutron diffraction and nanoindentation methods. *Int. J. Adv. Manuf. Technol.* **2017**, *93*, 3733–3747. [CrossRef]
10. Mehra, S.; Dhanda, P.; Khanna, R.; Goyat, N.S.; Verma, S. Effect of tool on tensile strength in single and double sided friction stir welding. *Int. J. Sci. Eng. Res.* **2012**, *3*, 1–6.
11. Hassan, K.A.; Prangnell, P.B.; Norman, A.F.; Price, D.A.; Williams, S.W. Effect of welding parameters on nugget zone microstructure and properties in high strength aluminium alloy friction stir welds. *Sci. Technol. Weld. Join.* **2003**, *8*, 257–268. [CrossRef]
12. Meng, X.; Gao, S.; Ma, L.; Li, Z.; Yue, Y.; Xiao, H. Effects of rotational velocity on microstructures and mechanical properties of surface compensation friction stir welded 6005A-T6 aluminum alloy. *Eng. Rev.* **2016**, *36*, 107–113.

13. Sakala, R.S.; Renangi, S.; Indira, R.M. Experimental study of double sided friction stir welding of AA 6061 plates using hexagonal tool tip. *Int. J. Res. Advent Technol.* **2018**, *6*, 32–37.

14. Asgar, K.; Abdul, S.M.K.; Bharat, K. Fabrication of a butt joint using friction stir welding (FSW)—A non consumable tool to generate heat. *Int. J. Res. Sci.Innov.* **2016**, *3*, 53–55.

15. Woo, W.; Feng, Z.; Wang, X.L.; Brown, D.W.; Clausen, B.; An, K.; Choo, H.; Hubbard, C.R.; David, S.A. In situ neutron diffraction measurements of temperature and stresses during friction stir welding of 6061-T6 aluminium alloy. *Sci. Technol. Weld. Join.* **2007**, *12*, 298–303. [CrossRef]

16. Simar, A.; Bréchet, Y.; De Meester, B.; Denquin, A.; Gallais, C.; Pardoen, T. Integrated modeling of friction stir welding of 6xxx series Al alloys: Process, microstructure and properties. *Prog. Mater. Sci.* **2012**, *57*, 95–183. [CrossRef]

17. Liu, H.; Yang, S.; Xie, C.; Zhang, Q.; Cao, Y. Microstructure characterization and mechanism of fatigue crack initiation near pores for 6005A CMT welded joint. *Mater. Sci. Eng. A* **2017**, *707*, 22–29. [CrossRef]

18. Dong, P.; Sun, D.; Wang, B.; Zhang, Y.; Li, H. Microstructure, microhardness and corrosion susceptibility of friction stir welded AlMgSiCu alloy. *Mater. Des.* **2013**, *54*, 760–765. [CrossRef]

19. Feng, Z.; Wang, X.L.; David, S.A.; Sklad, P.S. Modeling of residual stresses and property distributions in friction stir welds of aluminum alloy 6061-T6. *Sci. Technol. Weld. Join.* **2013**, *12*, 348–356. [CrossRef]

20. Woo, W.; Feng, Z.; Wang, X.L.; David, S. Neutron diffraction measurements of residual stresses in friction stir welding: A review. *Sci. Technol. Weld. Join.* **2011**, *16*, 23–32. [CrossRef]

21. Malopheyev, S.; Vysotskiy, I.; Kulitskiy, V.; Mironov, S.; Kaibyshev, R. Optimization of processing-microstructure-properties relationship in friction-stir welded 6061-T6 aluminum alloy. *Mater. Sci. Eng. A* **2016**, *662*, 136–143. [CrossRef]

22. Deplus, K.; Simar, A.; Van Haver, W.; De Meester, B. Residual stresses in aluminium alloy friction stir welds. *Int. J. Adv. Manuf. Technol.* **2011**, *56*, 493–504. [CrossRef]

23. Simar, A.; Bréchet, Y.; De Meester, B.; Denquin, A.; Pardoen, T. Sequential modeling of local precipitation, strength and strain hardening in friction stir welds of an aluminum alloy 6005A-T6. *Acta Mater.* **2007**, *55*, 6133–6143. [CrossRef]

24. Tao, W.; Yong, Z.; Xuemei, L.; Matsuda, K. Special grain boundaries in the nugget zone of friction stir welded AA6061-T6 under various welding parameters. *Mater. Sci. Eng. A* **2016**, *671*, 7–16. [CrossRef]

25. Węglowski, M.S.; Sędek, P.; Hamilton, C. The effect of process parameters on residual stress in a Friction stir processed cast aluminium alloy. *Eng. Trans.* **2016**, *64*, 301–309.

26. Prime, M.; Gnaupel-Herold, T.; Baumann, J.; Lederich, R.; Bowden, D.; Sebring, R. Residual stress measurements in a thick, dissimilar aluminum alloy friction stir weld. *Acta Mater.* **2006**, *54*, 4013–4021. [CrossRef]

27. Sepe, R.; Armentani, E.; Di Lascio, P.; Citarella, R. Crack Growth Behavior of Welded Stiffened Panel. *Procedia Eng.* **2015**, *109*, 473–483. [CrossRef]

28. Citarella, R.; Carlone, P.; Lepore, M.A.; Sepe, R. Hybrid technique to assess the fatigue performance of multiple cracked FSW joints. *Eng. Fract. Mech.* **2016**, *162*, 38–50. [CrossRef]

29. Citarella, R.; Carlone, P.; Sepe, R.; Lepore, M.A. DBEM crack propagation in friction stir welded aluminum joints. *Adv. Eng. Softw.* **2016**, *101*, 50–59. [CrossRef]

30. Patel, V.; Li, W.; Wang, G.; Wang, F.; Vairis, A.; Niu, P. Friction Stir Welding of Dissimilar Aluminum Alloy Combinations: State-of-the-Art. *Metals* **2019**, *9*, 270. [CrossRef]

31. Sonne, M.R.; Carlone, P.; Hattel, J.H. Assessment of the Contour Method for 2-D Cross Sectional Residual Stress Measurements of Friction Stir Welded Parts of AA2024-T3—Numerical and Experimental Comparison. *Metals* **2017**, *7*, 508. [CrossRef]

32. Fadaeifard, F.; Matori, K.A.; Aziz, S.A.; Zolkarnain, L.; Rahim, M.A.Z.B.A.; Rahim, M.A. Effect of the Welding Speed on the Macrostructure, Microstructure and Mechanical Properties of AA6061-T6 Friction Stir Butt Welds. *Metals* **2017**, *7*, 48. [CrossRef]

33. Li, S.; Chen, Y.; Kang, J.; Amirkhiz, B.S.; Nadeau, F. Effect of Revolutionary Pitch on Interface Microstructure and Mechanical Behavior of Friction Stir Lap Welds of AA6082-T6 to Galvanized DP800. *Metals* **2018**, *8*, 925. [CrossRef]

34. Niu, P.; Li, W.; Zhang, Z.; Yang, X. Global and local constitutive behaviors of friction stir welded AA2024 joints. *J. Mater. Sci. Technol.* **2017**, *33*, 987–990. [CrossRef]

35. Hutchings, M.T.; Withers, P.J.; Holden, T.M.; Lorentzen, T. *Introduction to the Characterization of Residual Stress by Neutron Diffraction*, 1st ed.; Taylor and Francis: London, UK, 2005.

36. Liu, X.; Wimpory, R.C.; Gong, H.; Liu, Y.; Chen, D.; Liu, Y.; Wu, Y.; Li, C. The Determination of Residual Stress in Quenched and Cold-Compressed 7050 Aluminum Alloy T-Section Forgings by the Contour Method and Neutron Diffraction. *J. Mater. Eng. Perform.* **2018**, *27*, 6049–6057. [CrossRef]

37. Wimpory, R.; Mikula, P.; Šaroun, J.; Poeste, T.; Li, J.; Hofmann, M.; Schneider, R. Efficiency Boost of the Materials Science Diffractometer E3 at BENSC: One Order of Magnitude Due to a Horizontally and Vertically Focusing Monochromator. *Neutron News* **2008**, *19*, 16–19. [CrossRef]

38. Mishra, R.; Ma, Z.; Mishra, R. Friction stir welding and processing. *Mater. Sci. Eng. R Rep.* **2005**, *50*, 1–78. [CrossRef]

39. Qian, J.W.; Li, J.L.; Xiong, J.T.; Zhang, F.S.; Li, W.Y.; Lin, X. Periodic variation of the torque and its relations to interfacial sticking and slipping during friction stir welding. *Sci. Technol. Weld. Join.* **2012**, *17*, 338–341. [CrossRef]

40. Randau, C.; Garbe, U.; Brokmeier, H.G. StressTextureCalculator: A software tool to extract texture, strain and microstructure information from area-detector measurements. *J. Appl. Crystallogr.* **2011**, *44*, 641–646. [CrossRef]

41. Gnäupel-Herold, T. ISODEC: Software for calculating diffraction elastic constants. *J. Appl. Crystallogr.* **2012**, *45*, 573–574. [CrossRef]

Article

Interface Behavior and Impact Properties of Dissimilar Al/Steel Keyhole-Free FSSW Joints

Zhongke Zhang [1], Yang Yu [2,*], Huaxia Zhao [3] and Xijing Wang [1]

[1] School of Materials Science and Engineering, Lanzhou University of Technology, Lanzhou 730050, China; zhangzke@lut.cn (Z.Z.); wangxj@lut.cn (X.W.)
[2] School of Materials Science and Engineering, Tongji University, Shanghai 201804, China
[3] Aeronautical Key Laboratory for Welding and Joining Technologies, AVIC Manufacturing Technology Institute AECC, Beijing 100024, China; zhhq@sina.com
* Correspondence: y.yu@tongji.edu.cn or yuyangmr@163.com; Tel.: +86-021-6958-1508

Received: 28 April 2019; Accepted: 13 June 2019; Published: 18 June 2019

Abstract: This work systematically investigates the interface behavior and impact properties of the keyhole-free friction stir spot welding (FSSW) of a dissimilar metal AA6082-T4 Al alloy and DP600 galvanized steel. The keyhole is eliminated by pin retraction technology. The welding process is in accordance with the welding temperature curve and the maximum temperature of the periphery of the shoulder, measured at about 500 °C. The transition layers were formed at the interface, in which the Al, Fe, and Zn elements form an inhomogeneous diffusion. A cloud cluster-like mechanical mixing of the Al and steel components is formed in the stirring zone. The impact toughness of the specimen with a welding parameter of 1000 rpm is the best. To a certain extent, the factors affecting the impact energy are not the maximum impact load but the maximum impact deformation. The maximum impact deformation directly reflects the post-crack propagation energy, which significantly affects its impact toughness. In addition, the impact fracture showed a mixed ductile and brittle fracture mode with a brittle–ductile transition zone. Most of the impact energy was absorbed by the ductile fracture.

Keywords: FSSW; dissimilar metals; interface behavior; impact properties

---

1. Introduction

Friction stir spot welding (FSSW) is a type of solid-state welding process that combines friction stir welding (FSW) and spot welding. FSSW is widely used for the connection of similar Al alloys [1] and Mg alloys [2] and dissimilar Al/steel [3], Al/Cu [4], Al/Mg [5] and Mg/steel [6] materials in the transportation and aerospace industries. In order to realize lightweight automobiles, Al and Mg alloys are widely used as substitutes for steel, which can effectively reduce the overall weight of the automobile body, and achieve energy conservation and environmental protection. However, the melting points of Al and steel are quite different, which makes them difficult to weld.

Compared with solid state welding, it is easier to produce gas pockets and intermetallic compounds (IMCs) in dissimilar Al/steel joints during fusion welding, which results in a decline in the mechanical properties of joints. Similarly, there are some inevitable defects in FSW and FSSW, such as keyholes [7], hook defects [8–10], and IMCs [11–13]. These defects seriously affect the forming quality, tensile-shear strength, impact toughness, and fatigue life of welded joints. Therefore, a deeper understanding of dissimilar Al/steel FSSW is necessary. In order to overcome the above welding defects, keyhole-free FSSW techniques have been developed based on traditional FSSW technology. The remaining keyhole is the weakest area of the weld, and the joint first form cracks at the edge of the keyhole. Kubit et al. [7] found that keyholes and alclads are the main weld defects that worsen the joint quality in the FSSW process. Refilled friction stir spot welding (RFSSW) can fill the keyhole to avoid structural defects. In addition, FSSW joints can be divided into four zones: the weld nugget zone

(WNZ), thermo-mechanically affected zone (TMAZ), heat-affected zone (HAZ), and base metal (BM). Reimann et al. [14] found partial recrystallization in the stir zone for the first time using the stop-action technique. Chen [10] and Cao [15] et al. observed fine grains and shear texture in the stir zone and proved that the microstructure's evolution was related to recrystallization in the RFSSW of the 6061-T6 alloy.

The hook defect is formed by the plunge of the tool pin, causing a large extrusion deformation of the material during FSW. Garg and Bhattacharya [9] found that hook formation is related to the length of the tool pin. A pinless tool can obscure hook formation and improve joint strength. Chen et al. [10] investigated the microstructure and mechanical properties of the RFSSW of a 6061 Al alloy to transformation-induced plasticity (TRIP) steel. They proved that RFSSW increases the joint's strength by 56.33% compared with that of conventional FSSW joints. In contrast, they thought that the hook structure generated in the regular FSSW step is indispensable for a strong joint.

IMCs, as brittle phases, are also one of the main causes of joint cracking. Hsieh et al. [11] obtained dissimilar lap joints for a low carbon steel (SS400) sheet on a 6061-T6 Al alloy sheet, which were achieved by FSSW with a welding tool that had an independent tool pin and shoulder. They found that two IMC layers, Fe_2Al_5 and Fe_4Al_{13}, were formed at this surface, and the failure load was related to the thickness of the IMC layer. Bozzi et al. [12] also studied the IMCs of joints of 6016 Al alloy to IF-steel performed by FSSW and reached the same conclusion. They noted, at the same time, that the presence of IMCs depends on the welding conditions. Dong et al. [13] studied dissimilar lap joints of a Novelist AC 170 PX Al alloy and 1.2 mm thick ST06 Z galvanized steel sheets with RFSSW. They found that the IMC layer of ZnO, which was as thin as 0.68 μm in the sleeve-plunging zone, was the weakest part of the structure. Although the diameter of the welded spot was 9 mm, the maximum tensile/shear fracture load was only 4.5 kN. They also predicted the material flow during the Refilled FSSW/sleeve plunging process by the distribution of stirred zinc coating.

A large number of numerical simulation analyses of FSSW have been carried out. The temperature field, the stress and strain fields, the welding force, the residual stress, the formability, the material flow, and grain size of welding have become the focus of numerical simulations of FSSW [16–24]. At the same time, many articles have reported the shear-tensile strength and fatigue properties of FSSW joints [25–28]. However, there are few publications that study the impact properties of spot-welding structures, especially keyhole-free FSSW joints. Several studies on impact toughness mainly focused on the friction stir butt-welded joints of thick plates [29–31]. In this work, the microstructure and interface behavior of dissimilar Al/steel keyhole-free FSSW joints are studied and analyzed in combination with the keyhole-free FSSW process. In order to determine the relationship between the microstructure and the mechanical properties of the keyhole-free FSSW joints, an impact test was carried out on the dissimilar Al/steel keyhole-free FSSW joint.

2. Experimental Procedures

2.1. Materials and Fabrication Process

AA6082-T4 Al alloy plates and DP600 galvanized steel plates were used in the experiment. Their dimensions were 150 mm × 50 mm × 2 mm and 150 mm × 50 mm × 1 mm, respectively. Tables 1 and 2 show the chemical composition of the 6082 Al alloy and DP600 galvanized steel.

Table 1. The chemical composition of the 6082 Al alloy (in wt%).

6082	Si	Fe	Cu	Mn	Mg	Cr	Zn	Ti	Other	Al
Content	0.7–1.3	0.50	0.10	0.4–1.0	0.6–1.2	0.25	0.20	0.10	0.15	Bal.

Table 2. The chemical composition of DP600 galvanized steel (in wt%).

DP600	C	Mn	Si	Al	Mo	Cr	Cu	S	P	Fe
Content	0.09	1.84	0.36	0.05	0.01	0.02	0.03	0.005	0.005	Bal.

The joints of the dissimilar 6082 Al alloy and the DP600 galvanized steel were welded by a retractile keyhole-free FSSW machine. The welding tool was made of a nickel-based super alloy with a tool shoulder of 20 mm in diameter and a tool pin of 5 mm in diameter and 2.1 mm in length. The pin can move up and down and rotate freely in the inner hole of the shoulder [32]. The spot-welded joints adopt the overlap form of positioning the steel plate on Al plate. This is different from another form of Al plate on the top of the joint [33], although in terms of the cost of the welding tool, it can improve the life of the welding tool. The overlapping form of steel on top can make the tool pin penetrate the steel plate and mix the steel plate and the aluminum plate thoroughly. At the same time, it increases the welding temperature and the thermoplastic deformation of the steel side, which can improve the quality of the dissimilar Al/steel spot-welding joint. The welding process of the dissimilar Al/steel retractile keyhole-free FSSW is shown in Figure 1. In the welding process, the pin and shoulder rotated simultaneously, and the rotational speeds of the welding tool ω_r were 800 rpm, 1000 rpm, and 1200 rpm, respectively. When the rotational speed of the welding tool reaches a steady state, it begins to plunge, and the plunge speed v_p was 5 mm/min (in Figure 1a). The shoulder plunge depth d_s was 0.3 mm after the tool shoulder reached the surface of the workpiece (in Figure 1b) [33,34]. Then, the welding tool was moved forward at a speed v_f of 3 mm/min, and the tool pin moved upward at a speed v_u of 5 mm/min (in Figure 1c). When the pin was lifted by 2.1 mm relative to the shoulder, the bottom of the pin was level with the bottom of the shoulder. At that point, the pin stopped moving upward, and the shoulder stopped moving forward. This is equivalent to the tool pin moving forward approximately 1.5 mm. With the stirring, retracting, and advancing of the stirring pin, the thermoplastic metal near the stirring pin could gradually fill the rear keyhole under the action of the shoulder. Then, the welding tool moved upward (in Figure 1d). When the welding tool was moved to a safe distance, the tool shoulder and tool pin simultaneously reset and the keyhole-free FSSW joint was obtained (in Figure 1e). At the same time, the thermocouples are used to measure the temperature of dissimilar Al/steel joints during keyhole-free FSSW. Figure 2 shows the schematic diagram of thermocouple distribution around the dissimilar Al/steel joint during keyhole-free FSSW. The welded structure adopts the overlap form, and the rotating direction is shown in Figure 2. Four thermocouples are evenly placed around the spot-welded joints on the steel plate, and the welding temperature is measured in real time through the four channels.

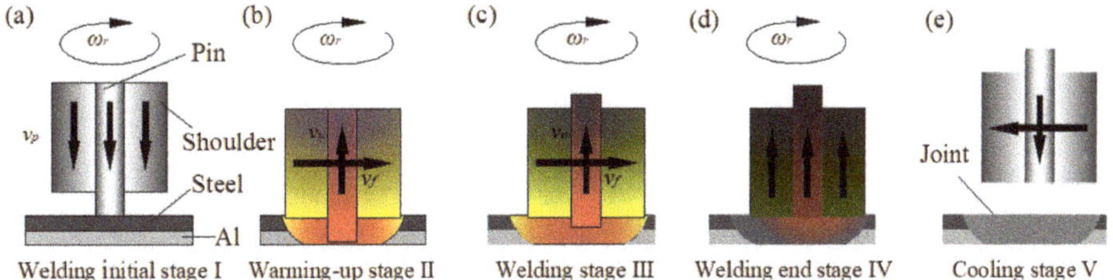

Figure 1. The welding process of the dissimilar Al/steel retractile keyhole-free friction stir spot welding (FSSW), (**a**) welding initial stage I, (**b**) warming-up stage II, (**c**) welding stage III, (**d**) welding end stage IV, (**e**) cooling stage V.

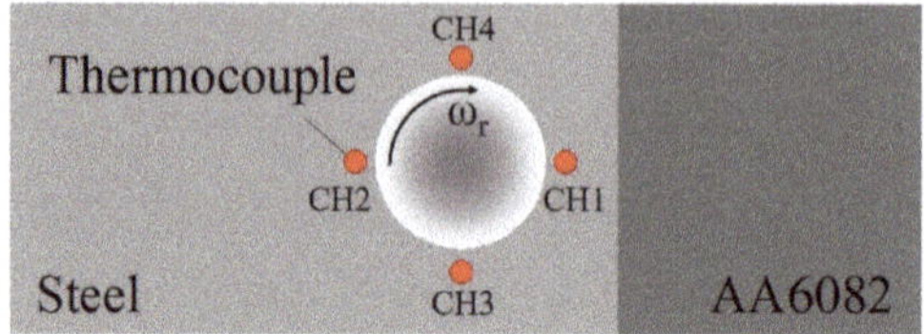

Figure 2. Schematic diagram of thermocouple distribution around the dissimilar Al/steel keyhole-free FSSW joint.

2.2. Microstructure Characterization

The cross-sectional specimens of the keyhole-free FSSW joints, with dimensions of 40 mm × 3 mm × 15 mm, were cut off by wire cutting. They were used for microscopic metallographic and interface analyses. The cross-sectional specimens were ground with sandpaper, polished with a polishing cloth, and then cleaned with anhydrous alcohol. The Al layer of the cross-sectional specimens was etched with Keller's etchant (2.5 mL HNO_3, 1.5 mL HCl, 1.0 mL HF, and 95 mL distilled water) for 20 s, and the steel layer was etched with a 4% nitric acid alcohol etchant for 4 s.

The microstructure of the keyhole-free FSSW joint was observed using an MeF3 large metallographic microscope (Leica Corporation, Wetzlar, Germany). The interfacial behavior, metal fluidity, and impact fracture of the keyhole-free FSSW joints were studied on a JSM-5600LV low-vacuum scanning electron microscope (SEM) (Japan Electron Optics Laboratory Company, Tokyo, Japan), and the interface element diffusion was measured with an X-ray energy-dispersive spectrometer (EDS) (Oxford Instruments, Oxford, UK). The IMCs of the keyhole-free FSSW joint were analyzed by X-ray diffractometry (XRD) (BRUKER-AXS Corporation, Billerica, Mass, Germany).

2.3. Impact Tests

The cantilever method was used to carry out impact experiments on the dissimilar Al/steel keyhole-free FSSW joints with different welding parameters. There were three specimens in each group of experiments. The cantilever impact test was performed on a CIEM-30D-CPC modified oscillographic impact testing machine (Tokyo Testing Machine MFG. CO. LTD, Tokyo, Japan). Figure 3 shows the cantilever impact schematic of the dissimilar Al/steel keyhole-free FSSW joints. One end of the sample was fixed on the impact fixture, and the other end was connected to an extension body with a long screw and nut, as shown in Figure 3. The extension body is separated from the impact fixture and can move freely. Aluminum and steel filler pieces of the same thickness and material are installed on both sides of the sample to keep the direction of force unchanged. The falling pendulum hammer hit the extension body horizontally, resulting in the fracture of the keyhole-free FSSW joint. The impact energy was calculated by the start and end angles before and after the pendulum impact on the dial. The oscilloscope recorded and displayed the impact load-displacement curve.

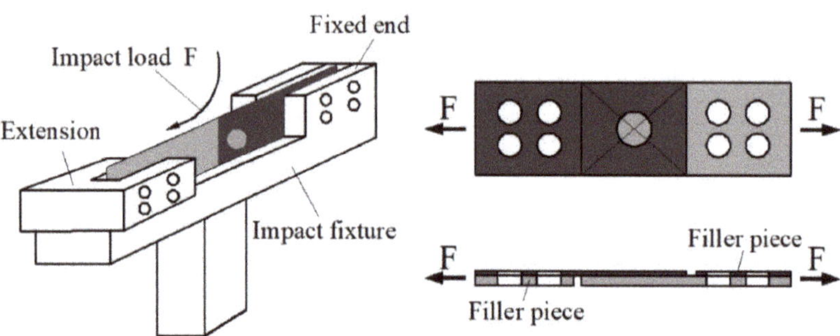

Figure 3. Cantilever impact schematic of the dissimilar Al/steel keyhole-free FSSW joints.

3. Results and Discussion

3.1. Microstructure and Interface Behavior

3.1.1. Welding Temperature Curve

Figure 4 shows the real-time temperature–time curves and histogram of the maximum welding temperature of the dissimilar Al/steel keyhole-free FSSW process. It can be seen that the temperature variation curves of the four thermocouple channels around the spot-welded joints are slightly different, but the curves are identical in shape, which shows that the temperature variation is in good agreement with the welding stages. According to the temperature curve, the welding process can be divided into five stages as follows, as shown in Figure 4a.

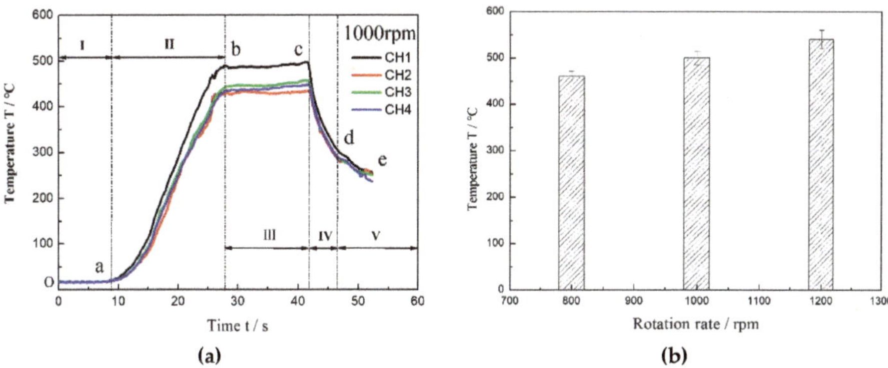

Figure 4. The real-time temperature–time curves (**a**) and histogram of maximum welding temperature (**b**) of dissimilar Al/steel keyhole-free FSSW process.

The welding initial stage (I): The point O is the starting point of welding, and the temperature of the sample is about 20 °C at room temperature. Starting from point O, the welding tool begins to rotate and plunge in stage Oa, but it is not in contact with the workpiece, so the sample temperature has not changed.

The warming-up stage (II): The temperature of the sample begins to rise rapidly in the ab stage. At this time, the tool pin begins to plunge into the workpiece, and the shoulder then contacts the surface of the steel plate. The temperature of the workpiece rises sharply due to friction heat generated by friction action. The higher heating rate $v_c = dT/dt$ indicates that the temperature rises very fast in this stage, and the maximum temperature reaches about 500 °C.

The welding stage (III): this stage, bc, is the stable stage of welding, and the temperature is still maintained at a maximum of about 500 °C and is almost constant. At this time, when the tool pin is retracted back, the contact between the shoulder and the workpiece is still stable. The friction heat is almost invariant, and the welding temperature is maintained near 500 °C.

The welding end stage (IV): The welding tool is lifted in stage cd, and the shoulder begins to detach from the workpiece surface. The workpiece temperature drops rapidly. The cooling rate $v_d = dT/dt$ is almost the same as the heating rate v_c. This indicates that the friction heat between the shoulder and the workpiece surface is an important factor of the temperature rise in the FSSW welding process.

The cooling stage (V): This stage, de, is the cooling period of the specimen. At this stage, the FSSW welding process has ended and the workpiece temperature begins to decrease naturally. However, the cooling mode is mainly the heat conduction between the workpiece and the worktable and the heat convection and radiation between the specimen and the air. Therefore, the cooling rate is relatively slow, and the time is relatively long.

With the agitation of the mixing head and the increase of the welding temperature, the top steel plate and the bottom aluminum plate begin to deform and gradually soften under the action of the external force and friction heat. When the peak temperature of the workpiece on the outer side of the shoulder in the stable stage of welding reaches 500 °C, the temperature inside the joint will be higher during the dissimilar Al/steel FSSW process. At this time, the welding temperature is close to the melting point $T_m = 660$ °C of the Al alloy, which makes the Al alloy completely reach the thermoplastic state. With the stirring of the tool pin at such a high temperature, a plastic mixing of Al and steel components can be formed, which will lead to element diffusion, transition layers, and new phases. However, this mixing is only in the stirring area, as it is not mixed sufficiently in the action area of the shoulder. Instead, the bonding between metals is formed under the action of the shoulder extrusion force, similar to the rolling or extrusion process. At the same time, as can be found in Figure 4b, the maximum welding temperature increases gradually with the increase of rotational speed. Although a high welding temperature can increase the mixing and diffusion of the dissimilar Al/steel joint, too high a temperature can also lead to defects, such as excessive transition layer thickness and brittle phase formation.

3.1.2. Microstructure

Figure 5 shows the sample of the keyhole-free FSSW joint of the dissimilar 6082 Al alloy and DP600 galvanized steel. It can be found that the keyhole was eliminated by the pin retraction technology in the welding process. Figure 6 shows the macroscopic morphology of the cross-section of the dissimilar Al/steel keyhole-free FSSW joint. The cross-sectional specimens can be divided into four zones: WNZ, TMAZ, HAZ, and BM. There is no keyhole on the cross-sectional specimens. A continuous inhomogeneous interface was formed at the Al/steel interface due to the effect of the stirring of the pin and the extrusion and friction of the shoulder of the tool. The walking path of the pin is shown by the blue arrow. The thickness of the steel plate was thinned from 1 mm to 0.48 mm, as shown in the thin area. There is a gap at the edge of the interface of the keyhole-free FSSW joint, which was the weak link in the spot-welded joints.

Figure 5. Sample of a dissimilar Al/steel keyhole-free FSSW joint.

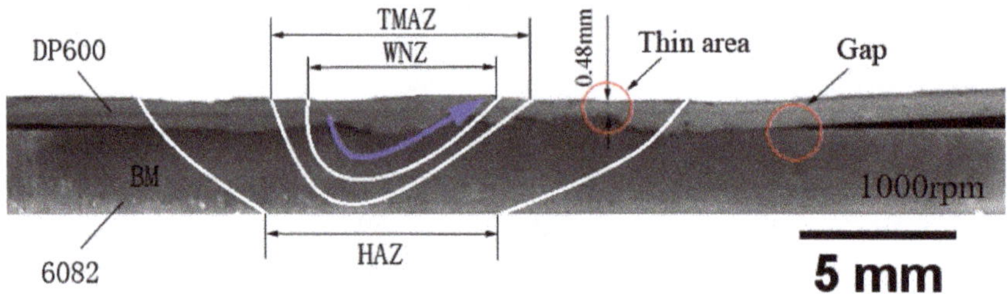

Figure 6. Macroscopic morphology of the cross-section of a dissimilar Al/steel keyhole-free FSSW joint.

Figures 7 and 8 show all regional microstructures of the steel and Al alloy in the keyhole-free FSSW joint, respectively. Figures 7a and 8a show the fine equiaxed grains of the WNZ, which formed due to the recrystallization caused by the large plastic deformation and the friction heat [35]. The grain size of the WNZ decreases from 20 µm of BM to 1 µm. Figures 7b and 8b show the elongated texture of the TMAZ, which was caused by large thermoplastic deformation [36]. Compared with the BM, the microstructure of the HAZ is slightly larger than that of the BM, as a result of the friction heat, as shown in Figure 7c,d and Figure 8c,d.

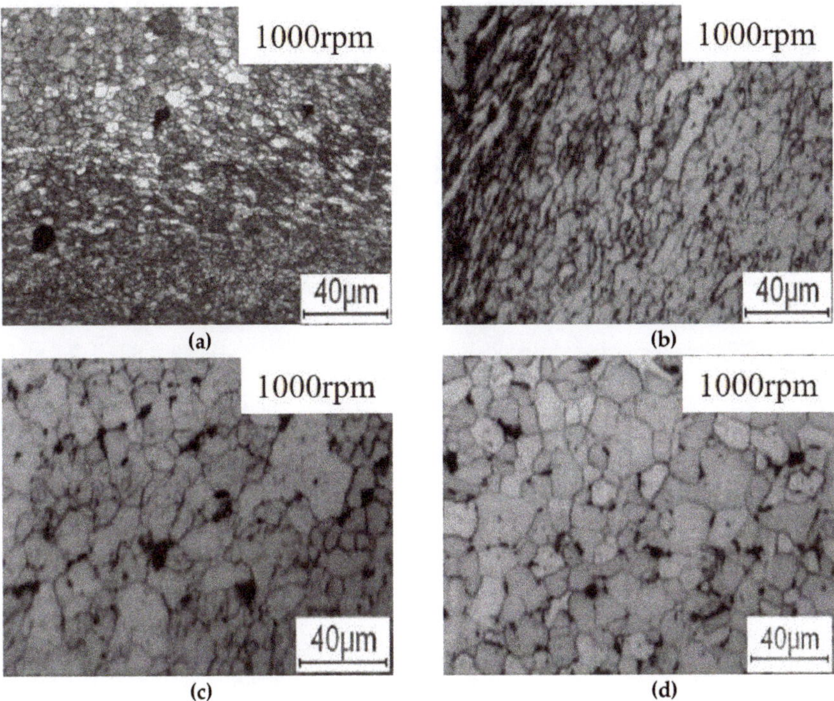

Figure 7. The microstructure of the steel of the keyhole-free FSSW joint: (**a**) weld nugget zone (WNZ), (**b**) thermo-mechanically affected zone (TMAZ), (**c**) heat-affected zone (HAZ), and (**d**) base metal (BM).

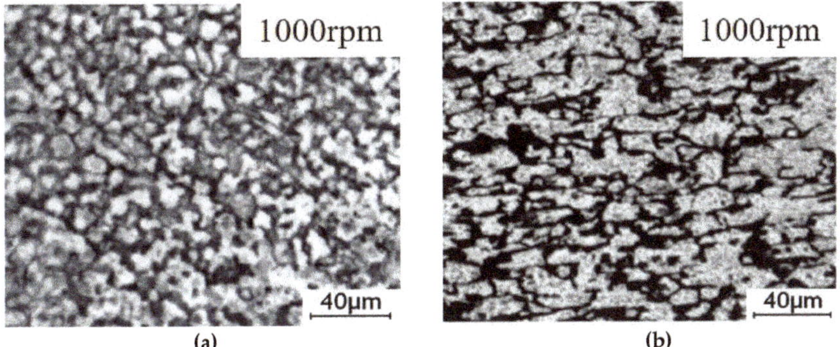

Figure 8. *Cont.*

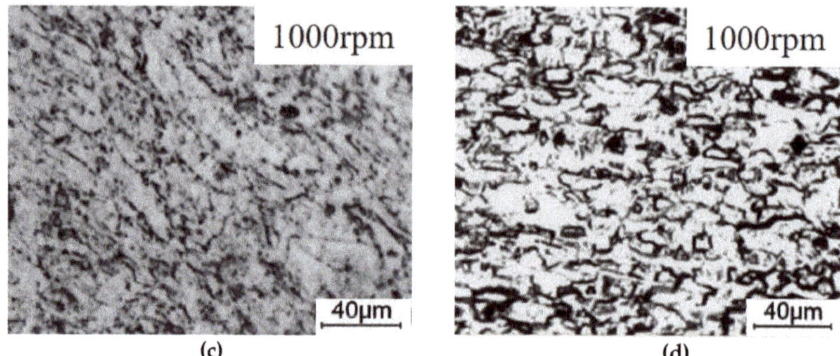

Figure 8. The microstructure of the Al alloy of the keyhole-free FSSW joint: (**a**) WNZ, (**b**) TMAZ, (**c**) HAZ, and (**d**) BM.

Figure 8d shows the microstructure of the 6082 Al alloy of BM zone, which has a typical basal texture. The grains of WNZ experienced a high temperature and strain rate under the combined action of the stirring pin and axial shoulder friction force, which caused the recrystallization of the grain and the redistribution of the enhancement phase and finally obtained the dynamic recrystallization grain, as shown in Figure 8a. The grain size of the WNZ decreases from 30 µm of BM to 5 µm. The width of the TMAZ is narrower, and the zone is partially plastic deformation due to the indirect agitation of the stirring pin. The temperature in the TMAZ is very close to that in the WNZ, so the grains of the recrystallization are shown in Figure 8b. The HAZ is only affected by the thermal cycle, so only the grain growth can be seen in Figure 8c.

Compared to Figures 7 and 8, the deformation of steel is more intense than that of Al. As shown in Figure 7b, the grains of steel at the TMAZ, located underneath the pin, exhibited greater elongation directly beneath the full penetration of the pin than that shown in Figure 8b. This phenomenon is caused by the difference of the compressive stirring force. The stirring force at the TMAZ in Figure 7b is higher than that in Figure 8b, so the grain deformation of steel is more serious than that of Al.

3.1.3. Interface Behavior

Figure 9 shows the micromorphology and EDS of the interface of the keyhole-free FSSW joint at the center and edge areas. In Figure 9a, the metal interlayer with discontinuous length and uneven thickness can be observed at the Al/steel interface. The metal interlayer is a transition layer composed of IMCs [11], which can be observed by local magnification of the metal interlayer region, as shown in the upper left corner in Figure 9a. The elements and their content were quantitatively analyzed by spot scanning the IMC layer, as shown in the upper right corner in Figure 9a. The IMC layer mainly contained Al, Fe, and Zn elements, and their contents were approximately 71.9%, 20.5%, and 7.1%, respectively. It can be concluded that the IMCs were mainly composed of Al, Fe, and Zn elements. There was an IMC layer at the interface at the edge of the keyhole-free FSSW joint, as shown in Figure 9b. It can be seen that the IMCs were not filled sufficiently at the interface gap, as shown in the lower right corner in Figure 8b. The interfacial bonding was poor and some of them became unbound [37,38]. The element diffusion was qualitatively analyzed by line scanning at the interface, as shown in the EDS at the interface in Figure 9b. It can be observed that the Fe, Al, and Mg elements linearly attenuated at the interface. However, the Zn element increased significantly at the interface. The zinc coating of DP600 galvanized steel participated in the interfacial reaction, resulting in the diffusion and increase in Zn in the interfacial layer [13].

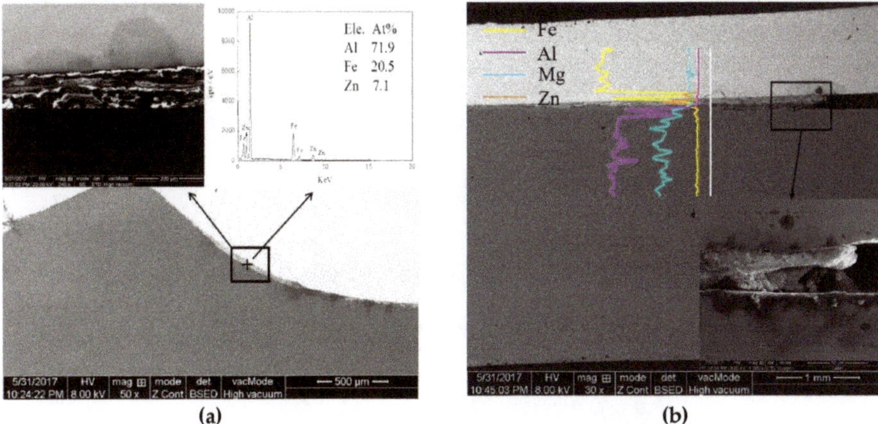

Figure 9. Micromorphology and energy-dispersive spectrometer (EDS) of the interface of the keyhole-free FSSW joint: (**a**) center area and (**b**) edge area.

The transition layer was further magnified and analyzed by EDS in Figure 10. Figure 10a shows the microstructure and EDS from the spot scans of the interface layer in the keyhole-free FSSW joint in the center area. The color of the interface transition layer is different from that of the matrix, and the thickness was approximately 15 μm [11]. The elements of the three atlases at points 1, 2, and 3 were quantitatively analyzed, as shown in the lower left corner of Figure 10a. The elements of the three spots in 1, 2, and 3 were mainly Al, Fe, and Zn. As the boundary of the interface became closer, the content of Zn increased, and the content of Fe decreased [13]. This indicates that the composition of the interface layer changed, and the transformed IMCs were also different [13,37–40]. Figure 10b shows the microstructure and EDS of the line-scanning analysis of the interface layer of the keyhole-free FSSW joint at the edge area. It can be found that the Fe, Al, Mg and Zn elements were staggered in the interface layer. This composition difference led to different IMCs, which had a non-uniform composition and content [13,37–40].

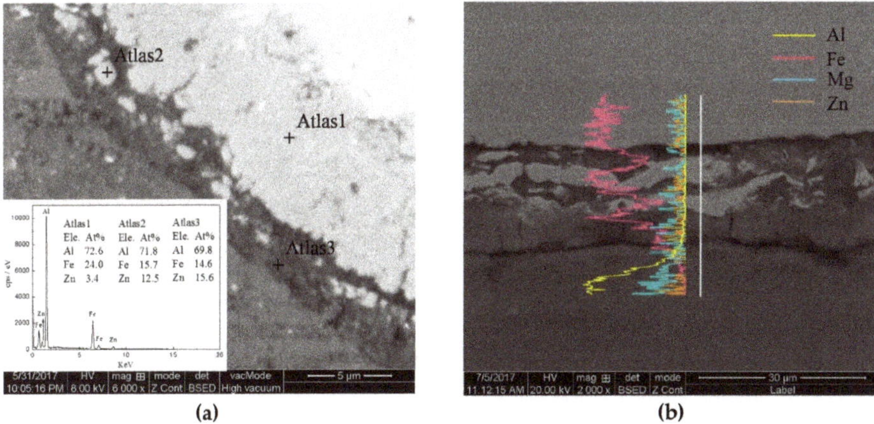

Figure 10. Micromorphology and EDS of the transition layer of the keyhole-free FSSW joint: (**a**) center area and (**b**) edge area.

Figure 11 shows the micromorphology and EDS of the Al/steel interface in WNZ. The cloud cluster-like chaotic microstructure at the Al/steel interface can be found in the stirring zone. The elements of Fe and Al are distributed alternately with streamlined lines [10]. Meanwhile, the elements of Mg, Cu, Si and Zn were also unevenly distributed. This indicates the stratified intersecting and inhomogeneous mixing of the Al alloy and steel plates [10,41]. The stirring of the pin was uneven for the material in the welding process. The material only reached the thermoplastic state during the short welding time and fast cooling rate, which led to the mechanical mixing of Al and steel components.

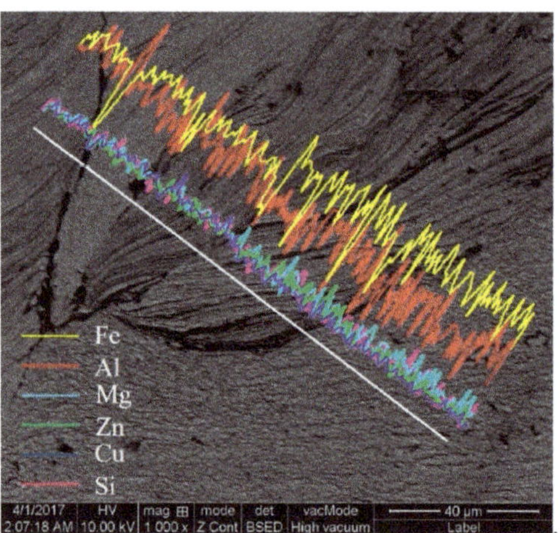

Figure 11. Micromorphology and EDS of the cloud cluster-like microstructure.

Figure 12 shows the XRD spectrum of the cross-section of the dissimilar Al/steel keyhole-free FSSW joint. It can be found that IMC FeAl$_3$ is formed at the Al/steel interface, as analyzed by the above elemental diffusion. Fereiduni et al. [38] also obtained an IMC with a similar atomic ratio at the interface layer. Due to the diffusion of the Zn element in the galvanized coating, the Zn element dissolved into the Al matrix to form phase AlZn$_x$, and form IMC FeAl$_3$Zn$_x$ with IMC FeAl$_3$ [13,37,38]. The Zn element diffused to the lattice space, and its elemental content was rather unstable [4]. At the same time, The IMC FeAl$_3$ formed under different welding parameters is the same, but the content is different. With the increase of rotation speed, the welding temperature becomes higher, and the stirring of joints becomes more sufficient; the content of IMC FeAl$_3$ and the thickness of the transition layers formed at a high temperature also increase.

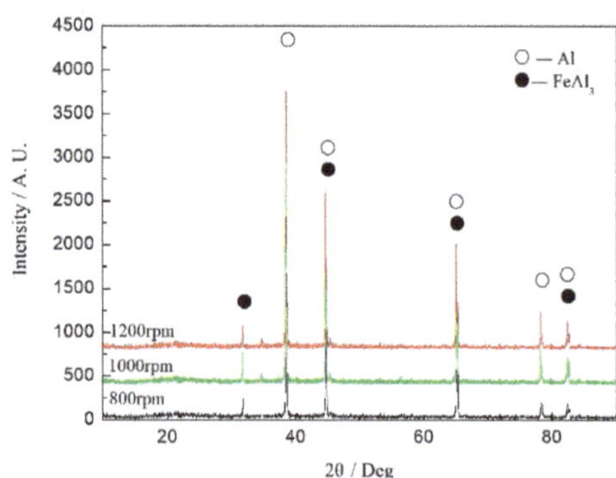

Figure 12. XRD spectrum of a cross-section of the dissimilar Al/steel keyhole-free FSSW joint.

3.2. Impact Properties

3.2.1. Impact Energy

The impact tests on the keyhole-free FSSW specimens with welding parameters of 800 rpm, 1000 rpm, and 1200 rpm were carried out to measure the impact toughness of the dissimilar Al/steel keyhole-free FSSW joint. The calculation formula for impact energy is as follows:

$$E = P \times D \times (cos\varphi - cos\alpha) \tag{1}$$

where E is the impact energy, P is the weight of the pendulum hammer, D is the distance between the center of the shaft and the center of gravity of the pendulum hammer, α is the starting angle of the pendulum hammer, and φ is the end angle of the pendulum hammer.

Parameters P, D and α are the fixed constants given by the equipment, and their values are 26.63 kgf, 0.6340 m and 60°, respectively. The impact test parameters and the impact energy are shown in Table 3. Table 3 shows that the impact toughness of the specimen with the welding parameter of 1000 rpm is the best, with an impact energy of approximately 42 J. However, the impact energy of the other two specimens is similar, approximately 32 J.

Table 3. The impact test parameters and the impact energy.

Rotation Speed ω/rpm	End Angle φ/°	Calculated Impact Energy E/J	Integral Impact Energy E/J
800	46	32.2	31.1
1000	41	42.1	48.7
1200	45.3	33.7	34.0

3.2.2. Load-Displacement Curves of Impact

Figure 13 shows the load-displacement curves and the histogram of the impact properties of the dissimilar Al/steel keyhole-free FSSW joint. The maximum load of the curve is the maximum impact load F_m, as the vertex of the dotted line in Figure 13a. The area integral of the left half of the dotted line is the crack generation energy E_i, as shown in the yellow area in Figure 13a. The area integral of the right half of the dotted line is the crack propagation energy E_p, as shown in the blue area in Figure 13a. The area integral of the load-displacement curves of impact is the impact energy $E = E_i + E_p$. It can be found in Figure 13a that the maximum impact loads of the three specimens are almost the same, approximately 55 kN. The specimen with a welding parameter of 1000 rpm had the largest impact deformation and the best impact toughness, which illustrates that the factor affecting the size of the impact energy is not the maximum impact load but the maximum impact deformation. In addition, the integral impact energy of the load-displacement curve approximately agrees with the calculated impact energy. At the same time, it can be found in Figure 13b that the maximum impact deformation directly reflects the crack propagation energy. The pre-crack generation energy is almost the same and cannot affect the maximum impact energy to a great extent. However, the post-crack propagation energy substantially affects the impact toughness. There is an internal relationship between the crack propagation energy and impact deformation. The greater the crack propagation energy is, the longer the crack propagation, the greater the impact deformation, the better the impact toughness, and vice versa.

3.2.3. Impact Fracture

This impact test is based on the lap-shear test, as show in Figure 3. Although the welding parameters and impact energy of each group of samples are different, the fracture morphology rules under different welding parameters are consistent. The joint fracture of the Al side, with welding parameters at 1000 rpm, was selected as the object of observation and analysis. Figure 14 shows the macroscopic morphology of impact fracture of the dissimilar Al/steel keyhole-free FSSW joint. It can be found that the impact fracture morphology was diverse and uneven [13]. The magnification figure of the white rectangle area in Figure 14 is shown at the bottom of Figure 14. According to the welding process and connection mode, the impact fracture morphology is divided into the three zones noted as the non-action zone (NAZ), shoulder action zone (SAZ), and the stir zone (TSZ). The NAZ is weakly connected, and the fracture shows a shallow dimple-like morphology [11,42–45], which is not affected by the plunge force of the shoulder. This is the edge of the keyhole-free FSSW joint shown in Figure 8b, where the IMCs were not completely filled. Therefore, the cracks develop from the gaps and propagate into the weld [37,42–45]. SAZ has a deeper dimple and stronger connection [42–45], which is affected by the dual effects of the friction heat and the plunge force of the shoulder. That is, the interface junction area of the keyhole-free FSSW joint in Figure 9b is in this zone. TSZ has the strongest connection strength and is affected by the stirring and heat effect of the pin, i.e., the cloud cluster-like microstructure of the keyhole-free FSSW joint shown in Figure 11.

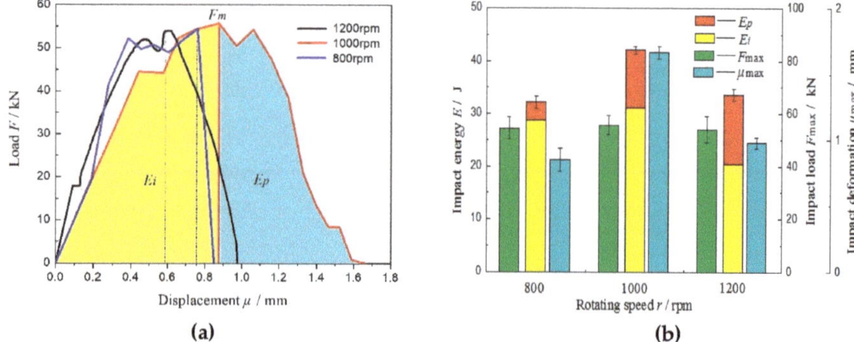

Figure 13. Load-displacement curves (**a**) and histogram of impact properties (**b**) of the dissimilar Al/steel keyhole-free FSSW joint.

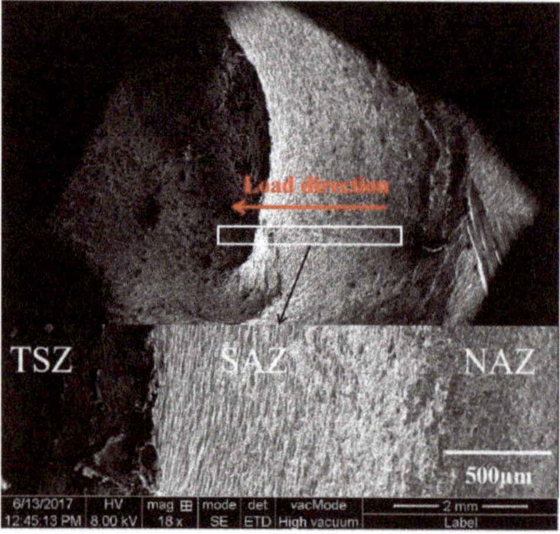

Figure 14. Macroscopic morphology of impact fracture of dissimilar Al/steel keyhole-free FSSW joint.

Figure 15 shows the microscopic fracture morphology for the TSZ of a dissimilar Al/steel keyhole-free FSSW joint. Figure 15a shows the whole picture of the fracture morphology of the TSZ. It can be found that the crack started at a point in the brittle phase and expanded outward along the blue line in the TSZ, as shown in Figure 15a [10]. After the crack was generated, it first extended along the brittle phase. During the large deformation during impact, the large impact energy causes the brittle phase to crack, as shown in the upper left corner of Figure 15a. When there is no brittle phase in the crack extension zone, the crack can continue to expand along the plastic phase [11]. Figure 15b shows the brittle–ductile transition zone of the impact fracture in the TSZ. This zone illustrates the transition of the fracture mode from a brittle fracture to a ductile fracture. Further magnification of the brittle and ductile areas is shown in Figure 15c,d, respectively. Figure 15c shows a brittle cleavage fracture, whereas Figure 15d shows a ductile fracture with typical shear dimples [40,42]. It can be inferred that the impact fracture mode of the keyhole-free FSSW joint is the mixed ductile and brittle fracture mode [40,42]. Most of the impact energy is absorbed by the ductile fracture.

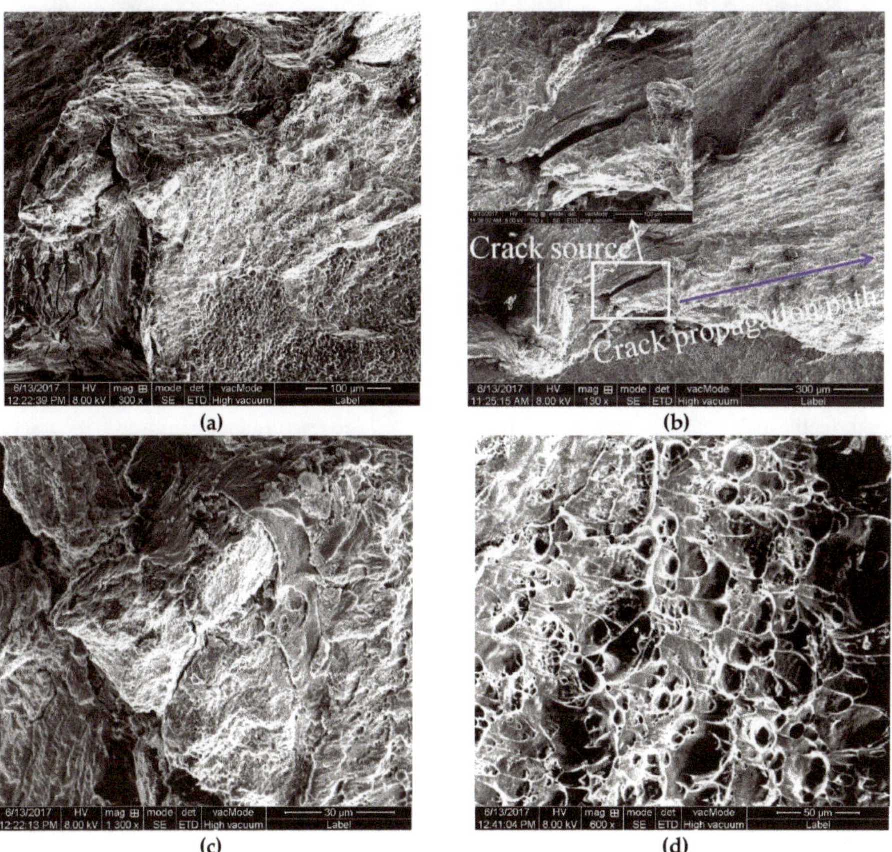

(a)

(b)

(c)

(d)

Figure 15. Microscopic fracture morphology of TSZ of dissimilar Al/steel keyhole-free FSSW joint: (**a**) whole picture, (**b**) brittle–ductile transition, (**c**) brittle cleavage fracture, and (**d**) ductile fracture.

Figure 16 shows the EDS surface-scanning results of the impact fracture in the TSZ. The elements were unevenly distributed on the fracture surface, as shown in Figure 16. Therefore, many brittle phases existed at the fracture surface, as discussed above, such as $AlZn_x$, $FeAl_3$, and $FeAl_3Zn_x$ [13,37,38]. This is another important cause of crack formation.

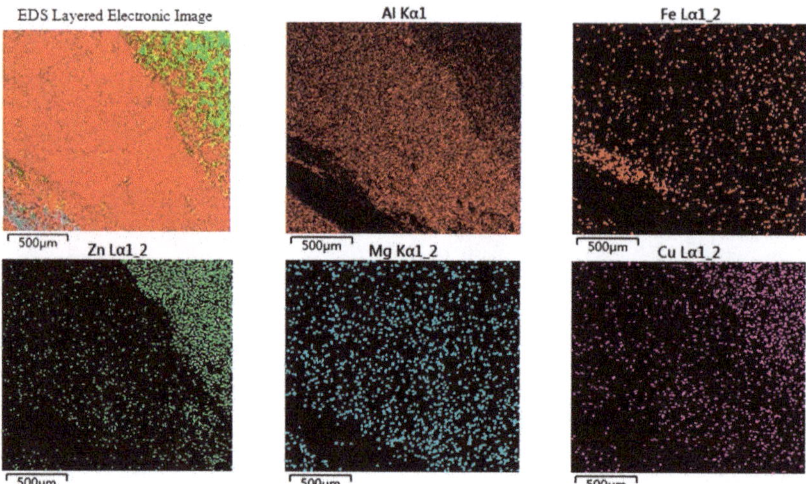

Figure 16. EDS of the surface-scanning analysis of the impact fracture in STZ.

From the above analysis, it can be seen that during the impact fracture of dissimilar Al/steel joints, the crack first starts from the weak zone of the joint, such as brittle phase and the gap. At the same time, the crack first propagates along the brittle zone, forming a dissociation fracture, and then propagates along the ductile zone, forming a ductile fracture. The intermetallic transition layer provides the environment for brittle crack growth, while the Al and steel matrix provide the buffer for ductile crack growth. Therefore, the impact fracture of the dissimilar Al/steel keyhole-free FSSW joint is the mixed ductile and brittle fracture mode, in which the plastic fracture mode improves the fracture impact energy.

4. Conclusions

The keyhole-free spot-welding joints of the dissimilar 6082 Al alloy and DP600 galvanized steel were successfully fabricated by retractile keyhole-free FSSW. The keyhole was eliminated by pin retraction technology. The interface behavior and impact performance of the dissimilar Al/steel keyhole-free FSSW joints were studied and analyzed in combination with the keyhole-free FSSW process. Important conclusions are as follows:

1. The maximum temperature of the periphery of the shoulder measured is about 500 °C, and fine equiaxed crystals are formed by dynamic recrystallization in WNZ, simultaneously accompanied by a cloud cluster-like mechanical mixing chaotic microstructure in TSZ. The Al/steel interface forms a transition layer composed of IMC $FeAl_3$, which is formed at the Al/steel interface, and there is diffusion of Al and Fe elements on the transition layer. When the welding parameters are 1000 rpm, the thickness of transition layer is approximately 15 μm. The grain size of WNZ on the steel side decreases from 20 μm of BM to 1 μm, and the grain size of WNZ on steel side decreases from 30 μm of BM to 5 μm.

2. The impact toughness of the specimen with a welding parameter of 1000 rpm is the best, and the impact energy is approximately 42 J. The brittle IMCs are the crack source, and mixed ductile and brittle fractures with brittle–ductile transition zones are formed, in which the ductile fracture improves impact deformation and absorbs most of impact energy. To a certain extent, the maximum impact deformation directly reflects the post-crack propagation energy, which significantly affects the impact toughness. The greater the crack propagation energy, the longer the crack propagation, the greater the impact deformation, and the better the impact toughness, and vice versa.

Author Contributions: Z.Z., Y.Y. and X.W. designed and planned the experiment. Y.Y. made the tests and wrote the paper. Z.Z. modified the paper. H.Z. provided the funding acquisition.

Funding: This research was funded by the Aviation Science Fund of China (No. 201611U2001) and Major Science and Technology Project of Gansu Province (No. 18ZD2GC013).

Acknowledgments: This work was funded by the Aviation Science Fund of China (No. 201611U2001) and Major Science and Technology Project of Gansu Province (No. 18ZD2GC013).

Conflicts of Interest: The authors declare no conflict of interest.

References

1. Piccini, J.M.; Svoboda, H.G. Effect of the Tool Penetration Depth in Friction Stir Spot Welding (FSSW) of Dissimilar Aluminum Alloys. *Procedia Mater. Sci.* **2015**, *8*, 868–877. [CrossRef]
2. Xu, R.; Ni, D.; Yang, Q.; Liu, C.; Ma, Z. Pinless Friction Stir Spot Welding of Mg-3Al-1Zn Alloy with Zn Interlayer. *J. Mater. Sci. Technol.* **2016**, *32*, 76–88. [CrossRef]
3. Piccini, J.M.; Svoboda, H.G. Effect of pin length on Friction Stir Spot Welding (FSSW) of dissimilar Aluminum-steel joints. *Procedia Mater. Sci.* **2015**, *9*, 504–513. [CrossRef]
4. Boucherit, A.; Avettand-Fènoël, M.-N.; Taillard, R. Effect of a Zn interlayer on dissimilar FSSW of Al and Cu. *Mater. Des.* **2017**, *124*, 87–99. [CrossRef]
5. Choi, D.-H.; Ahn, B.-W.; Lee, C.-Y.; Yeon, Y.-M.; Song, K.; Jung, S.-B. Formation of intermetallic compounds in Al and Mg alloy interface during friction stir spot welding. *Intermetallics* **2011**, *19*, 125–130. [CrossRef]
6. Shen, Z.; Ding, Y.; Chen, J.; Gerlich, A. Comparison of fatigue behavior in Mg/Mg similar and Mg/steel dissimilar refill friction stir spot welds. *Int. J. Fatigue* **2016**, *92*, 78–86. [CrossRef]
7. Kubit, A.; Kluz, R.; Trzepieciński, T.; Wydrzyński, D.; Bochnowski, W. Analysis of the mechanical properties and of micrographs of refill friction stir spot welded 7075-T6 aluminium sheets. *Arch. Civ. Mech. Eng.* **2018**, *18*, 235–244. [CrossRef]
8. Xu, R.; Ni, D.; Yang, Q.; Liu, C.; Ma, Z. Influencing mechanism of Zn interlayer addition on hook defects of friction stir spot welded Mg–Al–Zn alloy joints. *Mater. Des.* **2015**, *69*, 163–169. [CrossRef]
9. Garg, A.; Bhattacharya, A. Strength and failure analysis of similar and dissimilar friction stir spot welds: Influence of different tools and pin geometries. *Mater. Des.* **2017**, *127*, 272–286. [CrossRef]
10. Chen, K.; Liu, X.; Ni, J. Keyhole refilled friction stir spot welding of aluminum alloy to advanced high strength steel. *J. Mater. Process. Technol.* **2017**, *249*, 452–462. [CrossRef]
11. Hsieh, M.-J.; Lee, R.-T.; Chiou, Y.-C. Friction stir spot fusion welding of low-carbon steel to aluminum alloy. *J. Mater. Process. Technol.* **2017**, *240*, 118–125. [CrossRef]
12. Bozzi, S.; Helbert-Etter, A.; Baudin, T.; Criqui, B.; Kerbiguet, J. Intermetallic compounds in Al 6016/IF-steel friction stir spot welds. *Mater. Sci. Eng. A* **2010**, *527*, 4505–4509. [CrossRef]
13. Dong, H.; Chen, S.; Song, Y.; Guo, X.; Zhang, X.; Sun, Z. Refilled friction stir spot welding of aluminum alloy to galvanized steel sheets. *Mater. Des.* **2016**, *94*, 457–466. [CrossRef]
14. Reimann, M.; Goebel, J.; Dos Santos, J.F. Microstructure and mechanical properties of keyhole repair welds in AA 7075-T651 using refill friction stir spot welding. *Mater. Des.* **2017**, *132*, 283–294. [CrossRef]
15. Cao, J.; Wang, M.; Kong, L.; Zhao, H.; Chai, P. Microstructure, texture and mechanical properties during refill friction stir spot welding of 6061-T6 alloy. *Mater. Charact.* **2017**, *128*, 54–62. [CrossRef]
16. Malik, V.; Sanjeev, N.; Hebbar, H.S.; Kailas, S.V. Time Efficient Simulations of Plunge and Dwell Phase of FSW and its Significance in FSSW. *Procedia Mater. Sci.* **2014**, *5*, 630–639. [CrossRef]
17. Malik, V.; Sanjeev, N.; Hebbar, H.S.; Kailas, S.V. Finite Element Simulation of Exit Hole Filling for Friction Stir Spot Welding–A Modified Technique to Apply Practically. *Procedia Eng.* **2014**, *97*, 1265–1273. [CrossRef]
18. D'Urso, G. Thermo-mechanical characterization of friction stir spot welded AA6060 sheets: Experimental and FEM analysis. *J. Manuf. Process.* **2015**, *17*, 108–119. [CrossRef]
19. Malik, V.; Sanjeev, N.; Hebbar, H.S.; Kailas, S.V. Investigations on the Effect of Various Tool Pin Profiles in Friction Stir Welding Using Finite Element Simulations. *Procedia Eng.* **2014**, *97*, 1060–1068. [CrossRef]
20. Siddharth, S.; Senthilkumar, T.; Chandrasekar, M. Development of processing windows for friction stir spot welding of aluminium Al5052/copper C27200 dissimilar materials. *Trans. Nonferrous Met. Soc. China* **2017**, *27*, 1273–1284. [CrossRef]

21. Fanelli, P.; Vivio, F.; Vullo, V. Experimental and numerical characterization of Friction Stir Spot Welded joints. *Eng. Fract. Mech.* **2012**, *81*, 17–25. [CrossRef]

22. Mandal, S.; Rice, J.; Elmustafa, A. Experimental and numerical investigation of the plunge stage in friction stir welding. *J. Mater. Process. Technol.* **2008**, *203*, 411–419. [CrossRef]

23. Su, H.; Wu, C.S.; Bachmann, M.; Rethmeier, M. Numerical modeling for the effect of pin profiles on thermal and material flow characteristics in friction stir welding. *Mater. Des.* **2015**, *77*, 114–125. [CrossRef]

24. He, X.; Gu, F.; Ball, A. A review of numerical analysis of friction stir welding. *Prog. Mater. Sci.* **2014**, *65*, 1–66. [CrossRef]

25. Malafaia, A.; Milan, M.; Oliveira, M.; Spinelli, D. Evaluation of dynamic defect detection in FSSW welded joints under fatigue tests. *Procedia Eng.* **2010**, *2*, 1823–1828. [CrossRef]

26. Venukumar, S.; Muthukumaran, S.; Yalagi, S.G.; Kailas, S.V. Failure modes and fatigue behavior of conventional and refilled friction stir spot welds in AA 6061-T6 sheets. *Int. J. Fatigue* **2014**, *61*, 93–100. [CrossRef]

27. Joy-A-Ka, S.; Ogawa, Y.; Sugeta, A.; Sun, Y.; Fujii, H. Fatigue Fracture Mechanism on Friction Stir Spot Welded Joints Using 300 MPa-class Automobile Steel Sheets under Constant and Variable Force Amplitude. *Procedia Mater. Sci.* **2014**, *3*, 537–543. [CrossRef]

28. Uematsu, Y.; Tokaji, K.; Tozaki, Y.; Nakashimac, Y. Fatigue behaviour of dissimilar friction stir spot weld between A6061 and SPCC welded by a scrolled groove shoulder tool. *Procedia Eng.* **2010**, *2*, 193–201. [CrossRef]

29. Cui, H.; Xie, G.; Luo, Z.; Ma, J.; Wang, G.; Misra, R. The microstructural evolution and impact toughness of nugget zone in friction stir welded X100 pipeline steel. *J. Alloy. Compd.* **2016**, *681*, 426–433. [CrossRef]

30. Xie, G.; Cui, H.; Luo, Z.; Misra, R.; Wang, G. Asymmetric distribution of microstructure and impact toughness in stir zone during friction stir processed a high strength pipeline steel. *Mater. Sci. Eng. A* **2017**, *704*, 401–411. [CrossRef]

31. Sahu, P.K.; Pal, S. Mechanical properties of dissimilar thickness aluminium alloy weld by single/double pass FSW. *J. Mater. Process. Technol.* **2017**, *243*, 442–455. [CrossRef]

32. Ding, J.; Oelgoetz, P. Auto-adjustable tool for friction stir welding. U.S. Patent 5,893,507A, 13 April 1999.

33. Piccini, J.M.; Svoboda, H.G. Tool geometry optimization in friction stir spot welding of Al-steel joints. *J. Manuf. Process.* **2017**, *26*, 142–154. [CrossRef]

34. Xie, G.; Cui, H.; Luo, Z.; Yu, W.; Ma, J.; Wang, G. Effect of Rotation Rate on Microstructure and Mechanical Properties of Friction Stir Spot Welded DP780 Steel. *J. Mater. Sci. Technol.* **2016**, *32*, 326–332. [CrossRef]

35. Wang, L.; Xie, L.; Lv, Y.; Zhang, L.-C.; Chen, L.; Meng, Q.; Qu, J.; Zhang, D.; Lu, W. Microstructure evolution and superelastic behavior in Ti-35Nb-2Ta-3Zr alloy processed by friction stir processing. *Acta Mater.* **2017**, *131*, 499–510. [CrossRef]

36. Cho, H.-H.; Kang, S.H.; Kim, S.-H.; Oh, K.H.; Kim, H.J.; Chang, W.-S.; Han, H.N. Microstructural evolution in friction stir welding of high-strength linepipe steel. *Mater. Des.* **2012**, *34*, 258–267. [CrossRef]

37. Ding, Y.; Shen, Z.; Gerlich, A. Refill friction stir spot welding of dissimilar aluminum alloy and AlSi coated steel. *J. Manuf. Process.* **2017**, *30*, 353–360. [CrossRef]

38. Fereiduni, E.; Movahedi, M.; Kokabi, A. Aluminum/steel joints made by an alternative friction stir spot welding process. *J. Mater. Process. Technol.* **2015**, *224*, 1–10. [CrossRef]

39. Rao, H.M.; Yuan, W.; Badarinarayan, H. Effect of process parameters on mechanical properties of friction stir spot welded magnesium to aluminum alloys. *Mater. Des.* **2015**, *66*, 235–245. [CrossRef]

40. Zhao, Y.; Ding, Z.; Shen, C.; Chen, Y. Interfacial microstructure and properties of aluminum–magnesium AZ31B multi-pass friction stir processed composite plate. *Mater. Des.* **2016**, *94*, 240–252. [CrossRef]

41. Sarkar, R.; Pal, T.; Shome, M. Material flow and intermixing during friction stir spot welding of steel. *J. Mater. Process. Technol.* **2016**, *227*, 96–109. [CrossRef]

42. Jamili-Shirvan, Z.; Haddad-Sabzevar, M.; Vahdati-Khaki, J.; Chen, N.; Shi, Q.; Yao, K.-F. Microstructure characterization and mechanical properties of Ti-based bulk metallic glass joints prepared with friction stir spot welding process. *Mater. Des.* **2016**, *100*, 120–131. [CrossRef]

43. Shen, Z.; Yang, X.; Zhang, Z.; Cui, L.; Li, T. Microstructure and failure mechanisms of refill friction stir spot welded 7075-T6 aluminum alloy joints. *Mater. Des.* **2013**, *44*, 476–486. [CrossRef]

44. Pabandi, H.K.; Movahedi, M.; Kokabi, A.H. A new refill friction spot welding process for aluminum/polymer composite hybrid structures. *Compos. Struct.* **2017**, *174*, 59–69. [CrossRef]
45. Zheng, Q.X.; Feng, X.M.; Shen, Y.F.; Huang, G.Q.; Zhao, P.C. Dissimilar friction stir welding of 6061 Al to 316 stainless steel using Zn as a filler metal. *J. Alloys. Compds* **2016**, *686*, 693–701. [CrossRef]

Article

Numerical Simulation of Material Flow and Analysis of Welding Characteristics in Friction Stir Welding Process

Haitao Luo [1,2,*], Tingke Wu [1,3], Peng Wang [1,4], Fengqun Zhao [1,3], Haonan Wang [1,4] and Yuxin Li [1,4]

1   State Key Laboratory of Robotics, Shenyang Institute of Automation, Chinese Academy of Sciences, Shenyang 110016, China; wutingke@sia.cn (T.W.); 1710107@stu.neu.edu.cn (P.W.); zhaofengqun@sia.cn (F.Z.); 1770174@stu.neu.edu.cn (H.W.); liyuxin@stumail.neu.edu.cn (Y.L.)
2   Institutes for Robotics and Intelligent Manufacturing, Chinese Academy of Sciences, Shenyang 110016, China
3   School of Mechanical Engineering, Shenyang Ligong University, Shenyang 110159, China
4   Institute of Mechanical Engineering and Automation, Northeastern University, Shenyang 110819, China
*   Correspondence: luohaitao@sia.cn; Tel.: +86-131-3021-6191

Received: 5 May 2019; Accepted: 25 May 2019; Published: 28 May 2019

Abstract: Friction stir welding (FSW) material flow has an important influence on weld formation. The finite element model of the FSW process was established. The axial force and the spindle torque of the welding process were collected through experiments. The feasibility of the finite element model was verified by a data comparison. The temperature field of the welding process was analyzed hierarchically. It was found that the temperature on the advancing side is about 20 °C higher than that on the retreating side near the welding seam, but that the temperature difference between the two sides of the middle and lower layers was decreased. The particle tracking technique was used to study the material flow law in different areas of the weld seam. The results showed that part of the material inside the tool pin was squeezed to the bottom of the workpiece. The material on the upper surface tends to move downward under the influence of the shoulder extrusion, while the material on the lower part moves spirally upward under the influence of the tool pin. The material flow amount of the advancing side is higher than that of the retreating side. The law of material flow reveals the possible causes of the welding defects. It was found that the abnormal flow of materials at a low rotation speed and high welding speed is prone to holes and crack defects. The forming reasons and material flow differences in different regions are studied through the microstructure of the joint cross section. The feasibility of a finite element modeling and simulation analysis is further verified.

Keywords: material flow; finite element model; temperature field; microstructure; welding defects

1. Introduction

Friction stir welding (FSW) has achieved great success in joining aluminum alloys since its invention. Due to the characteristics of a high joint quality, small welding deformation, green welding process and no pollution, FSW technology is often applied to the welding of aluminum, magnesium and other light alloy materials [1–3]. High-strength aluminum alloys are used in the welding of aerospace products. Because of its good strength and hardness, 2XXX aluminum alloy is used for the welding of rocket storage tanks and aircraft wings [4,5]. However, FSW is a complicated thermal-mechanical coupling process, and the forming quality is mainly affected by the tool shape, rotation speed, welding speed and other process parameters [6–8]. If it is not properly controlled, it is easy to change the physical quantities, such as the heat generation and plastic flow state of the weld materials, which will lead to welding defects [9]. The material flow has an important influence on the quality of the weld forming. Studying the material flow will help to understand the welding process and prevent welding defects.

In FSW, there is a close contact between the tool and the material to be welded, so there is severe friction between the tool and the workpiece. During the welding process, the intense friction between the tool and the workpiece generates a large amount of heat. This heat causes the temperature of the material to rise, and the material softens at high temperatures [10–12]. The softened material undergoes severe plastic flow under the drive of the tool. At the same time, the shear deformation of the material also generates a part of the heat [13]. Along with the plastic flow of the material at a high temperature, the heat transfer inside the material is convective and conductive. The severe plastic flow of the material under the drive of external force causes heat convection, and the temperature distribution also affects the plastic flow of the material [14,15].

Franke et al. [16] examines the intermittent flow of material and its relation to defect formation. In addition, advances have been made in a force-based defect detection model that links changes in process forces to the formation and size of defects. Gratecap et al. [17] placed a steel powder mark on the butt to study the material flow. The results show that the material in the weld zone extrudes around the retreating side (RS) of the tool pin as it moves from the leading side to the trailing side. Huang et al. [18] observed the vertical flow behavior of the material by observing the distribution of Cu foil fragments and Al-Cu intermetallic compounds. The downward and upward flows encounter each other at the advancing side (AS) in the material depositing process, changing the morphology of the weld nugget zone. Dialami et al. [19] used the fourth-order Runge-Kutta (Rk4) integration method to calculate the particle trajectory. The effect of the input process parameters and pin shape on the weld quality was studied by a particle tracking method.

Li et al. [20] added a 0.1 mm thick bronze foil to the 7075-T651 aluminum alloy welding interface. The distribution characteristics of the horizontal and vertical cross-section bronze foil were analyzed by X-ray. They measured that the shear force at the front of the tool was stronger than the shear at the back of the tool pin. The metal is mainly sheared from the advancing side in front of the tool to the retreating side. Then, under the squeezing force of the rotary tool, the plastic material is pushed back to the advancing side at a slow speed. As a result, a temporary cavity is formed on the advancing side. The plasticized material is extruded into a cavity and filled therein. Figure 1 shows a schematic of the flow of material around the tool. Shanavas and Dhas [21] discussed the development of fuzzy models for predicting the weld quality. The effects of welding parameters, such as the tool geometry, tool rotation speed, welding speed and tool inclination angle, on the welding quality are studied. De Filippis et al. [22] established a simulation model to monitor, control and optimize the friction stir welding process. The correlation between the process parameters and mechanical properties can be identified. In order to improve the welding performance, the parameters of the joint are controlled in real time through experiments. Cisko et al. [23] conducted a large number of experiments and calculations on the fatigue behavior of aluminum-lithium alloy (AA2099) friction stir welding (FSW). The effect of the friction stir welding process on the fatigue life was studied by using a microstructure sensitive fatigue model.

Zhang et al. [24] proposed a conceptual model describing weld formation to study the effect of the Alclad layer on material flow and defect formation. He found that the top Alclad assembled at the shoulder/workpiece interface, thereby weakening the material flow in the shoulder-driven zone and favoring the formation of a void defect at high traveling speeds. Zhu et al. [25] established a three-dimensional coupled finite element model of friction stir welding defects based on the Euler-Lagrange coupling method. The results show that smaller gaps will be generated at higher welding speeds.

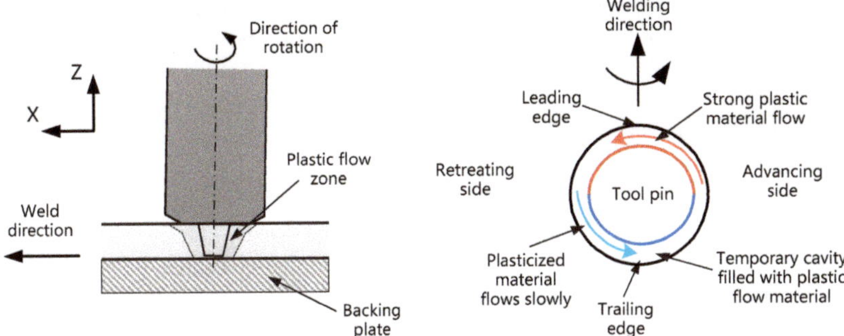

Figure 1. Schematic of the material movement around the friction stir welding (FWS) tool.

Although some scholars have conducted some research on material flow [26–28], the numerical model is still in the initial stage, and there are still many problems to be studied. According to the author's knowledge, there is no report about the process from the plunging process to the material staying behind the weld. This paper understands the frictional heat generation of the tool and the material by establishing the heat input model. A finite element model was constructed to simulate the AA2A14-T6 friction stir welding process. The axial force and torque of the welding process were collected by electromagnetic coupling technology and compared with the simulation data. The results verified the correctness of the finite element model.

The contribution of this article is to establish the thermogenic physical model of the FSW process and to establish the finite element model of FSW. The material flow and temperature fields of the welding process are studied by a numerical simulation. The numerical simulation results are further verified by using the microstructure of the weld. Through the stratified analysis of the temperature of the weld zone, it was found that the advancing side and the retreating side above the material showed obvious temperature differences, and this phenomenon was also found in the test [29]. A material flow analysis was carried out on the inside of the stirring pin via particle tracking technology. It was found that a small part of the internal material of the pin was extruded to the bottom of the weld and that most of the materials moved upward with the pin. A material flow analysis was performed on the advancing side and the retreating side of the welded area, respectively. The material flow exhibits different flow conditions on the advancing and retreating sides. At the same time, a similar situation was found through the microstructure of the joint. The analysis of the flow trend of the microstructure confirmed the correctness of the finite element analysis. Since the material flow is closely related to the weld formation, this paper can predict or provide a reference for future welding experiments by analyzing the defects of the finite element model.

2. Modeling and Acquisition Methods

2.1. Friction Stir Welding Heat Input Model

The heat input of friction stir welding makes the plastic flow of materials a complex process, which is accompanied by the coupling and interaction of the friction heat generation, metal plastic flow and structural transformation. Therefore, it is very difficult or even impossible to carry out a comprehensive analysis of all factors in the friction stir processing (FSP) [30]. Therefore, the heat input during the friction stir welding is studied.

It is difficult to establish the model within actual conditions, so the model studies the main factors of the heat production process and makes an ideal treatment for the less influential factors. Therefore, this model is an ideal model based on these assumptions. The following assumptions are made: (1) Ignoring latent heat of the phase change during the tissue structure transformation; (2) The thread on the surface of the tool pin is not considered; (3) Ignoring the welding inclination to simplify the model; (4) It is assumed that the friction work is all converted into friction heat.

The tool pin and tool shoulder dimensions are shown in Figure 2. The main variables used in this article are shown in Table 1. Assuming that the tool shoulder forging force acts evenly on the shoulder area, the frictional force received on the dA area is:

$$df = \frac{\mu F}{\pi r_1^2} r dr d\theta, \tag{1}$$

where F is the axial force, r_1 is the radius of the outer circumference of the tool, and μ is the coefficient of friction. The torque and power generated on the dA area are:

$$dM = \frac{\mu F}{\pi r_1^2} r^2 dr d\theta, \tag{2}$$

$$dp = dM\omega = \frac{\mu F}{\pi r_1^2} r^2 dr d\theta \omega, \tag{3}$$

where ω is the rotational angular speed. For the integration of both sides of Equation (3), the total heat production power of the shoulder zone is:

$$P_1 = \int_{r_2}^{r_1} \int_0^{2\pi} dp, \tag{4}$$

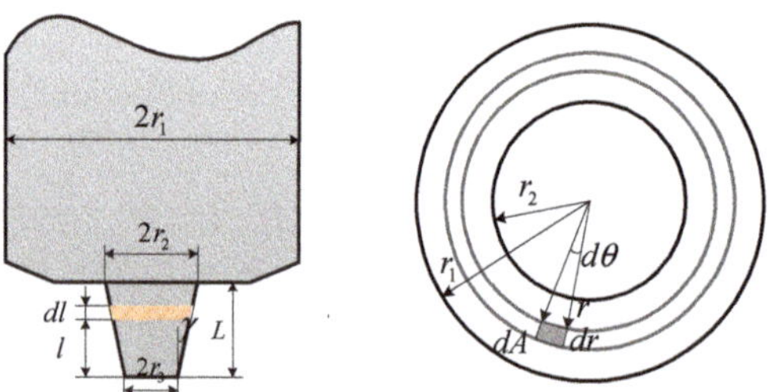

Figure 2. Schematic diagram of the pin and tool shoulder size.

Table 1. Variables used in this article.

Variables	Physical Meanings	Variables	Physical Meanings
F (N)	Axial force	σ_s (MPa)	Yield stress of the material
ω (rad)	Rotational angular speed	L (m)	Length of the tool pin
r_1 (m)	Radius of the outer circumference	σ'_{ij} (MPa)	Deviatoric stress component
r_2 (m)	Maximum radius of tool pin	$\dot{\bar{\varepsilon}}$ (s^{-1})	Strain rate component
r_3 (m)	Radius of the end of the tool pin	λ (/)	Large penalty factors
μ (/)	Coefficient of friction	F_i (N)	Arbitrary variables
γ (°)	Taper angle of the tool pin	$\bar{\sigma}$ (MPa)	Flow stress
τ (Pa)	Ultimate shear strength of the material	m (/)	The rate sensitivity
T (°C)	The temperature	$(\rho_i\ cm^{-2})$	Dislocation density

Therefore, the surface heat flux density of the tool shoulder area is:

$$q_1 = \frac{P_1}{\pi(r_1^2 - r_2^2)}, \tag{5}$$

The friction force on the pin side dl is:

$$df_2 = 2\pi\tau(r_3 + h\tan\gamma)dh, \tag{6}$$

$$\tau = \frac{\sigma_s}{\sqrt{3}}, \tag{7}$$

where r_3 is the radius of the end of the tool pin, γ is the taper angle of the tool pin, τ is the ultimate shear strength of the material, and σ_s is the yield stress of the material. The torque and power of dl are:

$$dM_2 = \frac{2\pi\sigma_s(r_3 + l\tan\gamma)^2 dl}{\sqrt{3}}, \tag{8}$$

$$dp_2 = \frac{2\pi\sigma_s\omega(r_3 + l\tan\gamma)^2 dl}{\sqrt{3}}, \tag{9}$$

By integrating Equation (9), the heat production power on the side of the pin is obtained.

$$P_2 = \int_0^L dp = \frac{4\pi^2 n\sigma_s L}{3\sqrt{3}}(3r_3^2 + 3r_3 L\tan\gamma + L^2\tan^2\gamma), \tag{10}$$

where L is the length of the tool pin. The heat generation at the bottom surface of the pin is similar to that at the tool shoulder. The heat generation is as follows:

$$P_3 = \int_0^{r_3} \frac{\mu F}{\pi r_1^2} 2\pi r dr \times \omega r = \frac{4\mu F\pi n r_3^3}{3r_1^2}, \tag{11}$$

The body heat flux density of the tool pin is:

$$q_2 = \frac{P_2 + P_3}{V}, \tag{12}$$

$$V = \frac{\pi L(r_2^2 + r_2 r_3 + r_3^2)}{3}, \tag{13}$$

2.2. Finite Element Model

2.2.1. Geometric Model and Boundary Conditions

The FSW process is a dynamic nonlinear analysis based on the Lagrange method. The finite element software DEFORM-3D (V10.2, Scientific Forming Technologies Corporation, Columbus, OH, USA) was used to simulate the whole process of friction stir welding. The material of the tool was W6, and the workpiece size was 150 mm × 100 mm × 6 mm for a 2A14-T6 lightweight aluminum alloy. The tool shoulder diameter of the tool is 16.3 mm, the maximum diameter of the tool pin is 8.15 mm, and the length of the tool pin is 5.65 mm.

The geometric model of the friction stir welding is shown in Figure 3. To make the simulation results more accurate, three sets of meshes are added to the workpiece weld area, and the minimum mesh is divided into 0.6 mm. Because the pin participates in the material flow of the welding process, the tool plays an important role in the flow of the material. However, considering the solution duration, the mesh size of the tool shoulder and the welding influence area are gradually increased, which ensures the calculation accuracy and improves the calculation efficiency [31]. A set of mesh window is also set on the tool, and the mesh window is set to move synchronously with the tool during the welding process.

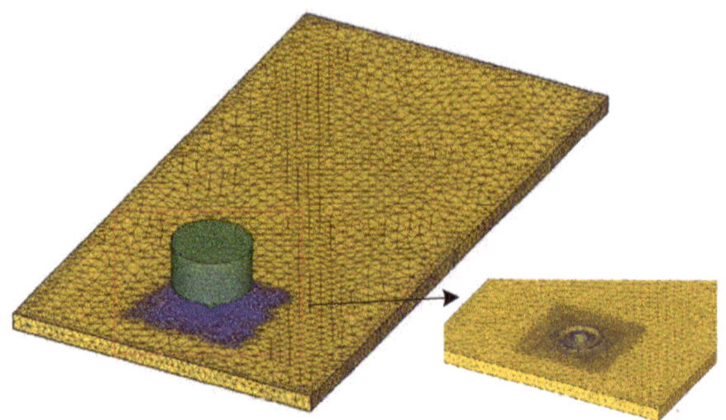

Figure 3. Geometric model in finite element model (FEM) simulation.

Due to the large plastic deformation of the material that is to be welded during the FSW process, the mesh adaptive re-division technique can be used to control the mesh distortion caused by the rotation and translation of the tool. The mesh distortion is controlled by the absolute minimum mesh size. The absolute mesh size can ensure the accuracy of the solution but also increase the solution time [32]. The total number of workpiece tetrahedral elements is divided into 68,762. The total number of mixing head tetrahedral meshes is 35,104. In the simulation process, since the tool material strength is much higher than the workpiece material, the tool is set as a rigid body, and the workpiece is set as a rigid plastic body. In order to simplify the model, the fixture and the backing plate are not built in the model, and the bottom and side degrees of freedom of the workpiece are fully constrained.

2.2.2. Finite Element Formula

A rigid viscoplastic model with a von-Miss yield criterion is used [33]. The finite element formula for rigid viscoplastic materials is based on the variational principle, where the allowable velocity should satisfy the conditions of compatibility and incompressibility, as shown in Equation (14):

$$\eta = \int_V E(\dot{\varepsilon}_{ij})dV - \int_{S_F} F_i u_i dS, \tag{14}$$

where F_i, V, S_F and $E(\dot{\varepsilon}_{ij})$ are respectively the surface traction force, the workpiece volume, the force surface and the plastic deformation power function. A penalty function for incompressibility is added to eliminate the incompressibility constraint on the allowable velocity field. The actual velocity field can now be determined by the steady value of the variation equation [34], which is expressed as:

$$\delta\eta = \int_V \overline{\sigma}\delta\dot{\overline{\varepsilon}}dV + \lambda \int_V \dot{\varepsilon}_V \delta\dot{\varepsilon}_V dV - \int_{S_F} F_i \delta u_i dS, \tag{15}$$

where $\dot{\overline{\varepsilon}}$, $\dot{\varepsilon}_{ij}$, and σ'_{ij} are the effective strain rate, strain rate component and deviatoric stress component. λ, and δu_i are, respectively, the large penalty factors and arbitrary variables. $\delta\dot{\overline{\varepsilon}}_V$ and $\delta\dot{\overline{\varepsilon}}$ represent the change in the strain rate derived from δu_i.

2.2.3. Material Model

A proper selection of material models is critical to an accurate solution in the simulation process. The material changes from a solid state to a viscous state, so it is necessary to define a wide range of strain, strain rate and flow stress values at temperatures. The flow stress is defined as a function of the strain, strain rate, and temperature, and Equation (16) is used to define the flow stress:

$$\overline{\sigma} = \overline{\sigma}(\varepsilon, \dot{\overline{\varepsilon}}, T),\tag{16}$$

Among them, $\overline{\sigma}$ is the flow stress, ε is the strain, and T is the temperature.

The flow stress is temperature-dependent and strain rate sensitive. Figure 4 shows the flow stress curve of the 2A14 aluminum alloy [35]. The flow stress data is imported into the numerical model. As can be seen from Figure 4, the material is sensitive to the temperature at high strain rates. As the temperature rises, the flow stress decreases. The main reason for this is that the thermal vibration energy of the crystal lattice becomes larger at a high temperature and the external force required for the dislocation movement decreases. The thermal properties of the 2A14-T6 workpiece and W6 tool steel are summarized in Table 2.

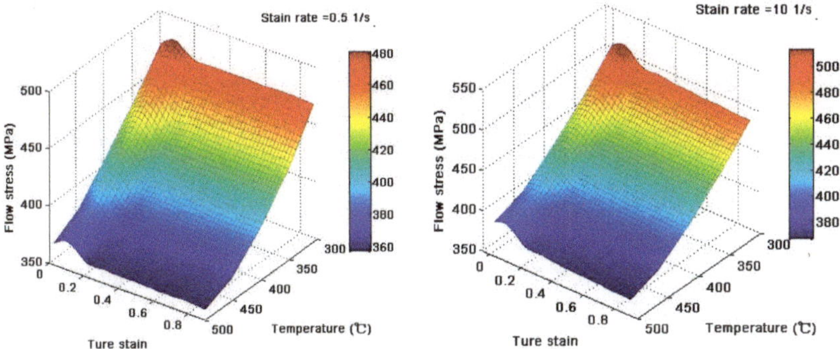

Figure 4. Flow stress curve of the 2A14 aluminum alloy.

Table 2. Thermal characteristics of the 2A14-T6 workpiece and W6 tool.

Properties	AA 2014-T6	Tool Steel W6
Heat capacity $(N/mm^2/°C)$	2.46	3.18
Young's modulus (N/mm^2)	69,300	230,000
Heat expansion coefficient $(\mu mm/mm/°C)$	21	11.9
Thermal conductivity $(N/s/°C)$	176	30.8
Poisson's ratio (/)	0.33	0.3

2.2.4. Dislocation Motion Model

The plastic deformation of the metal results from the dislocation movement of the material, so the dislocation density changes constantly in the process of friction stir welding, and the material flow changes based on this model. The dislocation motion is expressed as follows:

$$d\rho_i = (h - \dot{r}\rho_i)d\varepsilon,\tag{17}$$

$$h = h_0\left(\frac{\dot{\overline{\varepsilon}}}{\dot{\varepsilon}_0}\right)^m \cdot \exp\left(\frac{mQ}{RT}\right),\tag{18}$$

$$\dot{r} = \dot{r}_0\left(\frac{\dot{\overline{\varepsilon}}}{\dot{\varepsilon}_0}\right)^{-m} \cdot \exp\left(\frac{-mQ}{RT}\right),\tag{19}$$

where ρ_i is the dislocation density, h is the height of the action range of the dislocation stress field, $\dot{r}$ is the radius of the action range of the dislocation stress field, m is the rate sensitivity, and $\dot{\varepsilon}_0$ is the non-dynamic strain rate.

2.3. Test Acquisition Method

The welding material used in the test is an 2A14-T6 aluminum alloy plate with the same size and simulation. The 2A14 aluminum alloy is a high-strength aerospace aluminum alloy. The internal material composition is shown in Table 3 [36]. The tool shoulder diameter is 16.3 mm, the tool cone angle is 15°, and the overall size is consistent with the simulation model parameters. The welding process is completed on a self-developed high-precision friction stir welding robot. During the welding process, the spindle speed is 900 r/min, the welding speed is 100 mm/min, and the tool shoulder is plunged against the workpiece surface by 0.2 mm. The welding inclination angle is 2.5°, after the tool needle is pressed down to a certain amount, it will dwell and rotate for 10 s. The overall welding parameters are consistent with the simulation parameters.

Table 3. Chemical composition of 2A14 Aluminum Alloys.

Cu	Si	Mn	Mg	Fe	Zn	Ti	Ni	Al
3.9~4.8	0.6~1.2	0.4~1.0	0.4~0.8	≤0.7	≤0.3	≤0.15	≤0.1	Margin

The axial force and torque in the welding process are collected by electromagnetic coupling technology, and wireless power transmission is adopted, as shown in Figure 5. Most scholars use the method of setting the mechanical sensor at the bottom of the workpiece for data acquisition. However, since the measurable area of the sensor is limited, when the welding center deviates from the center of the sensor, the accuracy of the test method will be greatly reduced; thus, it is difficult to track and collect long welds.

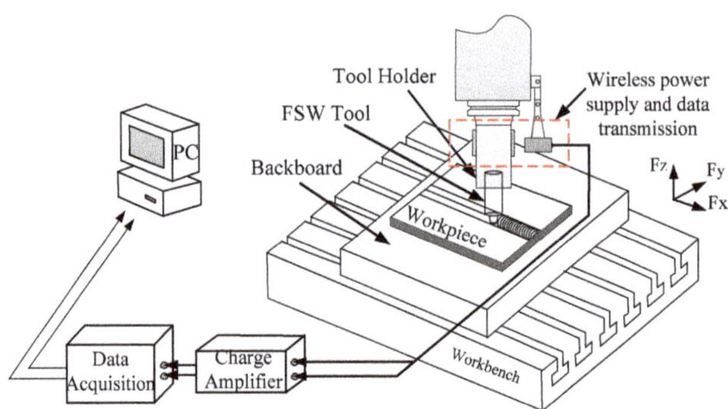

Figure 5. Signal acquisition method.

In the magnetic coupling resonant radio energy and signal transmission system, the primary side uses the amplitude shift keying (ASK) method to load the signal into the power transmission channel for a synchronous transmission, and uses the non-coherent demodulation method to extract and recover the signal on the secondary side. The schematic diagram of the acquisition method is shown in Figure 6. The power and signal synchronous transmission method can reduce the inverter link, reduce the loss and improve the quality of the advancing signal transmission.

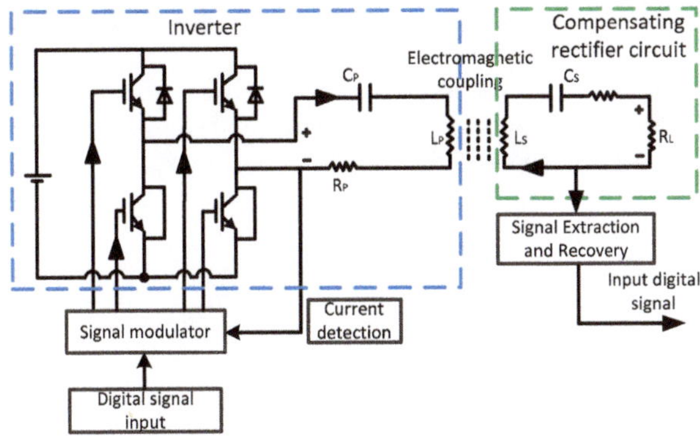

Figure 6. Schematic diagram of electromagnetic coupling signal transmission.

3. Results and Discuss

3.1. Verification of Finite Element Model

The axial force was tested throughout the FSW process. Figure 7 shows the comparison of the test results with the simulation results. It can be seen from the curve that the temperature gradually rises and reaches a peak in the first phase of the curve. In other words, the tool pin is just in contact with the workpiece, and some of the material is extruded due to plastic shear. The temperature rise softens the base material around the tool pin and promotes the pressing down of the tool. The material around the second stage tool gradually softens, and the external force of the tool plunged into the material increases slowly. In the third stage, as the tool shoulder comes into contact with the extruded material, the axial force begins to gradually increase, mainly due to the need for a greater axial force due to the tool shoulder squeezing the spilled material. At the peak of the axial force, the tool is prone to wear. Fracture can easily occur under extreme conditions. However, the axial force of the simulation output is slightly lower than the actual test results, but the overall trend of each step is in good agreement.

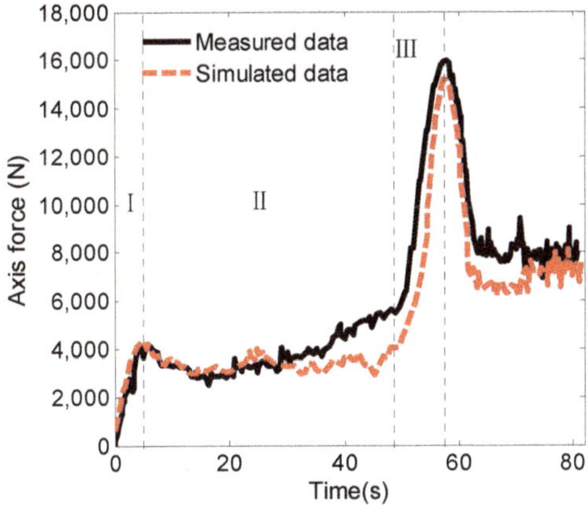

Figure 7. Axis force curve in the Z direction.

The spindle torque gives important information about the power consumption of the welding process, so torque is a decisive factor in the power requirements of the friction stir welding robot spindle. The change in the torque curve is gradual compared to the axial force, and the torque curve and the axial force curve have similar trends. As shown in Figure 8, the torque of the tool is relatively small at the beginning of the plunging. As the contact area between the tool pin and the material increases, the spindle torque also increases. During the dwelling phase of the tool, the torque is gradually reduced, and the torque variation during the welding phase is relatively flat. The simulated data and the experimental data have certain errors, and the simulated curve changes relatively gently during the plunging phase, but the overall values and trends are close to the experimental data, thus verifying that the finite element model is relatively accurate.

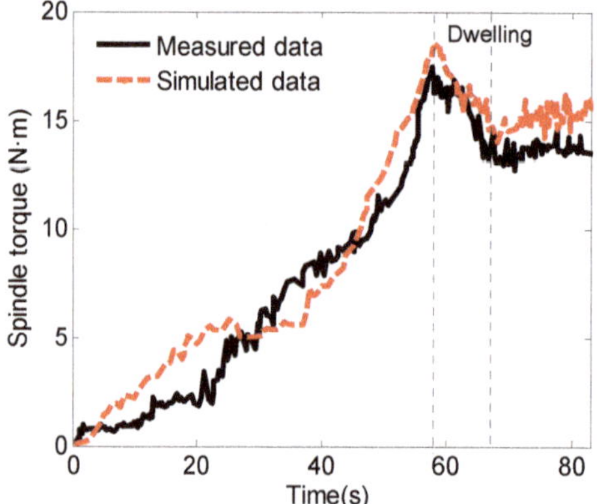

Figure 8. Spindle torque variation curve.

3.2. FSW Temperature Field Distribution

The simulation of the temperature field has a certain significance for the study of the friction stir welding process. The friction stir welding process is also the result of the joint action of the tool shoulder and the tool pin. The temperature field in the welding area is analyzed in three layers, in which the upper layer is the surface layer, the middle layer is 3.5 mm deep below the surface layer, and the bottom layer is the bottom surface of the workpiece. As shown in Figure 9, the temperature field in the surface layer presents the highest temperature, and one can observe that the temperature on the advancing side is significantly higher than that on the retreating side. The temperature on the advancing side is about 20 °C higher than that on the retreating side near the welding seam; furthermore, the experimental test can be referred to [26].

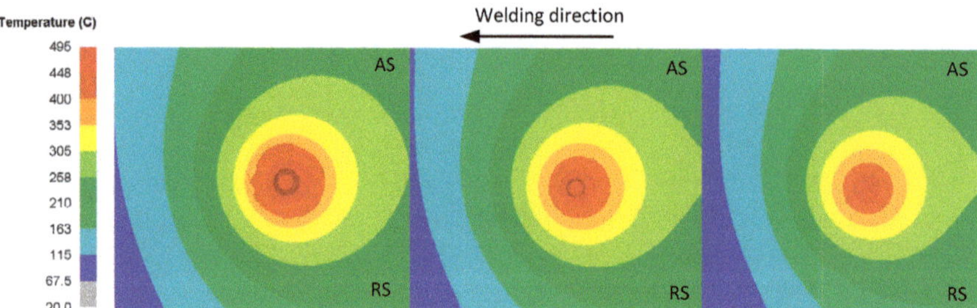

Figure 9. Temperature distribution of different thickness layers of the joint.

There is a thin layer of high temperature around the interface between the tool and the material that is to be welded. The material temperature in this area is about 500 °C. The rheological resistance of this area is extremely low, and the material is in a viscous state, which constitutes a viscous layer. The material temperature of the outer part of the viscous layer is about 450 °C, and the material is in a thermoplastic state, which constitutes a thermoplastic layer [37]. Since the tool pin is conical, the contact surface of the middle layer tool pin becomes smaller, and the temperature of the middle layer material, in which the temperature radiation area is also relatively reduced, decreases; furthermore, the plastic softening layer area of the material decreases. Similarly, in the welding area, the contact surface of the bottom tool pin is the smallest, the welding temperature of the tool pin at the bottom is relatively low, and the material flow speed also decreases, which is one of the reasons for the welding defects at the bottom.

The temperature profile of the temperature distribution is obtained from the upper and lower temperature gradients. The temperature field simulation of the friction stir welding process can determine the temperature distribution of different thickness layers. The temperature field directly affects the material flow, and the shearing action of the tool pin drives the plastic softening layer material flow. Therefore, the effect of the temperature on the material flow in different areas of the weld can be further studied.

3.3. Material Flow

The material flow occurs in the weld area during the friction stir welding. However, the plunging process is an important process of the friction stir welding. The material softens and plastically deforms during this process. With the increasing temperature in the welding zone, the material flows plastically. Studying the material flow is helpful for understanding the welding seam forming rule and for preventing welding defects. In this study, a particle tracking technique was used to study the material flow.

The plastic flow of materials in the simulation analysis process easily causes an excessive deformation of the cell shape, which leads to distortion. Therefore, it is guaranteed that the material can continue to operate after a large plastic deformation. DEFORM-3D will automatically re-divide the distorted mesh after the mesh distortion reaches a certain level, and will generate a new high-quality mesh. The mesh re-division technique can be used to reconstruct the plastically deformed mesh in time, so that it can be used to study the plastic flow of the materials.

3.3.1. Material Flow inside the Tool Pin

Before the tool pin is not plunged, as shown in Figure 10, three sets of tracking particles are arranged inside the tool pin, where P1–P8 particles are arranged in the center area of the tool pin, P1 is set on the surface of the workpiece, and the minimum diameter of the tool pin end is 4.6 mm. P9–P16 is set in the retreating side area 2 mm away from the axis, and the rest, P17–P24, is set in the symmetrical advancing side area. Where the plunging direction is the negative direction of the Z-axis, the spindle does a rotation movement around the Z axis, and the welding direction is the negative direction of the X-axis. Therefore, the top view, front view and side view are the positive Z direction, positive X direction and negative Y direction, respectively.

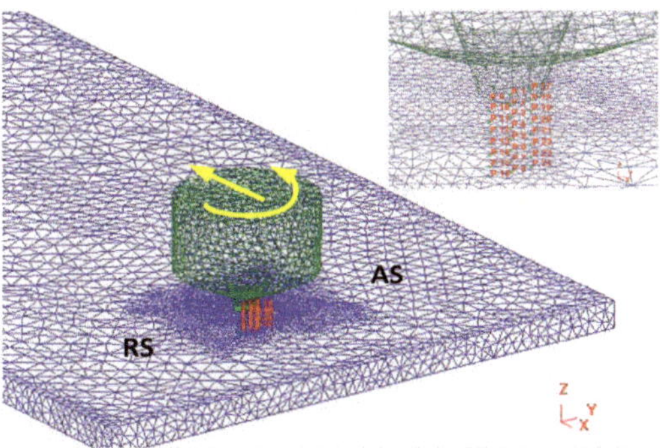

Figure 10. Setting of tracking particles inside the tool pin.

Figure 11 shows that during the pressing process, the particles (P1–P8) at the axis do not rotate around the tool pin but are squeezed to the bottom of the keyhole created by the tool pin. The reason for this phenomenon is that the linear velocity generated by the tool pin at the axial position is small, and the material fails to participate in the stirring action. After plunging onto the given amount, some of the particles inside the tool pin rotate to the top of the material, and some of the particles are distributed below the material.

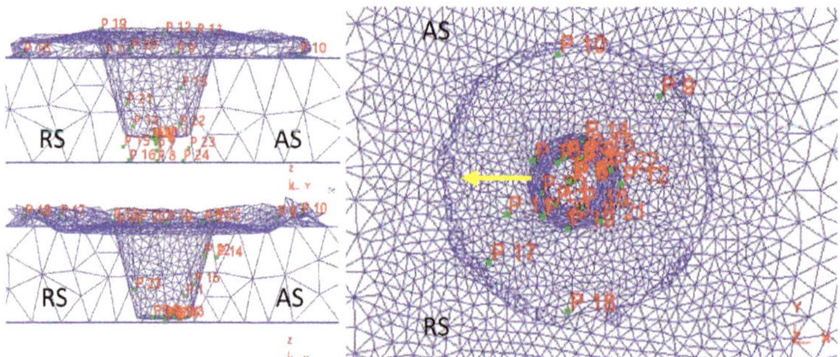

Figure 11. Tool needle internal particle plunging process.

Figure 12a,b shows that in the welding stage, the plastic material flows with the pin rotation motion. The particles undergo multiple rotational movements and are finally deposited behind the tool pin. A part of the material stays in the outer edge of the weld seam under the forging of the tool shoulder. A small number of particles follow the shoulder of the tool, and most of the material is deposited behind the weld zone, with a material distribution on both the retreating side and the advancing side. For example, both P10 and P18 particles flow around the tool shoulder. The P10 particles eventually stay on the retreating side of the weld, while the P18 particles stay on the advancing side of the weld. At the same time, due to the extrusion of the tool pin, a part of the material is deposited on the bottom of the tool pin.

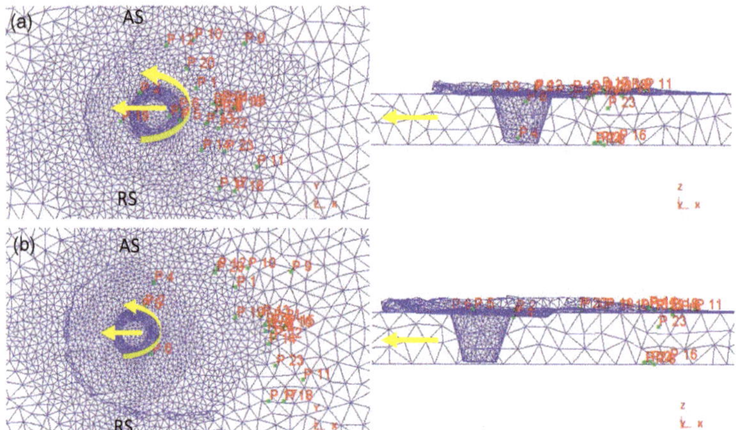

Figure 12. Material flow inside the tool needle during the welding phase. (**a**) Particle flow during welding; and (**b**) particle deposition behind the tool needle.

3.3.2. Material Flow on the Advancing Side

The tracking particles are arranged on the advancing side of the welding area. In order to improve the comprehensiveness of the simulation analysis, two sets of tracking particles (P1–P8 and P9–P16) are set. The distance from the axis of the tool pin is 3 mm, and the two groups of tracking particles are symmetrically distributed, as shown in Figure 13. It can be seen from the side view that they are located on both sides of the center line of the tool pin.

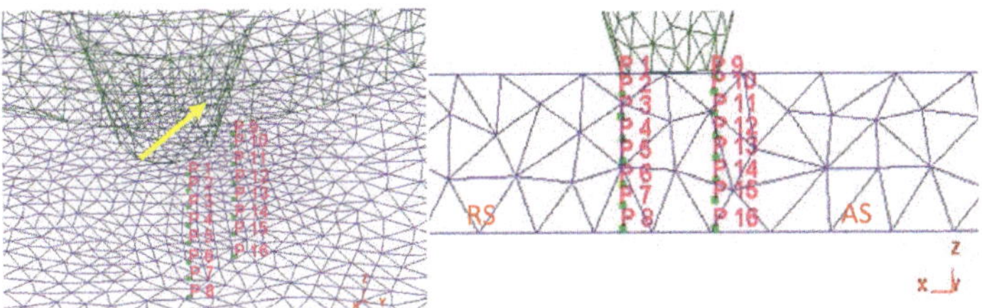

Figure 13. Advancing side tracking particle position setting.

During the plunging process, the particles on the advancing side differ from the flow of particles in the tool pin. During this stage, the particles are not rotated around the tool pin in the initial stage of pressing, but are squeezed by the tool pin to both sides of the tool pin, as shown in Figure 14. A plastic softening layer is formed, and the plastic softening zone becomes larger in the late stage of the plunging stage; the particles flow with the tool, the upper layer particles are spiraled to the upper side of the material and are finally moved to the advancing side behind the welding, and most of the other particles remain on the retreating side.

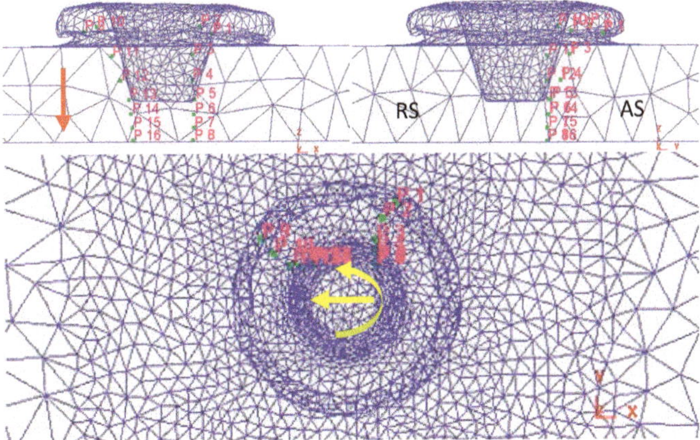

Figure 14. The trend of the particle flow during pressing.

Figure 15a,b shows that the particles P6, P7, and P8 are found behind the welding advancing side during the welding process. During the plunging process, the particles are plunged by the tool pin into the opposite direction of the welding, but the distance from the end of the tool pin becomes farther. The diameter of the end of the tool pin is small, and the effective influence area is smaller. Therefore, only a small part of the displacement of the three particles occurs during the welding process, and it cannot participate in the material flow.

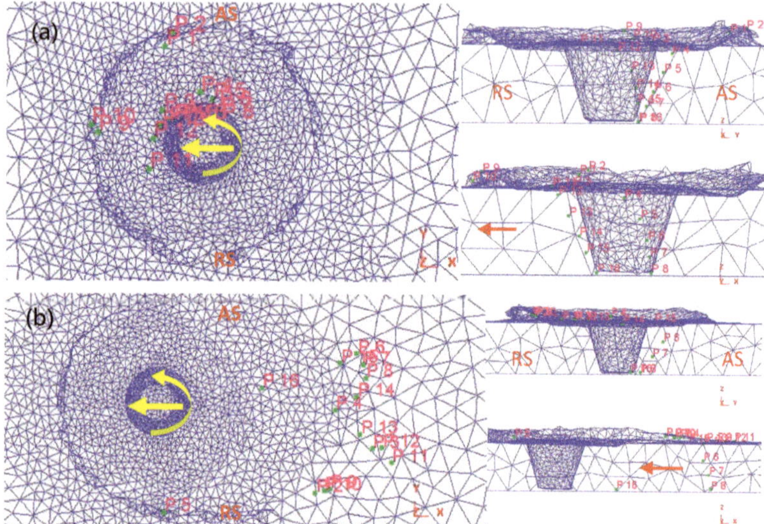

Figure 15. Material flow on the advancing side. (**a**) The tool is pressed down onto the given amount; and (**b**) the particles are deposited behind the tool.

Figure 15b shows that most of the tracking particles on the advancing side are rotated by the tool pin to the vicinity of the centerline of the weld bead behind the tool, and then gradually deposited at different positions between the center of the weld and the retreating side, which has also been observed in the experiment [38]. More particles were distributed near the retrograde side, and it was also confirmed that more materials flowed to the retreating side than to the advancing side. At the same time, partial backflow occurs due to the tool shoulder material, and finally the tool shoulder is forged to the rear of the tool.

3.3.3. Material Flow on the Retreating Side

Two sets of tracking particles are arranged on the retreating side, and the position is symmetrical to the advancing side. P1–P8 is close to the leading edge of the tool, and P9–P16 is close to the trailing edge of the tool. Since the rotation direction of the tool is counterclockwise, the particles in the initial stage of pressing extend outward under the extrusion action of the tool pin, as shown in Figure 16. As the pressing temperature gradually increases, the plastic softening layer in the weld zone becomes thicker, and the material plastically flows. The particles in the P1 group move backwards to the rear of the weld with the rotation of the tool pin.

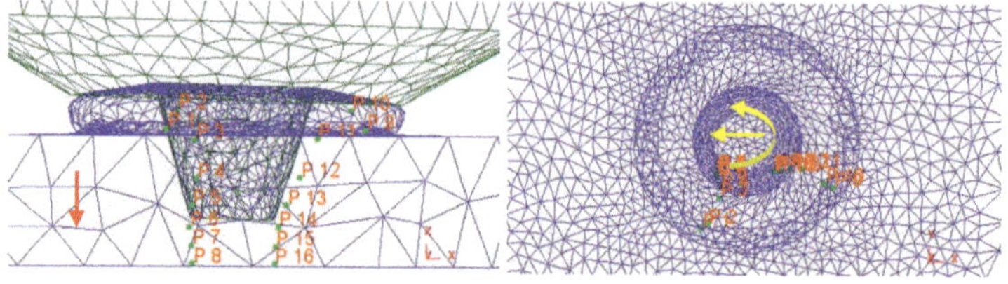

Figure 16. Particle migration during the plunging stage.

Figure 17a,b shows that the material flows under the action of the tool pin, whereas the material flow velocity on the retreating side is less than the advancing side, which also results in less material accumulating on the advancing side, which confirms the results of the literature [39].The trajectory of the marker material particles on the return side has two main parts. A part of the material is rotated around the tool pin for about half a week and then deposited onto the area near the advancing side of the tool. The other part of the material hardly rotates around the tool pin, but is gradually pushed down by the tool onto the area between the center line and the retreating side of the rear weld. A small number of particles on the tool shoulder is deposited in front of the retreating side after rotating around the tool for several weeks.

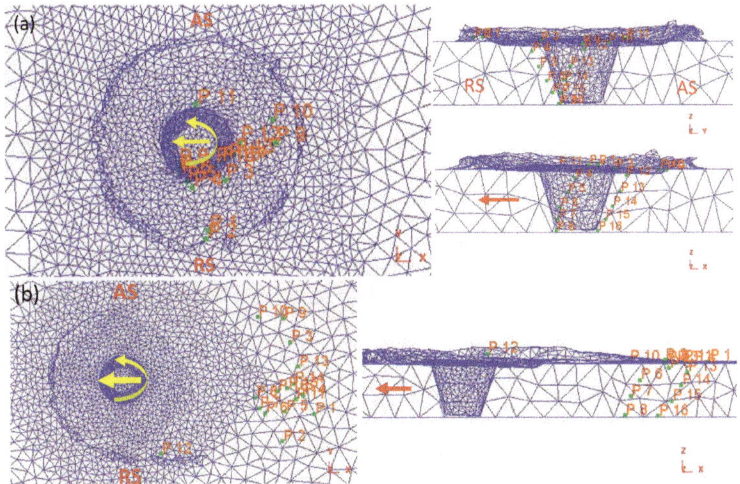

Figure 17. Material flow on the retreating side. (**a**) The tool is pressed down onto the given amount; and (**b**) the particles are deposited behind the tool.

3.4. Prediction of Welding Defects

Friction stir welding is a solid phase joining technique. Welding defects will also occur if improper welding parameters lead to an abnormal material flow. Two kinds of welding defects of the joint appeared in many simulation attempts.

The formation of hole defects is mainly due to insufficient heat input in the welding process, insufficient materials reaching the plastic softening state and an insufficient material flow, resulting in an incomplete closure in the welding interior [40]. When the spindle speed is low or the welding speed is too fast, the joint is prone to holes and crack defects. This type of weld defect is usually located in the middle and lower parts of the advancing side and near the weld surface. In the simulation process, the holes that were found after the welding speed increased by 2 times were formed in the lower parts of the advancing side, as shown in Figure 18b. When the rotation speed of the main shaft is further reduced by 0.5 times, a crack defect occurs after the welding, as shown in Figure 18a.

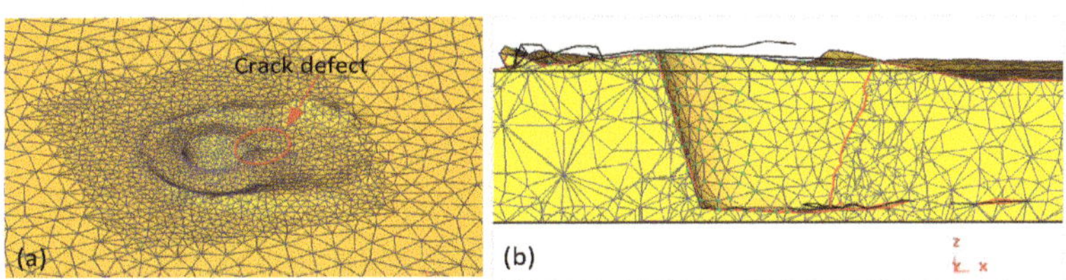

Figure 18. Simulation of the formation of weld defects. (**a**) Crack defects; and (**b**) hole defects.

According to the above research on the material flow, during the welding process the material flows from the advancing side to the retreating side of the material, and the material on the retreating side flows less to the advancing side. However, when the material flow is abnormal, the material on the rear side of the tool is difficult to fill in time, which will lead to the generation of holes. Therefore, in order to avoid the appearance of holes and cracks in the joint during the welding, it is very important to optimize the welding process parameters. A finite element analysis can effectively predict the welding defects. In this paper, the formation of welding defects is predicted by the law of material flow. Welding defects themselves are not expected to occur. This paper mainly studies the law of material flow, so the detailed study of welding defects will be carried out in the following work.

3.5. Effect of Material Flow on Microstructure

Figure 19 shows a comparison of the simulated temperature field and the metallographic structure of the joint cross section. It can be seen from the observation that the weld nugget zone (WNZ, zone I) in the cross section of the friction stir welding joint is connected with the thermo-mechanically affected zone (TMAZ, zone II). The heat affected zone (HAZ, zone III) is adjacent to the thermo-mechanically affected zone. In the simulation model, the temperature in the contact area with the tool is relatively high and gradually decreases to both sides.

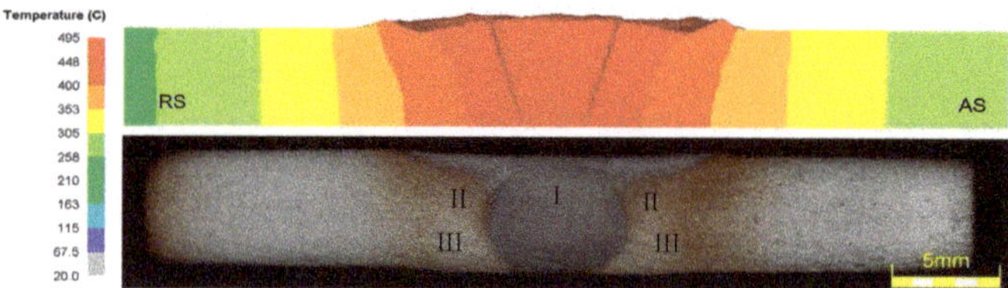

Figure 19. Cross-sectional comparison of the joints.

Due to the different material flows and microstructures in each region of the joint, the WNZ of the material forms a clear boundary with the TMAZ, and the material flow direction can be clearly seen from the formation of the microstructure in the TMAZ. As shown in Figure 20b, the microstructure in the WNZ is composed of fine recrystallized grains due to the intense stirring action of the pin. At the same time, it can be observed that the material in the TMAZ also has a plastic flow. The material under TMAZ near WAZ is rotated by the tool pin, and the material has an upward flow trend. Some of the material is squeezed into the WNZ, which is also found in the simulation. The material in TMAZ far away from the WNZ tends to move obliquely upward, and shows a similar rule on the advancing side, as shown in Figure 20a. However, it can be observed that the material flow slope on the advancing side is larger, which also confirms that the material flow on the advancing side is more intense.

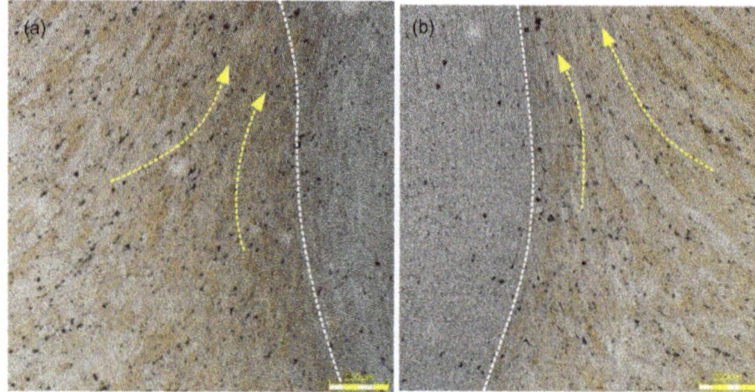

Figure 20. The boundary and material flow between the weld nugget zone (WNZ) and the thermo-mechanically affected zone (TMAZ). (**a**) The boundary area between the WNZ and the TMAZ on the advancing side; and (**b**) the boundary area between the WNZ and the TMAZ on the retreating side.

Figure 21a shows that the material flows obliquely to the weld nugget region above the TMAZ, and that the material in the tool shoulder action region has a slight flow trend. However, it can be seen from Figure 21b that a triangular region of material flow appears on the upper part of the advancing side due to the joint action of the tool pin and the tool shoulder. Here, the upper part of the weld has a downward flow trend under the extrusion of the tool shoulder. The simulation process also observed the action process of the tool shoulder. The action of the tool shoulder is more intense on the retreating side. The three-directional flows of the TMAZ material, tool shoulder action zone material and the WNZ material meet sharply. At the same time, if the material flow is abnormal in this area, weld defects will easily occur.

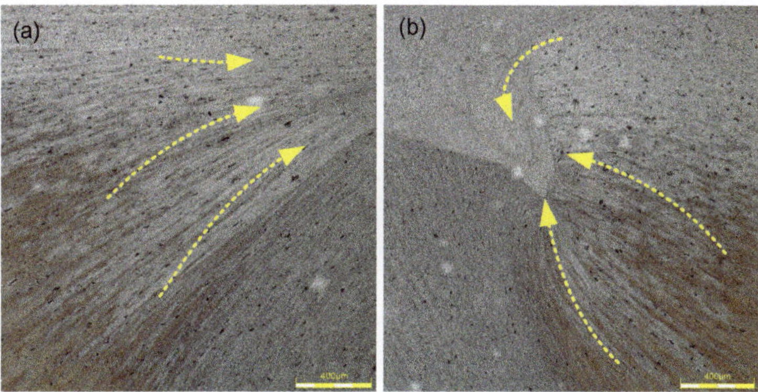

Figure 21. The material flow intersection of the welded joint. (**a**) The material intersection on the advancing side; and (**b**) the material intersection on the retreating side.

4. Conclusions

In this paper, the thermogenic physical model of the FSW process is established, and the finite element model of FSW is established. The material flow and temperature field in the welding process are studied by a numerical simulation. The finite element model is updated by the results of the axial force and torque tests. Moreover, the overall trend of the material flow in a finite element analysis is verified by the experiments on the weld cross section. The correctness of the finite element model analysis is verified by the microstructure formation and flow trend. Based on the results of this study, the following conclusions were made:

(1) The thermogenic physical model of the FSW process is established, and the finite element model is constructed to simulate the FSW process. The axial force and torque of the friction stir welding process were collected by electromagnetic coupling technology, and the data of the test and simulation were compared. The curve trend verified the correctness of the finite element model.

(2) Through a simulation analysis, it was found that the temperature on the advancing side is about 20 °C higher than that on the retreating side near the welding seam, and that the FSP temperature field has an important influence on the material flow field. By analyzing the temperature of the workpiece at different thickness layers, it was found that the temperature difference between the two sides of the middle and lower layers was relatively reduced.

(3) The material flow law in different areas of the weld was studied, and it was found that a small part of the inner material of the tool pin was extruded to the bottom of the workpiece. There is a large difference in the flow conditions of the upper and lower parts of the weld. The material on the upper surface tends to move downward under the influence of the shoulder extrusion, while the material on the lower part moves spirally upward under the influence of the tool pin. The material flow amount of the advancing side is higher than that of the retreating side.

(4) The material on the advancing side finally stays behind the weld, and most of the particles are distributed on the retreating side. The degree of the material flow gradually decreases along the thickness direction. Most of the material on the retreating side is finally distributed behind the welding seam, and some of it flows to the advancing side.

(5) Through a simulation analysis, it was found that the abnormal material flow in the welding process is prone to welding defects under the conditions of a low rotating speed and high welding speed.

(6) Through the comparative analysis of the cross-section of the joint, it was also found that there is a significant difference in the flow of the material between the advancing side and the retreating material, and that the material flow in the different regions makes the tissue forming boundary distinct. The tensile state of the joint microstructure can observe the flow tendency of the material. The advancing side forms a multi-directional material flow intersection area under the tool shoulder, which easily forms welding defects when the material flow is abnormal.

Author Contributions: H.L. and T.W. developed the FEM of friction stir welding process and wrote the paper; H.L., T.W. and P.W. conceived and designed the experiments; T.W. and Y.L. analyzed the data and modified the paper. T.W., F.Z. and H.W. performed the experiments.

Funding: The authors would like to acknowledge the financial support from the National Natural Science Foundation of China (Grant No. 51505470), the Youth Innovation Promotion Association, CAS and 'JXS' innovation fund, SIA.

References

1. Wang, T.; Zou, Y.; Matsuda, K. Micro-structure and micro-textural studies of friction stir welded AA6061-T6 subjected to different rotation speeds. *Mater. Des.* **2016**, *90*, 13–21. [CrossRef]
2. Morita, T.; Yamanaka, M. Microstructural evolution and mechanical properties of friction-stir-welded Al-Mg-Si joint. *Mater. Sci. Eng. A* **2014**, *595*, 196–204. [CrossRef]
3. Guo, J.F.; Chen, H.C.; Sun, C.N.; Bi, G.; Sun, Z.; Wei, J. Friction stir welding of dissimilar materials between AA6061 and AA7075 Al alloys effects of process parameters. *Mater. Des.* **2014**, *56*, 185–192.
4. Zhang, H.; Wang, M.; Zhang, X.; Yang, G. Microstructural characteristics and mechanical properties of bobbin tool friction stir welded 2A14-T6 aluminum alloy. *Mater. Des.* **2015**, *65*, 559–566. [CrossRef]
5. Carlone, P.; Palazzo, G.S. Influence of process parameters on microstructure and mechanical properties in AA2024-T3 friction stir welding. *Metall. Microstruct. Anal.* **2013**, *2*, 213–222. [CrossRef]
6. Su, H.; Wu, C.S.; Bachmann, M.; Rethmeier, M. Numerical modeling for the effect of pin profiles on thermal and material flow characteristics in friction stir welding. *Mater. Des.* **2015**, *77*, 114–125. [CrossRef]
7. Tutunchilar, S.; Haghpanahi, M.; Besharati Givi, M.K.; Asadi, P.; Bahemmat, P. Simulation of material flow in friction stir processing of a cast Al-Si alloy. *Mater. Des.* **2012**, *40*, 415–426. [CrossRef]
8. Chen, Z.W.; Pasang, T.; Qi, Y. Shear flow and formation of nugget zone during friction stir welding of aluminium alloy 5083-O. *Mater. Sci. Eng. A* **2008**, *474*, 312–316. [CrossRef]
9. Zhu, Y.; Chen, G.; Chen, Q.; Zhang, G.; Shi, Q. Simulation of material plastic flow driven by non-uniform friction force during friction stir welding and related defect prediction. *Mater. Des.* **2016**, *108*, 400–410.
10. Assidi, M.; Fourment, L.; Guerdoux, S.; Nelson, T. Friction model for friction stir welding process simulation: Calibrations from welding experiments. *Int. J. Mach. Tools. Manuf.* **2010**, *50*, 143–155. [CrossRef]
11. Zhang, Z.; Wu, Q. Numerical studies of tool diameter on strain rates, temperature rises and grain sizes in friction stir welding. *J. Mech. Sci. Technol.* **2015**, *29*, 4121–4128. [CrossRef]
12. Chowdhury, S.M.; Chen, D.L.; Bhole, S.D.; Cao, X. Tensile properties of a friction stir welded magnesium alloy: Effect of pin tool thread orientation and weld pitch. *Mater. Sci. Eng. A* **2010**, *527*, 6064–6075. [CrossRef]
13. Yang, C.L.; Wu, C.S.; Lv, X.Q. Numerical analysis of mass transfer and material mixing in friction stir welding of aluminum/magnesium alloys. *J. Manuf. Process.* **2018**, *32*, 380–394. [CrossRef]
14. Tongne, A.; Desrayaud, C.; Jahazi, M.; Feulvarch, E. On material flow in Friction Stir Welded Al alloys. *J. Mater. Process. Tech.* **2017**, *239*, 284–296. [CrossRef]

15. Mironov, S.; Masaki, K.; Sato, Y.S.; Kokawa, H. Relation between material flow and abnormal grain growth in friction-stir welds. *Scripta Mater.* **2012**, *67*, 983–986. [CrossRef]

16. Franke, D.J.; Zinn, M.R.; Pfefferkorn, F.E. Intermittent Flow of Material and Force-Based Defect Detection During Friction Stir Welding of Aluminum Alloys. *Frict. Stir Weld. Process. X* **2019**, 149–160.

17. Gratecap, F.; Girard, M.; Marya, S.; Racineux, G. Exploring material flow in friction stir welding: Tool eccentricity and formation of banded structures. *Int. J. Mater. Form.* **2012**, *5*, 99–107. [CrossRef]

18. Huang, Y.; Wang, Y.; Wan, L.; Liu, H.; Shen, J.; dos Santos, J.F.; Zhou, L.; Feng, J. Material-flow behavior during friction-stir welding of 6082-T6 aluminum alloy. *Int. J. Adv. Manuf. Tech.* **2016**, *87*, 1115–1123. [CrossRef]

19. Dialami, N.; Chiumenti, M.; Cervera, M.; Agelet de Saracibar, C.; Ponthot, J.P.; Bussetta, P. Numerical Simulation and Visualization of Material Flow in Friction Stir Welding via Particle Tracing. *Numer. Simul. Coupl. Probl. Eng.* **2014**, *33*, 157–169.

20. Li, W.-Y.; Li, J.-F.; Zhang, Z.-H.; Gao, D.-L.; Chao, Y.-J. Metal flow during friction stir welding of 7075-T651 Aluminum alloy. *Exp. Mech.* **2013**, *53*, 1573–1582. [CrossRef]

21. Shanavas, S.; Dhas, J.E.R. Quality Prediction of Friction Stir Weld Joints on AA 5052 H32 Aluminium Alloy Using Fuzzy Logic Technique. *Mater. Today Proc.* **2018**, *5*, 12124–12132. [CrossRef]

22. De Filippis, L.A.C.; Serio, L.M.; Facchini, F.; Mummolo, G.; Ludovico, A.D. Prediction of the Vickers Microhardness and Ultimate Tensile Strength of AA5754 H111 Friction Stir Welding Butt Joints Using Artificial Neural Network. *Materials* **2016**, *9*, 915. [CrossRef] [PubMed]

23. Cisko, A.B.; Jordon, J.B.; Avery, D.Z.; Liu, T.; Brewer, L.N.; Allison, P.G.; Carino, R.L.; Hammi, Y.; Rushing, T.W.; Garcia, L. Experiments and Modeling of Fatigue Behavior of Friction Stir Welded Aluminum Lithium Alloy. *Metals* **2019**, *9*, 293. [CrossRef]

24. Zhang, Z.; Xiao, B.L.; Wang, D.; Ma, Z.Y. Effect of Alclad Layer on Material Flow and Defect Formation in Friction-Stir-Welded 2024 Aluminum Alloy. *Metall. Mater. Trans. A* **2011**, *42*, 1717–1726. [CrossRef]

25. Zhu, Z.; Wang, M.; Zhang, H.; Zhang, X.; Yu, T.; Wu, Z. A finite element model to simulate defect formation during friction stir welding. *Metals* **2017**, *7*, 256. [CrossRef]

26. Edwards, P.D.; Ramulu, M. Material flow during friction stir welding of Ti-6Al-4V. *J. Mater. Process. Technol.* **2015**, *218*, 107–115. [CrossRef]

27. Mao, Y.; Ke, L.; Liu, F.; Chen, Y.; Xing, L. Effect of tool pin-tip profiles on material flow and mechanical properties of friction stir welding thick AA7075-T6 alloy joints. *Int. J. Adv. Manuf. Technol.* **2017**, *88*, 949–960. [CrossRef]

28. Ji, S.D.; Shi, Q.Y.; Zhang, L.G.; Zou, A.L.; Gao, S.S.; Zan, L.V. Numerical simulation of material flow behaviour of friction stir welding influenced by rotational tool geometry. *Comp. Mater. Sci.* **2012**, *63*, 218–226. [CrossRef]

29. Luo, H.; Wu, T.; Fu, J.; Wang, W.; Chen, N.; Wang, H. Welding Characteristics Analysis and Application on Spacecraft of Friction Stir Welded 2A14-T6 Aluminum Alloy. *Materials* **2019**, *12*, 480. [CrossRef] [PubMed]

30. Sun, Z.; Wu, C.S. A numerical model of pin thread effect on material flow and heat generation in shear layer during friction stir welding. *J. Manuf. Process.* **2018**, *36*, 10–21. [CrossRef]

31. Buffa, G.; Hua, J.; Shivpuri, R.; Fratini, L. A continuum based FEM model for friction stir welding-model development. *Mater. Sci. Eng. A* **2006**, *419*, 389–396. [CrossRef]

32. Jain, R.; Pal, S.K.; Singh, S.B. A study on the variation of forces and temperature in a friction stir welding process: A finite element approach. *J. Manuf. Process.* **2016**, *23*, 278–286. [CrossRef]

33. Leal, R.M.; Leitao, C.; Loureiro, A.; Rodrigues, D.M.; Vilaca, P. Material flow in heterogeneous friction stir welding of thin aluminium sheets: Effect of shoulder geometry. *Mater. Sci. Eng. A* **2008**, *498*, 384–391. [CrossRef]

34. Jain, R.; Pal, S.K.; Singh, S.B. Finite element simulation of pin shape influence on material flow, forces in friction stir welding. *Int. J. Adv. Manuf. Technol.* **2018**, *94*, 1781–1797. [CrossRef]

35. Zhou, Z.X.; Li, Y.L.; Suo, T. Microstructure and Dynamic Mechanical Properties of 2A14 Aluminium Alloy Sheet. *Mech. Sci. Technol. Aerosp. Eng.* **2009**, *28*, 1464–1467.

36. GB/T 3880-1997. *Aluminium and Aluminium Alloy Rolled Sheet*; Standardization Administration of the People's Republic of China: Beijing, China, 1998.

37. AlRoubaiy, A.O.; Nabat, S.M.; Batako, A.D.L. Experimental and theoretical analysis of friction stir welding of Al-Cu joints. *Int. J. Adv. Manuf. Technol.* **2014**, *71*, 1631–1642. [CrossRef]

38. Masaki, K.; Saito, H.; Nezaki, K.; Kitamoto, S.; Sato, Y.S.; Kokawa, H. Material Flow and Microstructure Evolution in Corner Friction Stir Welding of 5083 Al Alloy Using AdStir Technique. *Frict. Stir Weld. Process. X* **2019**, 181–188. [CrossRef]

39. Zhang, H.J.; Liu, H.J. Characteristics and Formation Mechanisms of Welding Defects in Underwater Friction Stir Welded Aluminum Alloy. *Metallogr. Microstruct. Anal.* **2012**, *1*, 269–281. [CrossRef]
40. Kumar, K.; Kailas, S.V. The role of friction stir welding tool on material flow and weld formation. *Mater. Sci. Eng. A* **2008**, *485*, 367–374. [CrossRef]

Article

Effect of Forced Air Cooling on the Microstructures, Tensile Strength, and Hardness Distribution of Dissimilar Friction Stir Welded AA5A06-AA6061 Joints

Guangjian Peng [1,2], Qi Yan [1], Jiangjiang Hu [1,2], Peijian Chen [3], Zhitong Chen [4,*] and Taihua Zhang [5,*]

[1] College of Mechanical Engineering, Zhejiang University of Technology, Hangzhou 310014, China; penggj@zjut.edu.cn (G.P.); yanqi@zjut.edu.cn (Q.Y.); jiangjianghu@zjut.edu.cn (J.H.)
[2] Key Laboratory of E&M (Zhejiang University of Technology), Ministry of Education & Zhejiang Province, Hangzhou 310014, China
[3] State Key Laboratory for Geomechanics and Deep Underground Engineering, School of Mechanics and Civil Engineering, China University of Mining and Technology, Xuzhou 221116, China; chenpeijian@cumt.edu.cn
[4] Department of Mechanical and Aerospace Engineering, The George Washington University, Washington, DC 20052, USA
[5] Institute of Solid Mechanics, Beihang University, Beijing 100191, China
*   Correspondence: zhitongchen@gwu.edu (Z.C.); zhangth66@buaa.edu.cn (T.Z.); Tel.: +1-202-834-8151 (Z.C.); +86-186-6710-2163 (T.Z.)

Received: 29 January 2019; Accepted: 4 March 2019; Published: 7 March 2019

Abstract: Friction stir welding (FSW) is a promising welding method for welding dissimilar materials without using welding flux. In the present work, 5A06-H112 and 6061-T651 aluminium alloys were successfully welded by friction stir welding with forced air cooling (FAC) and natural cooling (NC). Nanoindentation tests and microstructure characterisations revealed that forced air cooling, which can accelerate the cooling process and suppress the coarsening of grains and the dissolution of precipitate phases, contributes to strengthening and narrowing the weakest area of the joint. The tensile strength of joints with FAC were commonly improved by 10% compared to those with NC. Scanning electron microscopy (SEM) images of the fracture surface elucidated that FSW with FAC tended to increase the number and reduce the size of the dimples. These results demonstrated the advantages of FSW with FAC in welding heat-sensitive materials and provide fresh insight into welding industries.

Keywords: friction stir welding; aluminium alloys; forced air cooling; microstructures; tensile strength; hardness distribution

---

1. Introduction

Aluminium alloys have been widely used in the aviation, aerospace, shipping, and transportation industries because of their advantages of high specific strength, good fatigue performance, and strong corrosion resistance [1–3]. However, the welding of aluminium alloys presents great challenges. Traditional welding methods are not suitable for aluminium alloys, particularly dissimilar aluminium alloys, since they are prone to generate defects such as porosity and crack during solidification [4,5]. Friction stir welding (FSW)—which is a cost-effective and environmentally friendly solid-state method developed by the Welding Institute in 1991 [6]—is a promising welding method that avoids the above-mentioned issues. According to the microstructures and thermal effect, FSW joints can be typically divided into four zones: nugget zone (NZ), thermo-mechanical affected zone (TMAZ), heat affected zone (HAZ), and base material (BM).

Several researchers have investigated the friction stir welding of aluminium alloys and found that FSW could significantly improve the strength of welded joints compared with traditional welding methods [7–13]. However, due to the large amount of heat generated by friction and plastic deformation, the welded joint shows an obvious thermal softening effect, leading to the strength of the HAZ being lower than the BM [14,15]. Taking this into account, a forced cooling medium was applied during the welding process to reduce the thermal softening and improve the strength of the joint. Zhang et al. [16] pointed out that the tensile strength and corrosion resistance of FSW AA2014 joints were enhanced via water cooling. Sharma et al. [17] studied the effects of air, water, and liquid nitrogen cooling on AA7039 FSW, and suggested that water cooling was more helpful in improving the mechanical properties of the joints. Benavides et al. [18] investigated the mechanical properties of FSW AA2024 joints cooled by liquid nitrogen, and revealed that (1) the grain size in the NZ reduced, (2) the softening area of the joint narrowed, and (3) the hardness of the TMAZ and HAZ increased. Heirani et al. [19] focused on the influence of underwater FSW on the microstructure and mechanical properties of AA5083. They found that the HAZ of the joint disappeared, and the strength and hardness were improved. Other studies also indicated that a forced cooling medium is beneficial for improving the mechanical performances of FSW aluminium alloy joints [20–23].

The 5A06 is a corrosion-resistant non-heat treatable aluminium alloy (Al-Mg alloy), while 6061 is a high-strength heat-treatable aluminium alloy (Al-Mg-Si alloy) [24]. In the present work, a forced air cooling system that is cheap and easy to use was added to the commercial friction stir welding machine. The main purpose was to understand the effect of forced air cooling on the microstructure and mechanical properties of FSW AA5A06-AA6061 joints. The microstructures, tensile properties, hardness distribution, and fracture features of the joints welded in different conditions were systematically investigated and discussed.

2. Materials and Methods

Rolled AA5A06-H112 and AA6061-T651 plates with a thickness of 5 mm were employed as base materials, and the chemical compositions obtained by energy dispersive spectroscopy (EDS, Carl Zeiss SMT Pte Ltd., Oberkochen, Germany) are listed in Table 1. The plates were cut into small rectangular plates measuring 165 mm × 5 mm. These rectangular plates were welded in butt joint configuration using FSW by placing AA5A06 on the advancing side and AA6061 on the retreating side. A welding machine HT-JL10X12/2H (Shanghai Aerospace Equipments Manufacturer Co., Ltd., Shanghai, China) with a forced air cooling system was employed to carry out FSW. Both natural cooling (NC) and forced air cooling (FAC) conditions were considered in the FSW process. For FAC, forced air was blown on the welded area through a nozzle, as shown in Figure 1a. The rectangular nozzle with a size of 10 mm × 2 mm was placed 20 mm behind the tool and about 20 mm above the surface of the materials. The pressure of the forced air was 0.5 MPa, and the blowing direction was along the welding direction and had an intersection angle of 30° with the surface of the materials. The tapered left-hand threaded cylindrical tool used for FSW was made of "H13 steel", and the dimensions are shown in Figure 1b. During the welding process, the tool was plunged into the butt surface of the two base materials, with a tilt angle of 2.8° and a depth of 4.96 mm. Three different tool rotational speeds (RS)—i.e., 600, 900, and 1200 rpm—and two welding transverse speeds (TS)—i.e., 100 and 200 mm/min—were applied. It has been reported that the left-hand thread pin tool rotating clockwise generates better FSW joints [25]. Thus, a clockwise rotation direction of the tool was used in the experiments. As shown in Figure 1c, the welded AA5A06-AA6061 plates were processed into dumbbell-shaped specimens for uniaxial tensile tests and small rectangular specimens for nanoindentation and microstructure characterisation.

Table 1. Chemical composition of AA5A06 and AA6061 (wt. %).

Materials	Si	Fe	Cu	Mn	Mg	Cr	Zn	Ti	Al
AA5A06	0.40	0.26	0.06	0.86	5.37	-	0.10	0.11	92.84
AA6061	0.79	0.70	0.35	0.08	1.46	0.17	0.08	0.21	96.16

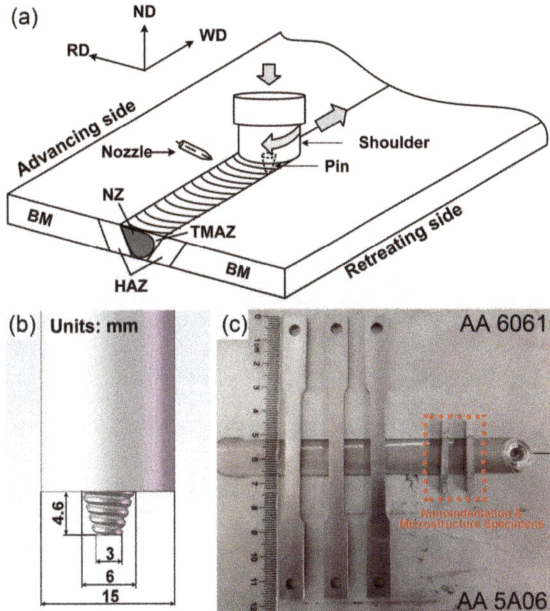

Figure 1. (**a**) Schematic of friction stir welding (FSW) with forced air cooling, (**b**) geometry of the tool, (**c**) specimens for uniaxial tensile tests, nanoindentation, and microstructure characterisation.

For microstructure characterisation, the side surface of the specimen was polished to a mirror finish using an automatic polishing machine (Shenyang Kejing Auto-instrument Co. Ltd., Shenyang, China) with 0.3 μm alumina powder. The surfaces were anodic coated and then observed using a polarizing microscope (Carl Zeiss AG Co. Ltd., Guangzhou, China). To obtain the hardness distribution in the joint area, nanoindentation tests were implemented in an Agilent Nano Indenter G200 system with a modified Berkovich indenter (Agilent Technologies, Oak Ridge, TN, USA). The nanoindentation hardness mapping method was used and an array consisting of 56 × 10 indents was carried out on the side surface of each specimen, as shown in Figure 1c. For each nanoindentation test, the maximum indentation load was 120 mN, and the thermal drift was controlled within 0.05 nm/s.

For uniaxial tensile tests, the dumbbell-shaped specimens had 1 mm of thickness machined away from both the top and bottom surface to ensure they had a uniform cross section. A material testing system, MTS CMT4204 (MTS System Co. Ltd., Shanghai, China) was employed to conduct the tensile tests. All the specimens were stretched to break at a tensile strain rate of about 0.04 s^{-1} at room temperature. The average value of the ultimate tensile strength was used for analysis. Scanning electron microscopy (SEM, Carl Zeiss SMT Pte Ltd., Oberkochen, Germany) was employed to analyse the microstructures of the fracture surface.

3. Results and Discussion

3.1. Microstructures

Defect-free joints were obtained for all welding parameters and conditions. Figure 2a shows a representative cross section of the joint, obtained at the ratio of rotational speed to transverse speed (R/T ratio) of 1200/100 r/mm with FAC. As shown in Figure 2, "onion rings" and interlaced ripple structures, which indicate a good mixing of materials, were observed in the NZ. The formation of an "onion ring" is attributed to the thermal softening of the materials, and the stirring action, extrusion, and transverse movement of the threaded tool [26]. Comparison of the "onion rings" in Figure 2b,c revealed that the spacing between the ripples decreases with an increase in the rotational speed of the tool, which agrees with Rodriguez's report [27]. There are two main reasons for this: (1) the higher heat input generated by the higher rotational speed makes the materials softer and easier to flow under mechanical stirring, and (2) higher rotational speed provides a higher stirring force, which mixes these two materials sufficiently to form narrower ripples.

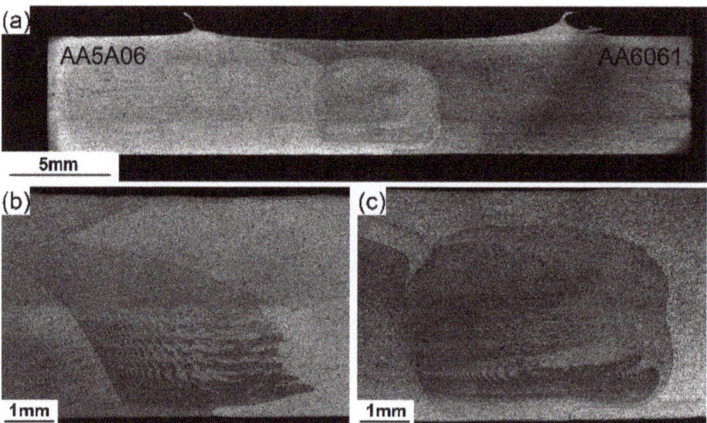

Figure 2. Macroscopic features of the joints with forced air cooling: (**a**) cross-section of a joint welded at a rotational speed to transverse speed (R/T) ratio of 1200/100 r/mm, (**b**) NZ of a joint welded at an R/T ratio of 600/100 r/mm, and (**c**) NZ of a joint welded at an R/T ratio of 1200/100 r/mm.

Figure 3 shows typical microstructures of different regions of the FSW AA5A06-AA6061 joints welded at a rotational speed of 1200 rpm and a welding speed of 100 mm/min with NC and FAC. Rolled prolate grains were observed in the BM of both AA5A06 and AA6061, and their average grain size was 49 μm and 50 μm, respectively (Figure 3a,b). For both cooling conditions, the NZ exhibits dynamically recrystallised microstructures with fine equiaxed grains (Figure 3c,d), which are the result of thermal softening and large plastic deformation during the FSW process. The average grain sizes in the NZ were 14 μm and 8 μm for FSW with NC and FAC, respectively. Figure 3e,f shows the grain structures of the HAZ of the 5A06 side for FSW with NC and FAC, and their average grain sizes, which were 52 μm and 51 μm, respectively. It was found that the average grain size in the HAZ and the BM of AA5A06 were almost the same and independent of the cooling conditions. The reason for this was that AA5A06 is a non-heat treatable aluminium alloy, and the heat generated by the FSW had little effect on the grain structures. On the 6061 side, however, the average grain size of the HAZ for FSW with NC is 57 μm (Figure 3g), an increase of about 14% compared to the average grain size of the BM. This was due to the fact that AA6061 is a heat-treatable alloy, and is therefore sensitive to temperature variations. The elevated temperature generated during the FSW process induced the growing and coarsening of grains in the HAZ. With the aid of the FAC, the average grain size of the HAZ on the 6061 side was reduced to 53 μm (Figure 3h). The FSW with FAC treatment could accelerate the cooling process and reduce the affecting time of high temperatures, thus suppressing the coarsening and growth of the grains.

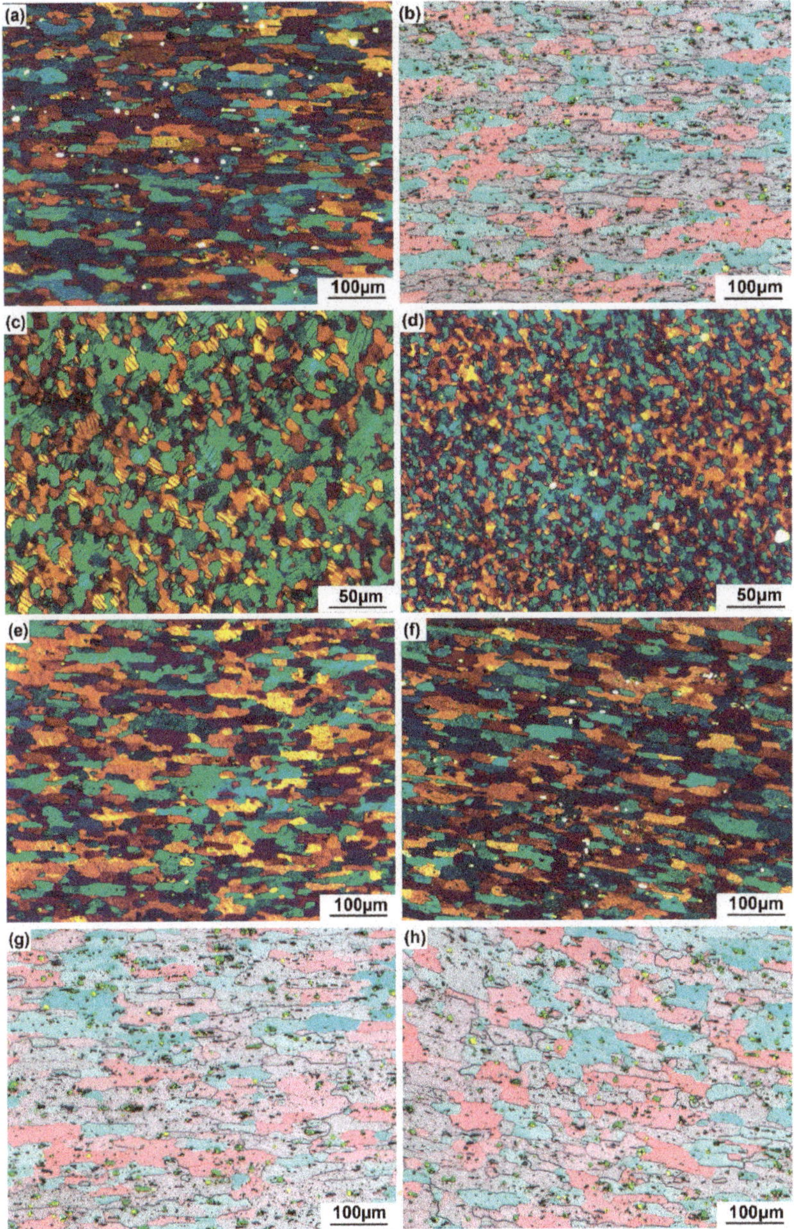

Figure 3. Microstructures of different regions of the FSW joints welded at an R/T ratio of 1200/100 r/mm with natural cooling (NC) and forced air cooling (FAC): (**a**) base material (BM) of 5A06, (**b**) BM of 6061, (**c**) nugget zone (NZ) (with NC), (**d**) NZ (with FAC), (**e**) heat affected zone (HAZ) of 5A06 (with NC), (**f**) HAZ of 5A06 (with FAC), (**g**) HAZ of 6061 (with NC), and (**h**) HAZ of 6061 (with FAC).

3.2. Uniaxial Tensile Properties

After the uniaxial tensile tests, it was found that the fracture occurred at the HAZ of the AA6061 side for all the FSW joints. This means that the ultimate tensile strength of AA6061 was weakened after welding, and the HAZ on the AA6061 side became the weakest area. Figure 4 shows the ultimate tensile strength of the base AA6061 and the FSW joints welded at various R/T ratios and under different cooling conditions. For FSW with NC, the joint welded at an R/T ratio of 600/200 r/mm had the maximum ultimate tensile strength of 198.27 MPa, which was 29% lower than that of the base AA6061. As the R/T ratio increased from 3 (600/200) r/mm to 12 (1200/100) r/mm, the ultimate tensile strength continued to decrease slowly, and the decrease was within 10%. It is evident from Figure 4 that there were two levels of decrease for the tensile strength of the FSW joints.

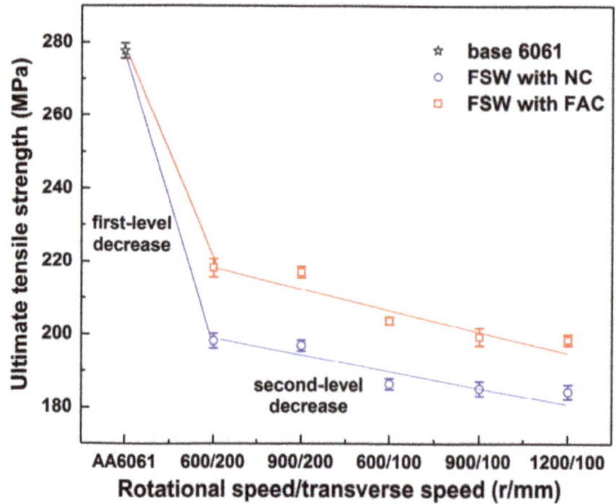

Figure 4. The variation in ultimate tensile strength with the ratio of rotational speed to welding speed. There were two levels of decrease for the tensile strength of the FSW joints. The ultimate tensile strength of the joints with FAC was generally 10% higher than for those with NC.

For the first-level decrease, the tensile strength of the FSW joint was largely weakened once the AA6061 was friction stir welded. As Peel et al. [28] pointed out, this is because the rolled AA6061 was kept in an extremely work-hardened state and had highly unstable microstructures. The recrystallisation caused by the temperature rising during FSW can readily destroy the hardened state and significantly weaken the mechanical properties. In addition, precipitation hardening is one of the strengthening mechanisms for heat-treatable AA6061 [29,30], and grain size also plays an important role according to the Hall-Petch relationship [31]. For AA6061, the main strengthening phase is the needle-shaped β'' phase. With the increase of the heat input, the fine needle-shaped β'' phase dissolved and grew to the coarsened rod-shaped β' phase and equilibrium β phase, which weakened the mechanical properties of the joint [32,33]. A comparison of Figure 3b,g reveals that the grains became coarser after welding. To sum up, the first-level decrease in tensile strength was attributed to the destruction of the hardened state (the primary reason), the dissolution and coarsening of the strengthening phases, and the coarsening of the grains. For the second-level decrease, the tensile strength of the FSW joint was slightly weakened with the increase of the R/T ratio. This was mainly because the higher R/T ratio generated a higher temperature, which led to more strengthening phases to be coarsened and causing grains to grow larger. The further dissolution of the precipitates and further coarsening of the grains could weaken the tensile strength with the increase of the R/T ratio, as shown in Figure 4.

With the aid of FAC, the cooling process was accelerated, and the affecting time of the high temperature was reduced. This was conducive to suppressing the coarsening of the grains and the dissolution of the precipitates in the HAZ of 6061, and thus improved the mechanical properties. As shown in Figure 4, the ultimate tensile strength of the joints with FAC was commonly improved by 10% compared to that with NC.

3.3. Nanoindentation Hardness Distribution

The nanoindentation hardness distribution on the cross-section of the FSW joints is shown in Figure 5. The hardness distribution on the AA5A06 side varied slightly for all the welding parameters. The maximum hardness appeared near the NZ and TMAZ, which was similar to the FSW AA5182 reported by Tronci et al. [31,34]. The reason for this was that AA5A06 is a non-heat treatable alloy, and therefore the temperature variation does not significantly affect the hardness. Near the NZ and TMAZ, the hardness was partially improved due to the refinement of the grains as a result of the stirring action. For the AA6061 side, a dramatic drop in hardness was observed in the HAZ due to the coarsening of the grains and the dissolution of the precipitates [35]. The HAZ of AA 6061 became the weakest area in the FSW joint. This intuitively explains why the fracture occurred in the HAZ of AA6061 during tensile tests. A comparison of Figure 5a,b or Figure 5c,d reveals that the hardness in the HAZ of AA6061 was improved and the weakest area was narrowed with the help of FAC. This is consistent with the results observed in the tensile test, namely that the strength was improved via FAC. The FSW with FAC could reduce the affecting time of high temperatures and suppress the coarsening of the grains and the dissolution of the precipitate phases in the HAZ of 6061, contributing to the improvement of the joint hardness.

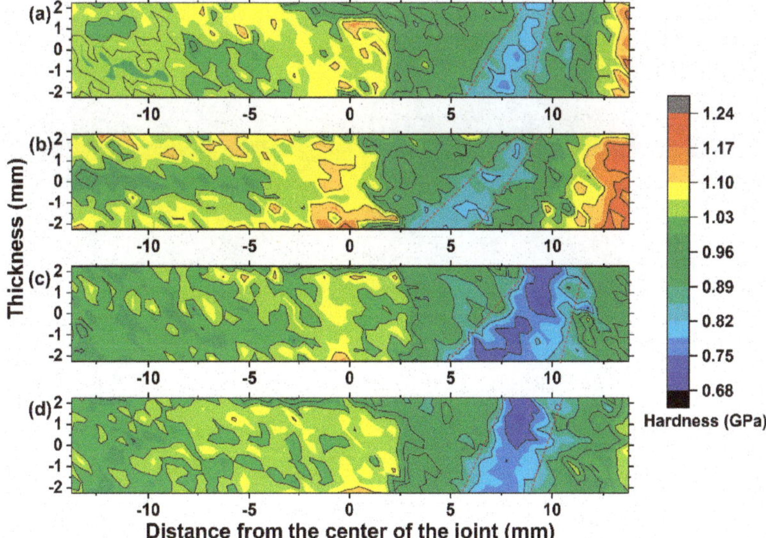

Figure 5. Nanoindentation hardness distribution on the cross-section of the FSW joint obtained in various welding conditions: (**a**) 600/200 r/mm with NC, (**b**) 600/200 r/mm with FAC, (**c**) 1200/100 r/mm with NC, and (**d**) 1200/100 r/mm with FAC. As the R/T ratio increased, the weakest area of the joint became wider. With the aid of FAC, the weakest area was narrowed.

3.4. Fracture Characteristics

Figure 6 shows the SEM images of the microstructures of the fracture surfaces. The dimples are observed in all fracture surfaces, indicating the ductile fracture mechanism of the joints. For FSW with NC, the fracture surface of the joint obtained at an R/T ratio of 1200/100 r/mm exhibited large and deep dimples (Figure 6a). When the R/T ratio reduced to 600/200 r/mm, the dimples became smaller (Figure 6b). For the fracture surfaces of the FSW joints with FAC, as shown in Figure 6c,d, the dimples continue to become smaller and shallower. It is understood that the size of the dimple is affected by the spacing of the precipitates, and a larger dimple leads to better plasticity and worse strength [36]. The quantity and size of the dimples can indirectly reflect the number and size of the precipitates. As mentioned above, precipitation hardening is the main strengthening mechanism for AA 6061. Typically, smaller and more uniformly distributed precipitates lead to better mechanical properties. Based on this point of view, it could be inferred from Figure 6 that the strength of joints with FAC is commonly higher than that of naturally cooled joints, which coincides well with the uniaxial tensile results.

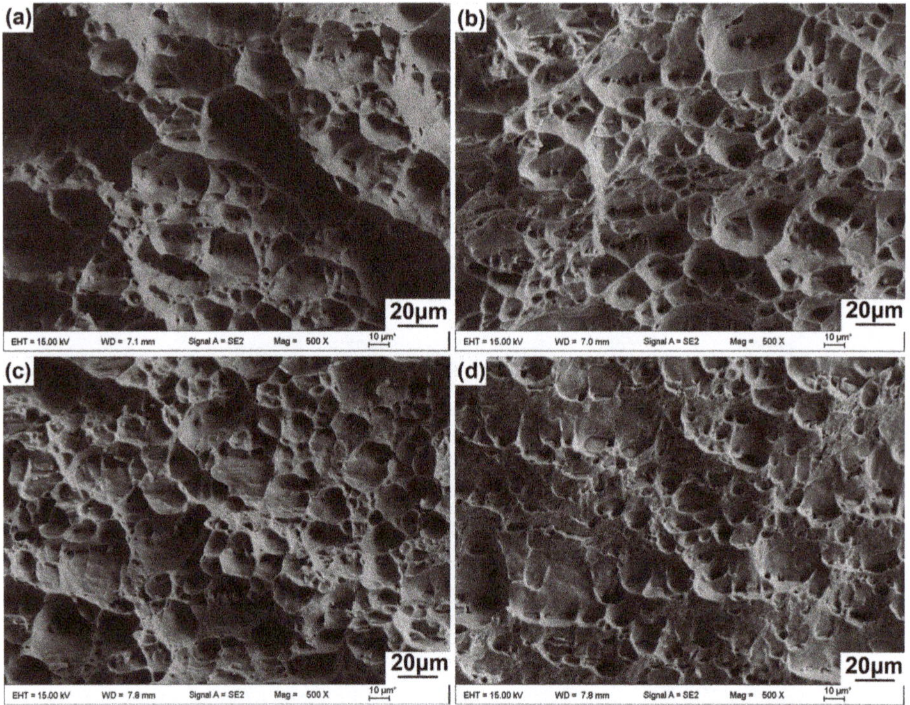

Figure 6. Scanning electron microscopy (SEM) images of the fracture surface of FSW specimens obtained at: (**a**) 1200/100 r/mm with NC, (**b**) 600/200 r/mm with NC; (**c**) 1200/100 r/mm with FAC, and (**d**) 600/200 r/mm with FAC. The fracture surfaces of the FSW joints with FAC tended to have smaller and shallower dimples than those with NC.

4. Conclusions

The 5A06 and 6061 aluminium alloys were successfully friction stir welded in both NC and FAC conditions. For all the welded joints, the HAZ of AA6061 was the weakest area and fracture occurred in this region during tensile tests. There were two levels of decrease in the tensile strength of the FSW joints. For the first-level decrease, the tensile strength of the FSW joint was largely weakened due to the destruction of the hardened state (the primary reason), the dissolution of the precipitates and the coarsening of the grains. For the second-level decrease, the tensile strength of the FSW joint was slightly weakened with the increase of the R/T ratio. This is attributed to the further dissolution of the precipitates and further coarsening of the grains, caused by the increase in temperature during FSW.

FAC can effectively accelerate the cooling process and reduce the affecting time of high temperature during FSW, suppressing the coarsening of the grains and the dissolution of the precipitates in HAZ of 6061, as well as generally improve the ultimate tensile strength by 10% compared with NC. The nanoindentation hardness contour maps also intuitively illustrate that FSW with forced air cooling is conducive to improving the hardness in the HAZ of AA 6061 and narrowing the weakest area.

Author Contributions: Funding acquisition, G.P. and T.Z.; Investigation, G.P., Q.Y., J.H., P.C. and Z.C.; Methodology, J.H. and P.C.; Supervision, G.P. and T.Z.; Writing—original draft, Q.Y.; Writing—review and editing, G.P. and Z.C.

Funding: The authors would like to gratefully acknowledge the support of the National Natural Science Foundation of China (Grant Nos. 11772302, 11727803 and 11672356) and Zhejiang Province Public Welfare Technology Application Research Project (2015C31074).

Conflicts of Interest: The authors declare no conflict of interest.

References

1. Dursun, T.; Soutis, C. Recent developments in advanced aircraft aluminium alloys. *Mater. Des.* **2014**, *56*, 862–871. [CrossRef]
2. Gharavi, F.; Matori, K.A.; Yunus, R.; Othman, N.K.; Fadaeifard, F. Corrosion Evaluation of Friction Stir Welded Lap Joints of AA6061-T6 Aluminum Alloy. *Trans. Nonferr. Metals Soc. Chin.* **2016**, *26*, 672–681. [CrossRef]
3. Hassan, K.A.A.; Prangnell, P.B.; Norman, A.F.; Price, D.A.; Williams, S.W. Effect of welding parameters on nugget zone microstructure and properties in high strength aluminium alloy friction stir welds. *Sci. Technol. Weld. Join.* **2003**, *8*, 257–268. [CrossRef]
4. Biradar, N.S. Investigation of hot cracking behavior in transverse mechanically arc oscillated autogenous AA2014 T6 TIG welds. *Metall. Mater. Trans. A* **2012**, *43*, 3179–3191. [CrossRef]
5. Ericsson, M.; Sandström, R. Influence of welding speed on the fatigue of friction stir welds, and comparison with MIG and TIG. *Int. J. Fatigue* **2003**, *25*, 1379–1387. [CrossRef]
6. Threadgill, P.L.; Leonard, A.J.; Shercliff, H.R.; Withers, P.J. Friction stir welding of aluminium alloys. *Metall. Rev.* **2009**, *54*, 49–93. [CrossRef]
7. Elangovan, K.; Balasubramanian, V.; Babu, S. Predicting tensile strength of friction stir welded AA6061 aluminium alloy joints by a mathematical model. *Mater. Des.* **2009**, *30*, 188–193. [CrossRef]
8. Kalemba-Rec, I.; Kopyscianski, M.; Miara, D.; Krasnowski, K. Effect of process parameters on mechanical properties of friction stir welded dissimilar 7075-T651 and 5083-H111 aluminum alloys. *Int. J. Adv. Manuf. Technol.* **2018**, *97*, 2767–2779. [CrossRef]
9. Rafiei, R.; Shamanian, M.; Fathi, M.H.; Khodabakhshi, F. Dissimilar friction-stir lap-welding of aluminum-magnesium (AA5052) and aluminum-copper (AA2024) alloys: Microstructural evolution and mechanical properties. *Int. J. Adv. Manuf. Technol.* **2018**, *94*, 3713–3730. [CrossRef]
10. Tamadon, A.; Pons, D.J.; Sued, K.; Clucas, D. Thermomechanical Grain Refinement in AA6082-T6 Thin Plates under Bobbin Friction Stir Welding. *Met.-Basel* **2018**, *8*. [CrossRef]
11. Zeng, X.H.; Xue, P.; Wang, D.; Ni, D.R.; Xiao, B.L.; Ma, Z.Y. Realising equal strength welding to parent metal in precipitation-hardened Al-Mg-Si alloy via low heat input friction stir welding. *Sci. Technol. Weld. Join.* **2018**, *23*, 478–486. [CrossRef]
12. Zeng, X.H.; Xue, P.; Wang, D.; Ni, D.R.; Xiao, B.L.; Wang, K.S.; Ma, Z.Y. Material flow and void defect formation in friction stir welding of aluminium alloys. *Sci. Technol. Weld. Join.* **2018**, *23*, 677–686. [CrossRef]
13. Zhang, Q.; Lang, L.; Zhang, Y.; Sun, G. Application of the modified critical voids expansion ratio criterion to the prediction of the forming limit of 6016-O aluminum alloy. *Int. J. Adv. Manuf. Technol.* **2018**, *98*, 2069–2082. [CrossRef]
14. Fratini, L.; Buffa, G.; Shivpuri, R. In-process heat treatments to improve FS-welded butt joints. *Int. J. Adv. Manuf. Technol.* **2009**, *43*, 664–670. [CrossRef]
15. Peng, G.; Ma, Y.; Hu, J.; Jiang, W.; Huan, Y.; Chen, Z.; Zhang, T. Nanoindentation Hardness Distribution and Strain Field and Fracture Evolution in Dissimilar Friction Stir-Welded AA 6061-AA 5A06 Aluminum Alloy Joints. *Adv. Mater. Sci. Eng.* **2018**, *2018*, 1–11. [CrossRef]

16. Zhang, Z.; Xiao, B.L.; Ma, Z.Y. Influence of water cooling on microstructure and mechanical properties of friction stir welded 2014Al-T6 joints. *Mater. Sci. Eng. A* **2014**, *614*, 6–15. [CrossRef]

17. Sharma, C.; Dwivedi, D.K.; Kumar, P. Influence of in-process cooling on tensile behaviour of friction stir welded joints of AA7039. *Mater. Sci. Eng. A* **2012**, *556*, 479–487. [CrossRef]

18. Benavides, S.; Li, Y.; Murr, L.E.; Brown, D.; Mcclure, J.C. Low-temperature friction-stir welding of 2024 aluminum. *Scr. Mater.* **1999**, *41*, 809–815. [CrossRef]

19. Heirani, F.; Abbasi, A.; Ardestani, M. Effects of processing parameters on microstructure and mechanical behaviors of underwater friction stir welding of Al5083 alloy. *J. Manuf. Process.* **2017**, *25*, 77–84. [CrossRef]

20. Zhang, H. Characteristics and formation mechanisms of welding defects in underwater friction stir welded aluminum alloy. *Metall. Microst. Anal.* **2012**, *1*, 269–281. [CrossRef]

21. Sinhmar, S.; Dwivedi, D.K. Enhancement of mechanical properties and corrosion resistance of friction stir welded joint of AA2014 using water cooling. *Mater. Sci. Eng. A* **2017**, *684*, 413–422. [CrossRef]

22. Sabari, S.S.; Malarvizhi, S.; Balasubramanian, V. Influences of tool traverse speed on tensile properties of air cooled and water cooled friction stir welded AA2519-T87 aluminium alloy joints. *J. Mater. Process. Technol.* **2016**, *237*, 286–300. [CrossRef]

23. Mofid, M.A.; Abdollah-Zadeh, A.; Gür, C.H. Submerged friction-stir welding (SFSW) underwater and under liquid nitrogen: An improved method to join Al alloys to Mg alloys. *Metall. Mater. Trans. A* **2012**, *43*, 5106–5114. [CrossRef]

24. Kim, J.R.; Ahn, E.Y.; Das, H.; Jeong, Y.H.; Hong, S.T.; Miles, M.; Lee, K.J. Effect of tool geometry and process parameters on mechanical properties of friction stir spot welded dissimilar aluminum alloys. *Int. J. Precis. Eng. Man.* **2017**, *18*, 445–452. [CrossRef]

25. Chowdhury, S.M.; Chen, D.L.; Bhole, S.D.; Cao, X. Tensile properties of a friction stir welded magnesium alloy: Effect of pin tool thread orientation and weld pitch. *Mater. Sci. Eng. A* **2010**, *527*, 6064–6075. [CrossRef]

26. Ouyang, J.H.; Kovacevic, R. Material flow and microstructure in the friction stir butt welds of the same and dissimilar aluminum alloys. *J. Mater. Eng. Perform.* **2002**, *11*, 51–63. [CrossRef]

27. Rodriguez, R.I.; Jordon, J.B.; Allison, P.G.; Rushing, T.; Garcia, L. Microstructure and mechanical properties of dissimilar friction stir welding of 6061-to-7050 aluminum alloys. *Mater. Des.* **2015**, *83*, 60–65. [CrossRef]

28. Peel, M.; Steuwer, A.; Preuss, M.; Withers, P.J. Microstructure, mechanical properties and residual stresses as a function of welding speed in aluminium AA5083 friction stir welds. *Acta. Mater.* **2003**, *51*, 4791–4801. [CrossRef]

29. MetalsBahemmat, P.; Haghpanahi, M.; Besharati, M.K.; Ahsanizadeh, S.; Rezaei, H. Study on mechanical, micro-, and macrostructural characteristics of dissimilar friction stir welding of AA6061-T6 and AA7075-T6. *Proc. Inst. Mech. Eng. B-J. Eng.* **2010**, *1*, 1–12.

30. Leitao, C.; Leal, R.M.; Rodrigues, D.M.; Loureiro, A.; Vilaca, P. Mechanical behaviour of similar and dissimilar AA5182-H111 and AA6016-T4 thin friction stir welds. *Mater. Des.* **2009**, *30*, 101–108. [CrossRef]

31. Tronci, A.; Mckenzie, R.; Leal, R.M.; Rodrigues, D.M. Microstructural and mechanical characterisation of 5XXX-H111 friction stir welded tailored blanks. *Sci. Technol. Weld. Join.* **2013**, *16*, 433–439. [CrossRef]

32. Lee, W.B.; Yeon, Y.M.; Jung, S.B. Mechanical Properties Related to Microstructural Variation of 6061 Al Alloy Joints by Friction Stir Welding. *Mater Trans.* **2004**, *45*, 1700–1705. [CrossRef]

33. Liu, F.J.; Fu, L.; Chen, H.Y. Effect of high rotational speed on temperature distribution, microstructure evolution, and mechanical properties of friction stir welded 6061-T6 thin plate joints. *Int. J. Mach. Tool. Manuf.* **2018**, *96*, 1823–1833. [CrossRef]

34. Leitao, C.; Emilio, B.; Chaparro, B.M.; Rodrigues, D.M. Formability of similar and dissimilar friction stir welded AA 5182-H111 and AA 6016-T4 tailored blanks. *Mater. Des.* **2009**, *30*, 3235–3242. [CrossRef]

35. Netto, N.; Tiryakioglu, M.; Eason, P.D. Characterization of Microstructural Refinement and Hardness Profile Resulting from Friction Stir Processing of 6061-T6 Aluminum Alloy Extrusions. *Met.-Basel* **2018**, *8*, 552. [CrossRef]

36. Liu, H.; Pan, Q.; Yu, L. Effect of friction stir welding parameters on microstructural characteristics and mechanical properties of 2219-T6 aluminum alloy joints. *Int. J. Mater. Form.* **2012**, *5*, 235–241. [CrossRef]

Article

Ultrasonic-Assisted Semi-Solid Forming Method and Microstructure Evolution of Aluminum/Copper Brazed Joints

Yin Liang [1,2], Jiruan Pan [3], Hua Zhang [2], Peng Huang [2], Jun Wang [1], Yuxin Shi [1] and Limin Chen [1,*]

[1] School of Information Engineering, Nanchang University, Nanchang 330031, China; liangyin@ncu.edu.cn (Y.L.); ncudsp20@126.com (J.W.); hr667sy@yeah.net (Y.S.)

[2] School of Mechatronics Engineering, Nanchang University, Nanchang 330031, China; robinkk4@gmail.com (H.Z.); ncuhp9090@163.com (P.H.)

[3] School of Mechanical Engineering, Tsinghua University, Beijing 100084, China; pjl-dme@mail.tsinghua.edu.cn

* Correspondence: chenlimin@ncu.edu.cn

Received: 31 December 2019; Accepted: 3 February 2020; Published: 6 February 2020

Abstract: Aluminum/Copper dissimilar metal connection devices have been widely used in equipment manufacturing. Ultrasonic vibration assisted semi-solid brazing technology is beneficial to improve brazing quality by using the good flow ability and strong deformation resistance of brazing alloy when it is in a semi-solid state. In this study, a new ultrasound-assisted high-frequency induction brazing method was used to braze aluminum and copper dissimilar metals under non-vacuum conditions with non-prefabrication of brazing filler metal. In a short time, uniform semi-solid brazed joint was obtained. Detailed investigations on the effects of aluminum-substrate side (Abbreviated as Al side) heat dissipation rate and ultrasonic vibration duration on the microstructure evolution and mechanical properties of semisolid weld were conducted. The results show that the welding seam obtained by the modification method increases the microstructure uniformity of brazing joint.

Keywords: ultrasonic vibration; dissimilar metal; semi-solid status; brazing; microstructure evolution

---

1. Introduction

Due to their excellent corrosion resistance, high mechanical strength, and excellent electric and heat transfer performance, Aluminum and copper and their alloys have been widely used in electric power, cooling, chemical industry and aerospace fields [1–5]. In the process of equipment manufacturing, the connection problem of aluminum and copper dissimilar metals is often faced.

Zn-Al alloy is a promising brazing filler alloy for Al/Cu brazing [6–9]. The main advantage of the filler metal is that zinc metal and Cu and Al metal all had a very good solubility, which form good metallurgical combination easily, and the same time the Zn-Al alloy has good toughness and corrosion resistance. The Zn-Al filler [10] is a medium temperature solder and the eutectic temperature is about 380 °C, which can basically adapt all Al alloy brazing [11] connection and can avoid stress cracks resulted from the high temperature brazing and substrate metals ablation. Zn alloy has good thermal and electrical conductivity, which meeting the needs of Al/Cu connector performance. With the advantages of above-mentioned, the Zn-Al filler brazing alloy of Al/Cu dissimilar metals is increasingly become a hot research area.

From what has been discussed above, it can be seen that in brazing welding, semi-solid filler metal has many advantages: reducing connection temperature; getting isometric crystals and uniform microstructure of the joint whose strength can be achieved or close to the strength of the substrate metals. In addition, semi-solid brazing can deposit a large amount of filler metal in a single-pass brazing, which avoids smoke and splash [12–17].

Existing research [12–14,18–20] shows that using semi-solid brazing can increase the uniformity of brazing seam joint structure, but the research on ultrasound-assisted brazing of Al/Cu dissimilar metals using non-prefabricated semi-solid Zn-22Al (Eutectic solder) as filler metal is rarely found.

Due to the heat and mass transfer characteristics [21], ultrasonic assisted method has been widely used in various kinds of metal welding [22] in recent years to improve the microstructure of the welding seam (including brazing seam).

So, a kind of non-vacuum ultrasound-assisted semi-solid Zn-22Al high-frequency induction brazing of Al/Cu is adopted in this paper, herein referred to as semi-solid-state brazing with ultrasonic-assisted unilateral nucleation (abbreviated as TSBUAUN). Then, the influence of aluminum side heat dissipation rate and ultrasonic duration on the microstructure evolution and physical properties of semi-solid forming brazed seam joint are studied, which are of great significance for promoting the industrial application of Al/Cu dissimilar metal connection.

The proposed TSBUAUN method can improve the uniformity of the brazing joint structure, and its forming mechanism can be extended to semi-solid brazing of other dissimilar metals, providing a method and theoretical support for new brazing technology and the design of dissimilar brazing solder.

2. Experimental Materials and Methods

The substrate materials used in the experiment are 6060 Al provided by Henan Jiayuan Aluminum company (Zhengzhou, China) and T1 copper (Cu) plate provided Zhongtian Copper Industry Co. LTD (Dongguan, China), and each size is 60 mm × 20 mm × 2 mm. The detailed composition of the substrate materials are shown in Table 1. The filler metal is Zn-22Al provided by Xinrui China (Shengzhou, China) with a thickness of 0.3 mm and a size of 7 mm × 20 mm. Before the experiment, the surface of substrate metals and filler metal were polished, and the oxide film was removed. After grinding, the substrate metals were put into acetone and then washed with water bath ultrasonic for 15 min to avoid the impact of grease and impurities on the joint structure on the material surface. Then, the cut composite filler metal is coated with flux. The substrate metals and the composite filler metal are fixed in accordance with the structure shown in Figure 1.

Table 1. Chemical compositions of 6061 Al alloy, T1 Cu alloy, and Zn–22Al alloy.

Material	Al	Fe	Cu	Si	Mn	Mg	Zn	Ti	P
6061 Al	Bal.	0.7	0.15–0.4	0.4–0.8	0.15	0.8–1.2	0.25	0.15	
T1 Cu		≤0.005	CuAg ≥ 99.95				≤0.005		$p < 0.001$
Zn-22Al	22			1.5			76.5		

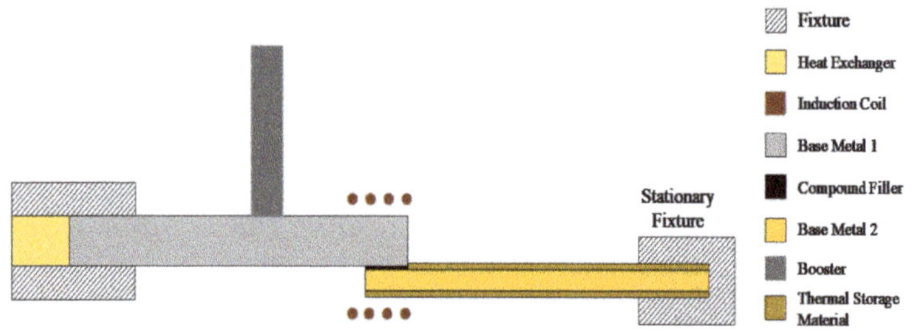

Figure 1. Schematic representation of bonding process for vibration brazing.

Brazing was carried out in atmospheric environment. The melting point of the brazing alloy Zn-22Al is about 450 °C. The use of pure cesium salt of medium temperature flux in the brazing process can protect the brazing materials and help to break the oxidation film on the metals surface.

The melting point of pure cesium salt is 445 °C, which plays a role in the vicinity of the filler metal liquid temperature line of 450 °C. The melting of pure cesium flux (Xinrui, China) before that of filler metal helps to protect the welding joint.

In the experiments, SP-25B model high frequency induction heating equipment which produced by Shuangping Supply Power Technologies Company (Shenzhen, China) was used as the brazing heat source. It consists of signal source, induction coil and cooling equipment. Meanwhile, thermocouple was used for temperature measurement.

In order to investigate the influence of Al side heat dissipation rate as while as ultrasonic duration on semi-solid forming and brazing joint microstructure [18,23–27], different experiments were designed. Firstly, under the condition of keeping the ultrasonic vibration duration unchanged, we carried out several groups of experiments under the condition of 2 K/s, 5 K/s, 7 K/s and 9 K/s (here K is the unit of thermodynamic temperature, named Kelvin, Abbreviated as K and K/s Represents the heat dissipation rate per second). Heat dissipation rate at Al side by adjusting the heat exchanger. Then, several groups of experiments under 0, 2, 4, 6, and 9 s ultrasonic vibration duration were carried out under the condition of the same heat dissipation rate on Al side.

The bonding cycle of heating, ultrasonic vibrating, solid/liquid ratio adjustment and Secondary ultrasonic dispersion and cooling are schematically illustrated in Figure 2.

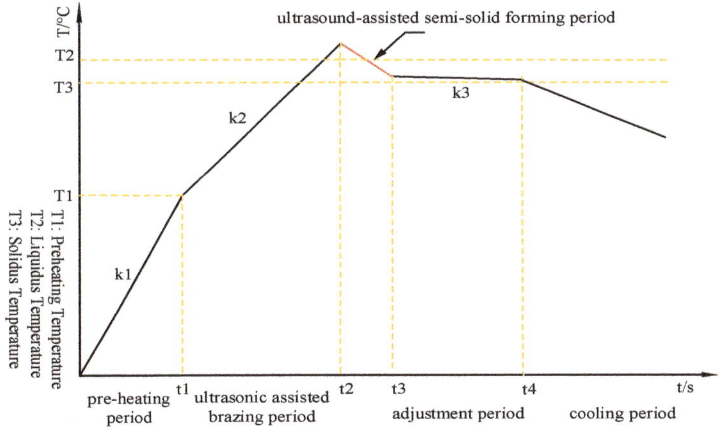

Figure 2. Sequence diagram of ultrasonic assisted one-side core semisolid brazing process.

After brazing, the samples were cut off along the width and each the section was polished with a series of metallographic sandpapers. The corrosion status of each polished specimen was observed under an optical microscope (OM) (GX71, Olympus corporation, Tokyo, Japan). The corrosion method was as follows: aluminum substrate material was corroded by 2% hydrofluoric acid, and copper substrate material was corroded by 3 g of $FeCl_3$ powder and 2 mL hydrochloric acid and 96 mL anhydrous ethanol. The corrosion time was about 5 s.

In order to further investigate the effect of semisolid process on microstructure and microstructure composition of brazed joints, the polished samples were observed by a scanning electron microscope (SEM, Quanta FEG 200, FEI Corporation, Hillsboro, OR, USA). Meanwhile, in order to explore the mechanism of the semi-solid process on the mechanical properties of brazed joints, the shear tensile strength (tensile rate of 1 mm/min) of each sample was measured on a universal mechanical tester.

In addition, experiments of common non-ultrasonic vibration brazing of Al-Cu were carried out to compare with the new method-TSBUAUN

3. Results and Discussion

3.1. Typical Interfacial Microstructure of the Al/Zn-22Al/Cu Brazed Joint

Optical microscope images of the interfacial microstructure of brazed joints by different brazing methods are shown in Figure 3a,b are the weld microstructure obtained by the common brazing process and ultrasonic assisted brazing process respectively.

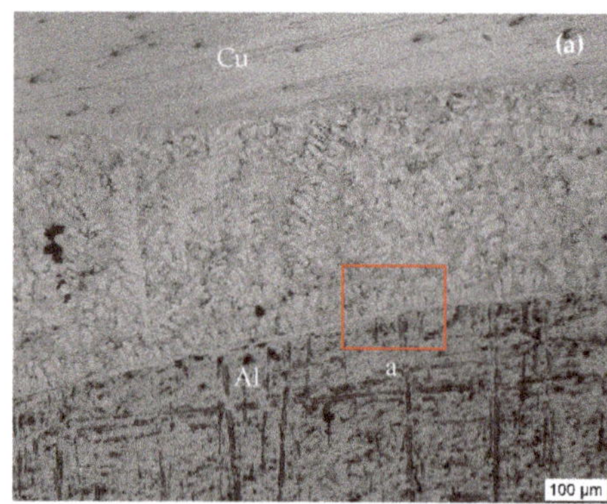

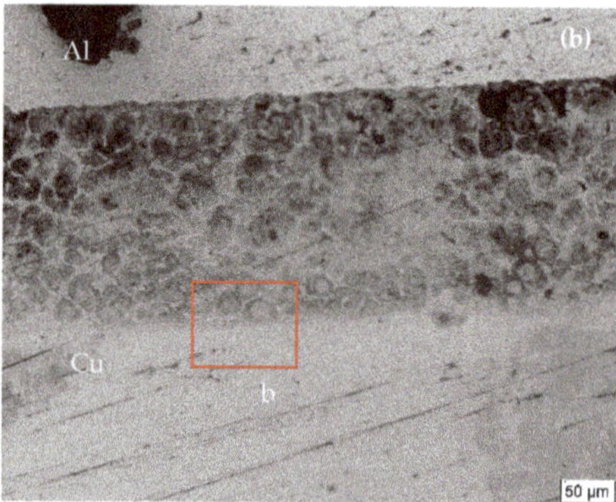

Figure 3. (**a**) brazing Al/Cu without ultrasonic; (**b**) semi-solid brazing Al/Cu with ultrasonic.

There are obvious differences in microstructure between the two kinds of brazing samples. The microstructure of brazed joint of common brazing sample in Figure 3a is mainly dendrite and fringe eutectic. The number of fringe eutectic tissues on the Cu- substrate- side (Abbreviated as Cu side) is larger than that on the Al side, which conforms to the diffusion law of Cu element.

In Figure 3b the microstructure of the TSBUAUN sample is mainly spherulite and fringe eutectic. The density of spherulite on the Al side is higher than that on the Cu side. Through graphic analysis of the corroded semi-solid specimen, it can be found that the core of the spherulite is more easily corroded than other parts of the semi-solid specimen.

Meanwhile, from TSBUAUN sample we can see there are obvious differences in corrosion resistance of spherulite cores on both sides of Cu and Al. Cu side spherulite core has better corrosion resistance than that of Al side. Under the condition of Al side supercooling, the weld formed dense α-Al columnar crystal and Large dendrites. Active heat dissipation on the Al side aggravated the imbalance of weld temperature and was beneficial to dendrite growth. The average length of dendrites can reach 200 to 300 microns under certain conditions.

Under the action of appropriate Al side heat dissipation rate and appropriate ultrasonic vibration duration semi-solid state in brazed joints was achieved. The diameter of semi-solid spherulites is mainly distributed between 25 and 40 microns, and the average size is about 35 microns.

By comparing Figure 3a,b, it was found that there was an obvious transition layer (about 10μm) on the Cu side of the common brazing sample, while the transition layer of the ultrasonic assisted brazing sample was very thin (0.5–0.8 μm).

The optical microstructure of the brazed joint of common brazing samples (Figure 4a), SEM microstructure (Figure 4b) and line-scan component analysis (Figure 4c) showed that the microstructure of common brazing was mainly composed of α-Al phase dendrite and Zn-Al eutectic. The rose-like intermetallic compounds are mainly distributed at grain boundaries. According to SEM/EDS results, the following two intermetallic compounds were identified: CuZn5-ε phase (light gray zone) and Al5Cu4Zn-T′ phase (dark gray zone).

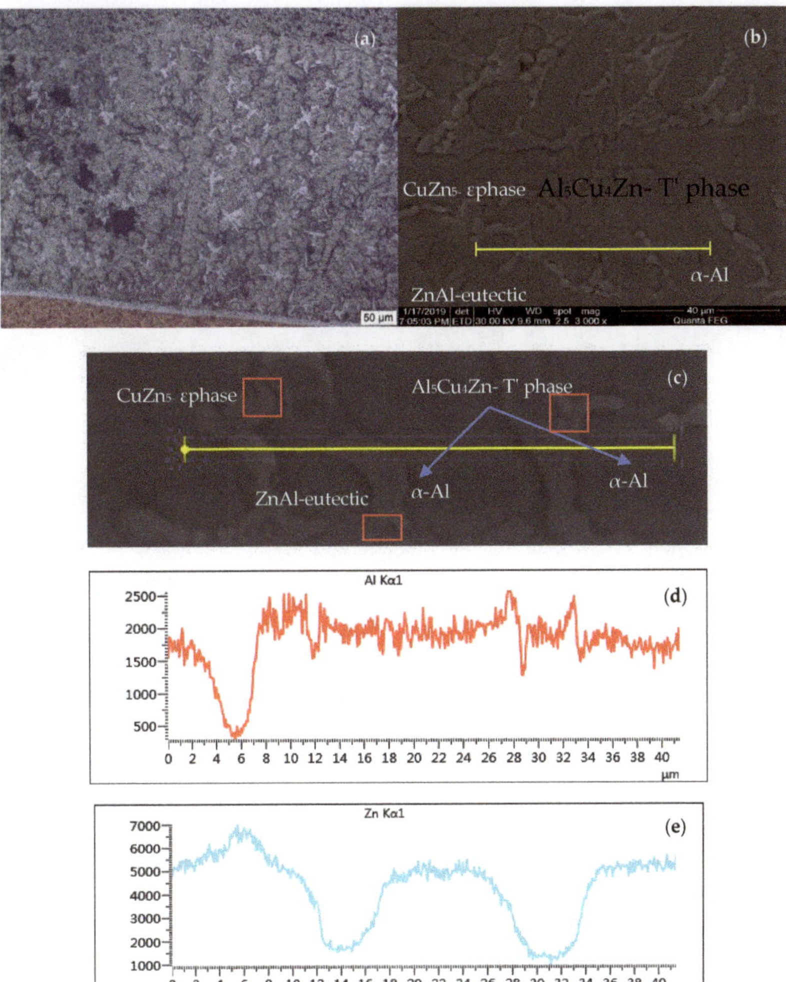

Figure 4. The dendrite microstructure OM diagram, SEM morphology diagram and line scan diagram of the brazed joint area of common sample: (**a**) OM morphology of dendrite in brazed zone; (**b**) SEM morphology of dendrite structure in brazed zone; (**c**) Linear scanning composition of dendrites in brazed joint microstructure; (**d**) Line scan for Al composition; (**e**) Line scan for Zn composition.

Under the ultrasonic-assisted semi-solid forming process, due to non-equilibrium solidification, striated eutectic was produced at the rapid cooling rate during the brazing process, and the number of eutectic tissues on the Cu side was significantly larger than that on the Al side. Due to the forced convection generated by ultrasonic waves, the dendrite-Al phase fragmentation was promoted, and the typical spherical structure composed of semi-solid particles was formed, as in Figure 5.

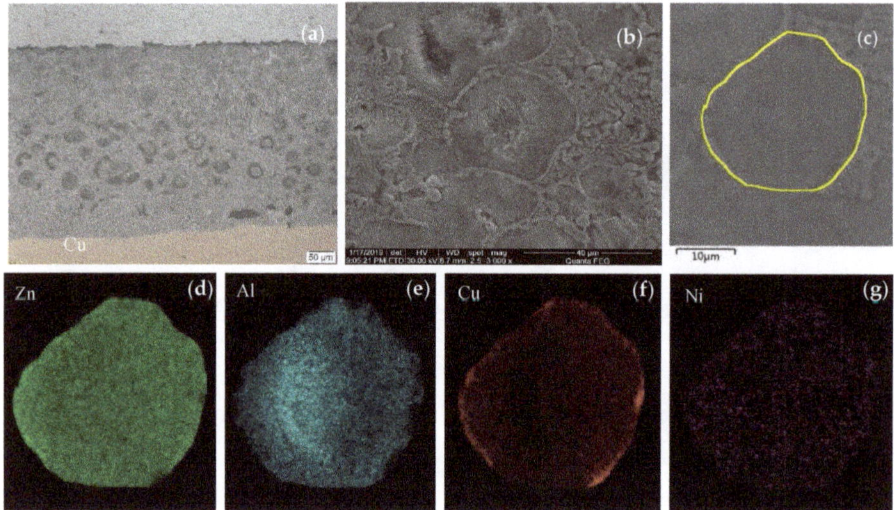

Figure 5. Spherulite OM morphology, SEM morphology, EDS layered image and composition distribution of ultrasonic-assisted semi-solid sample: (**a**) OM morphology of spherulite in the brazed joint zone; (**b**) SEM morphology of dendrite structure in the brazed joint zone; (**c**) EDS hierarchical images in spherulite; (**d**) Zn distribution in spherulite; (**e**) Al distribution in spherulite; (**f**) Cu distribution in spherulite; (**g**) Ni distribution in spherulite.

3.2. Influence of Aluminum Side Heat Dissipation Rate on Semi-Solid Forming and Microstructure Evolution of Brazed Joints

In order to explore the influence of different Al side heat dissipation rates on ultrasonic assisted brazing weld seams, we set four kinds of Al side heat dissipation rates of 2, 5, 7, and 9 k/s by adjusting the heat exchange equipment at Al plate side.

When the ultrasonic vibration duration is kept at 6 s, the microstructure appearance of each brazing seam sample obtained under different Al-side heat dissipation rates is shown in Figure 6.

As long as ultrasonic vibration is carried out, there are no large tree dendrites in the weld microstructure of brazing samples with different Al-side heat dissipation rate. Instead, there are rose-like crystals, columnar crystals and spherulites with relatively regular distribution, as can be seen in Figure 6.

The size of columnar crystal on Al side increases with the increase of heat dissipation rate on Al side. When the heat dissipation rate on the Al side increases to 9 K/s, some columnar crystals on the Al side have grown secondary crystal axes, as shown in region d1 of Figure 6d. Through careful observation of the weld microstructure in Figure 6d, we found that the columnar crystal size on the Cu side was much smaller than that firstborn on the Al side, but dendrite microstructure could be observed in the local area (d2). When the heat dissipation rate on the Al side increases to 9 K/s, the elongated columnar crystals generated in the weld joint are not conducive to the formation of semi-solid state spherulites.

In order to achieve a better semi-solid-state effect, we gradually reduced the heat dissipation rate on the Al side. When the rate dropped to 7 K/s, the size of primary columnar crystals on Al side decreases and there is no obvious dendrite tendency, meanwhile the size of primary columnar crystals on Al side decreased, and there was no obvious dendrite tendency. In Figure 6c, the dendrite degree of columnar crystals of C region presents an obvious gradient. Most columnar crystals distributed on the Cu side have secondary axis, but there is no fully developed dendrite structure in the weld seam.

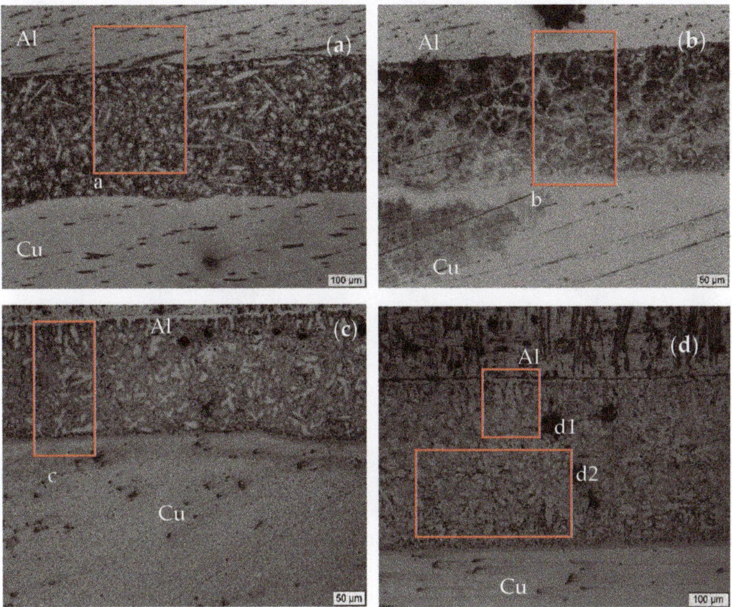

Figure 6. Microstructure of brazed joints under different Al-substrate-side heat dissipation rate: (**a**) Al-side heat dissipation rate 2 K/s; (**b**) Al-side heat dissipation rate 5 K/s; (**c**) Al-side heat dissipation rate 7 K/s; (**d**) Al-side heat dissipation rate 9 K/s.

It is concluded that the heat dissipation rate of 7 K/s on Al side is still not conducive to semi-solid spherulites forming. Therefore, the Al-side heat dissipation rate was gradually reduced. Under the combined action of ultrasonic vibration and the Al-side heat dissipation rate down to 5 K/s, the semi-solid state of weld microstructure was well realized, as shown in Figure 6b, where the spherulites in the weld seam have a good sphericity, and these ball spherulites diameters are mainly distributed between 20 and 35 microns. Meanwhile, the spherulite density on both sides of the substrate metal differs.

In order to comprehensively explore the effect of Al side heat dissipation rate on the microstructure of weld joint, we further reduce the heat dissipation rate of the Al side to the rate of 2 K/s. This time, in the welding joint seam, there were no thick columnar crystals or good spherulites on the Al side, but a large number of rose crystals formed (the copper-aluminum compounds are shown in a region of Figure 6a. We used Image software to conduct statistical analysis on the size of weld microstructure and crystal distribution obtained at different Al-side heat dissipation rates. The size of crystals and their distribution in weld microstructure were clearly demonstrated under different Al-side heat dissipation rates, as shown in Figure 7. Under the action of ultrasonic vibration, the influences of different Al-side heat dissipation rates on weld microstructure are quite different.

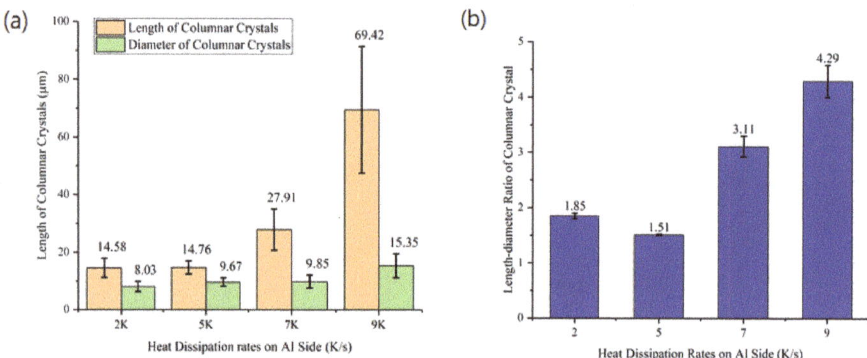

Figure 7. (**a**) columnar crystal length and diameter under four Al side heat dissipation rates (K/s) at 2 K/s, 5 K/s, 7 K/s and 9 K/s; (**b**) columnar crystal length-diameter ratio of Al-side heat dissipation rate (K/s) at 2 K/s, 5 K/s, 7 K/s and 9 K/s.

3.3. Effect of Ultrasonic Duration on Semi-Solid Forming and Microstructure Evolution of Corresponding Weld Seams

The Al-side heat dissipation rate and ultrasonic vibration are two key factors of semi-solid forming. The Al-side heat dissipation rate affects the morphology of the primary microstructure of the welding joint, while the appearance of weld joint microstructure is also related to ultrasonic vibration. We set a series of ultrasonic vibration duration (the first segment of ultrasound) to explore the mechanism of semi-solid forming under different ultrasonic duration, which are set as 0 s, 2 s, 4 s, 6 s and 9 s respectively in experiments. Under the optimal Al-side heat dissipation rate of 5 k/s, the metallography pictures of welding joints applied different ultrasonic vibration duration are shown in Figure 8.

By observing the metallographic diagrams in Figure 8, it is not difficult to find that, under suitable Al-side heat dissipation conditions at rate of 5 K/s, large dendrites grew from Al substrate material in the brazing process without adding ultrasonic vibration, as shown in Figure 8a. A Strong perturbation or agitation is required in the semi-solid forming process, and the weld joint microstructure of the sample with 2 s ultrasound no longer has large dendrite structure. In the selection area b in Figure 8, the coarse equiaxed crystal structure appears near the Al-substrate material, with each crystal diameter of 25~30 microns, and also find that a certain number of columnar crystal structure below the coarse equiaxed crystal structure, with a length of about 50 microns.

When Ultrasonic vibration duration is 2 s, it can't achieve better semi-solid structure forming, or even can't be called "semi-solid brazed joint". Figure 8c shows the brazed seam microstructure under 4 s ultrasonic vibration duration, which has basically achieved semi-solid state. and the process of primary equiaxed grain nucleation and being sheared can even be observed in the box selected area c.

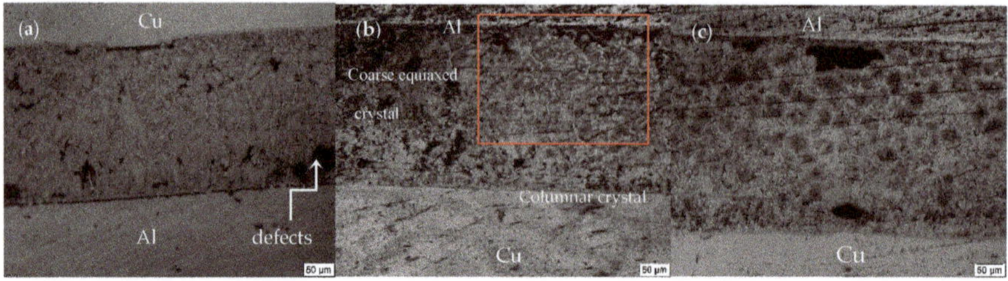

Figure 8. *Cont.*

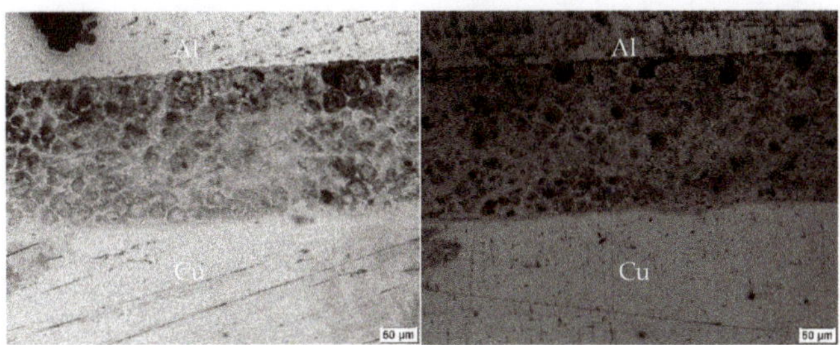

Figure 8. Microstructure of welds with different ultrasonic vibration duration: (**a**) 0 s; (**b**) 2 s; (**c**) 4 s; (**d**) 6 s; (**e**) 9 s.

At the Al-side heat dissipation rate of 5 K/s, the primary nucleation crystals of welding joint on the Al side, which shown in the selected area c have spherulite characteristics. Although ultrasonic vibration 4 s has basically achieved semi-solid welding seam, spherulite sphericity and density still have room for improvement. With the increase of ultrasonic vibration time to 6 s, the density of spherulite in weld increased greatly and the density of spherulite in the weld joint has been greatly improved, which means semi-solid forming effect is ideal, as shown in Figure 8, area d. However, the increase of ultrasonic vibration time will not always provide a positive effect for semi-solid forming, and more deformed spherulites appear in the weld seam of ultrasonic vibration sample of 9 s, as shown in area e Figure 8, which means that too long ultrasonic vibration time will have a negative impact on the semi-solid forming effect.

In order to better characterize the effect of ultrasonic vibration on the sphericity and distribution of semi-solid spherulite of the microstructure of joints, we used image software to analyze the sphericity and spherulite density of samples with different ultrasonic vibration durations, as shown in Figure 9.

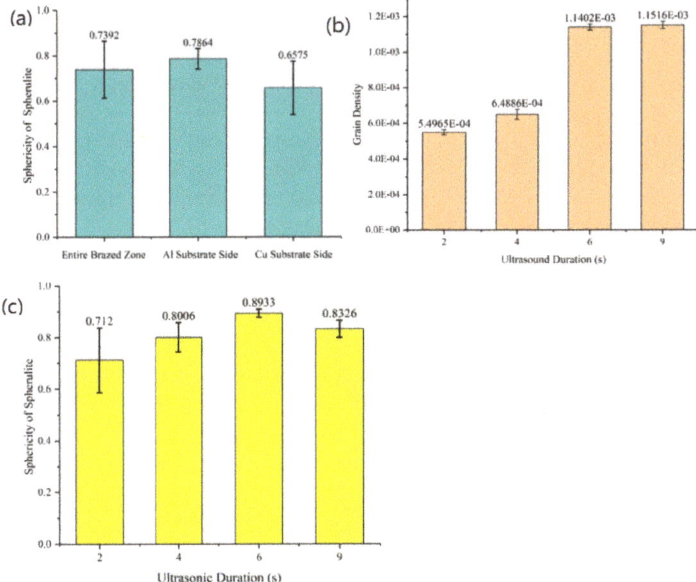

Figure 9. (**a**) spherulite sphericity of each weld seam area when ultrasonic vibration time is 2 s; (**b**) grain density under different ultrasonic duration of 2 s, 4 s, 6 s and 9 s; (**c**) spherulite sphericity under different ultrasonic duration of 2 s, 4 s, 6 s and 9 s.

Based on the above analyses, the aluminum side heat dissipation rate and ultrasonic duration had significant effects on semi-solid forming and the microstructure evolution of joints.

4. Conclusions

A reliable semi-solid state brazing joint of Copper connected to aluminum was successfully obtained using a new technique for non-vacuum ultrasound-assisted non-prefabricated semi-solid high-frequency induction brazing. The microstructure evolution and physical properties of Al/Zn-22Al/Cu joints were studied. Based on the achieved results, the conclusions are summarized as follows:

- Through a large number of cross tests, it was observed that when the ultrasonic time lasted 6 s and the heat dissipation degree on the aluminum side was 5 k/s, a better semi-solid forming effect was achieved, the spherulite density was large, and the weld structure was uniform.
- When the ultrasonic duration is 6 s, the effect of different heat dissipation rate on the microstructure of welding seam varies greatly. With the increase of heat dissipation rate on the Al side, the size of columnar crystal on the Al side keeps increasing. When the heat dissipation rate on the Al side reaches 9 K/s, some columnar crystals on the Al side have already grown secondary crystal axes. We gradually began to reduce the heat dissipation rate on the Al side from 9 K/s. When the supercooling rate drops to 7 K/s, the size of primary columnar crystals on Al side decreases and there is no obvious dendrite tendency. Further, 7 K/s Al side heat dissipation rate is still not conducive to semi-solid forming. Thus, we continue to reduce the Al side heat dissipation rate to 5 K/s. The joint action of ultrasonic vibration and Al side heat dissipation rate of 5 K/s achieves the perfect semi-solid weld structure. Further reduce the heat dissipation rate of Al side cooling equipment. There were no thick columnar crystals or spherulites in the Al side, but a large number of rose crystals were formed. So far, under a certain ultrasonic duration, the semi-solid forming effect is good when the heat dissipation rate of Al side is 5 K/s. The weld organization is uniform, and the shear strength of the weld point reaches a maximum of 65.3 Mpa.
- Under 5 K/s of the heat dissipation rate, the brazing process does not add in brazed joint ultrasonic vibration with Al parent metal as the basal bulky dendrite, add 2 s in ultrasonic sample brazing microstructure is no longer has a huge dendritic structure, and obtain the good effect of semi-solid forming can't even call the weld microstructure of semi-solid. With the increase of ultrasonic vibration time to 4 s, the weld structure of the sample has basically achieved semi-solid state, and the process of primary isometric crystal nucleation and shearing can even be observed in the box selection area. With the increase of ultrasonic vibration time to 6 s, the density of spherulite in the weld was greatly improved. The semi-solid forming effect is relatively ideal. However, the increase of ultrasonic vibration time will not always provide a positive effect for semi-solid forming, and the ultrasonic vibration time continues to increase. This means that the ultrasonic vibration time is too long to have a negative impact on the semi-solid forming effect.
- Compared with common brazing, the TSBUAUN brazing method changes the distribution and morphology of brazing intermetallic compounds, and the number of rose-like crystals is significantly reduced with fragmentation (thickness: 0.5–0.8 m). After fragmentation, rose-like intermetallic compounds become monolayer phase wrapped in the outer layer of α-Al phase. It is proven that the microstructure and intermetallic compound are changed by ultrasonic and the heat dissipation in the TSBUAUN process.

Author Contributions: Conceptualization, P.H. and J.W.; Formal analysis, Y.L.; Investigation, Y.L. and Y.S.; Project administration, H.Z.; Resources, J.P. and H.Z.; Supervision, J.P.; Validation, P.H.; Visualization, P.H. and Y.S.; Writing—original draft and revising, Y.L.; Writing—review and editing, all the authors; Writing—revising, L.C. and Y.L. All authors have read and agreed to the published version of the manuscript.

Funding: This project is supported by the National key R&D program (Grant No. 2018YFB1305305), and the National Natural Science Foundation of China (Grant No. U1731118).

Conflicts of Interest: The authors declare no conflict of interest.

References

1. Niu, Z.; Ye, J.; Huang, J.; Yang, H.; Yang, J.; Chen, S. Interfacial structure and properties of Cu/Al joints brazed with Zn-Al filler metals. *Mater. Charact.* **2018**, *138*, 78–88. [CrossRef]
2. Wang, X.-G.; Yan, F.J.; Li, X.-G.; Wang, C.-G. Induction diffusion brazing of copper to aluminum. *Sci. Technol. Weld. Join.* **2017**, *19*, 1–6.
3. Sahin, M. Effect of surface roughness on weldability in aluminum sheets joined by cold pressure welding. *Ind. Lubr. Tribol.* **2008**, *5*, 249–254. [CrossRef]
4. Saeid, T.; Abdollah-Zadeh, A.; Sazgari, B. Weldability and mechanical properties of dissimilar aluminum-copper lap joints made by friction stir welding. *J. Alloy. Compd.* **2010**, *2*, 652–655. [CrossRef]
5. Gulenc, B. Investigation of interface properties and weldability of aluminum and copper plates by explosive welding method. *Mater. Des.* **2008**, *1*, 275–278. [CrossRef]
6. Naka, M.; Hafez, K.M. Applying of ultrasonic waves on brazing of alumina to copper using Zn-Al filler alloy. *J. Mater. Sci.* **2003**, *16*, 3491–3494. [CrossRef]
7. Zhang, M.; Lin, Y.-B.; Jiang, H.-L. Effect of Al on Zn-Al Filler Metal Wettability on Pure Copper Surface. *Adv. Mater. Res.* **2012**, *538*, 196–199. [CrossRef]
8. Ji, H.; Chen, H.; Li, M. Effect of ultrasonic transmission rate on microstructure and properties of the ultrasonic-assisted brazing of Cu to alumina. *Ultrason. Sonochem.* **2017**, *3*, 491–495. [CrossRef]
9. Xu, H.; Xing, Q.; Zeng, Y. Semisolid stirring brazing of SiCp/A356 composites with Zn27Al filler metal in air. *Sci. Technol. Weld. Join.* **2011**, *6*, 483–487. [CrossRef]
10. Urena, A.; Gil, L.; Escriche, E.; Gómez de Salazar, J.M.; Escalera, M.D. High Temperature Soldering of Si C Particulate Aluminum Matrix Composites (series 2000) Using Zn-Al Filler Alloys. *Sci. Technol. Weld. Join.* **2001**, *6*, 1–11. [CrossRef]
11. Uremia, A.; de Salazar, J.G.; Escalera, M.D.; Fernandez, M.I. Study of the Braze-ability of Aluminum Matrix Composites. *Weld. J.* **1997**, *6*, 92–102.
12. Shi, L.; Yan, J.; Peng, B.; Han, Y. Deformation behavior of semi-solid Zn-Al alloy filler metal during compression. *Mater. Sci. Eng. A* **2011**, *22*, 7084–7092. [CrossRef]
13. Pola, A.; Tocci, M.; Kapranos, P. Microstructure and properties of semi-solid aluminum alloys: A literature review. *Metals* **2018**, *3*, 181. [CrossRef]
14. Xu, Z.; Li, Z.; Lei, X. Reduction of intermetallic compounds in ultrasonic-assisted semi-solid brazing of Al/Mg alloys. *Sci. Technol. Weld. Join.* **2018**, *24*, 1–8. [CrossRef]
15. Xu, Z.; Li, Z.; Peng, B.; Ma, Z.; Yan, J. Application of a new ultrasonic-assisted semi-solid brazing on dissimilar Al/Mg alloys. *Mater. Lett.* **2018**, *228*, 72–76. [CrossRef]
16. Xu, H.; Zhou, B.; Du, C.; Luo, Q.; Chen, H. Microstructure and properties of joint interface of semisolid stirring brazing of composites. *J. Mater. Sci. Technol.* **2012**, *12*, 1163–1168. [CrossRef]
17. Yu, X.; Xing, W.-Q.; Ding, M. Ultrasonic semi-solid coating soldering 6061 aluminum alloys with Sn-Pb-Zn alloys. *Ultrason. Sonochem.* **2016**, *31*, 216–221. [CrossRef]
18. Flemings, M.C. Behavior of metal alloys in the semi-solid state. *Metall. Mater. Trans. A* **1991**, *5*, 270–292.
19. Thadela, S.; Mandal, B.; Das, P.; Roy, H.; Lohar, A.K.; Samanta, S.K. Rheological behavior of semi-solid TiB2 reinforced Al composite. *Trans. Nonferrous Met. Soc. China* **2015**, *25*, 2827–2832. [CrossRef]
20. Wang, S.-C.; Qi, W.-J.; Zheng, K.-H.; Zhou, N.; Li, L. Microstructure and mechanical property of rheoforged A356 aluminum alloy in semisolid state. *Forg. Stamp. Technol.* **2011**, *36*, 127.
21. Gondrexon, N.; Cheze, L.; Jin, Y.; Legay, M.; Tissot, Q.; Tissot, Q.; Hengl, N.; Talansier, E. Intensification of heat and mass transfer by ultrasound: Application to heat exchangers and membrane separation processes. *Ultrason. Sonochem.* **2015**, *25*, 40–50. [CrossRef] [PubMed]
22. Krajewski, A.; Włosiński, W.; Chmielewski, T.; Kołodziejczak, P. Ultrasonic-vibration assisted arc-welding of aluminum alloys. *Bull. Pol. Acad. Sci. Tech. Sci.* **2012**, *60*, 841–852. [CrossRef]
23. Xia, C.-Z.; LI, Y.-J.; Wang, J.; Ma, H.-J. Microstructure and phase constitution near interface of Cu/ Al vacuum brazing. *Mater. Sci. Technol.* **2007**, *23*, 815–818. [CrossRef]
24. Yong, X.; Ji, H.; Li, M.; Kim, J. Ultrasound-induced equiaxial flower-like CuZn5/Al composite microstructure formation in Al/Zn-Al/Cu joint. *Mater. Sci. Eng. A* **2014**, *594*, 135–139.

25. Swallowe, G.M.; Field, J.E.; Rees, C.S.; Duckworth, A. A photographic study of the effect of ultrasound on solidification. *Acta Metall.* **1989**, *37*, 961–967. [CrossRef]

26. Zhang, Y.; Zhang, K.; Liu, G.; Xu, J.; Shi, L.; Cui, D.; Cui, B. The formation of rosette phase structural evolution during the reheating and semisolid casting of AlSi7Mg Alloy. *J. Mater. Process. Technol.* **2003**, *137*, 195–200. [CrossRef]

27. Sumitomo, T.; Stjohn, D.; Steiberg, T. The shear behavior of partially solidified Al-Si-Cu alloys. *Mater. Sci. Eng. A* **2000**, *286*, 18–29. [CrossRef]

MDPI

St. Alban-Anlage 66

4052 Basel

Switzerland

Tel. +41 61 683 77 34

Fax +41 61 302 89 18

www.mdpi.com

Metals Editorial Office

E-mail: metals@mdpi.com

www.mdpi.com/journal/metals